Green Energy and Environmental Systems

Green Energy and Environmental Systems

Edited by **Lucas Collins**

R CALLISTO REFERENCE

New York

Published by Callisto Reference,
106 Park Avenue, Suite 200,
New York, NY 10016, USA
www.callistoreference.com

Green Energy and Environmental Systems
Edited by Lucas Collins

International Standard Book Number: 978-1-63239-764-5 (Hardback)

Printed in the United States of America.

Contents

Preface

This book has been an outcome of determined endeavour from a group of educationists in the field. The primary objective was to involve a broad spectrum of professionals from diverse cultural background involved in the field for developing new researches. The book not only targets students but also scholars pursuing higher research for further enhancement of the theoretical and practical applications of the subject.

Green energy is derived from natural processes that produce small amount of pollution. Some of the resources used for this purpose are wind power, solar power, geothermal power, hydro power etc. Environmental systems include all the factors that affect the development, survival and evolution. This book unravels the recent studies in the field of green energy and environmental systems. It will also provide innovative topics for research which interested readers can take up. Scientists and students actively engaged in this field will find this book full of crucial and unexplored concepts. It extensively discusses the emerging technologies that increase energy efficiency. This text will serve as a valuable source of reference for engineers, environmentalists, ecologists, conservationists, professionals and students associated with this area.

It was an honour to edit such a profound book and also a challenging task to compile and examine all the relevant data for accuracy and originality. I wish to acknowledge the efforts of the contributors for submitting such brilliant and diverse chapters in the field and for endlessly working for the completion of the book. Last, but not the least; I thank my family for being a constant source of support in all my research endeavours.

Editor

Energy Analysis for the Compaction of Jerash Cohesive Soil

Talal Masoud, Hesham Alsharie, Ahmad Qasaimeh

Civil Engineering Department, Jerash University, Jerash, Jordan
Email: argg22@yahoo.com

Abstract

The aim of this research is to study the effect of compaction energy on Jerash cohesive soil. Qualitative and quantitative analyses of soil compaction energy with relation to unit weight and moisture content are conducted. These analyses spot the light on energy savings performed for soil compaction. The study shows that as the compaction energy increases; the unit weight of the Jerash cohesive soil increases and the optimum water content decreases. Generally, a soil with low moisture content is less vulnerable to compaction than a soil with high moisture content. But when the moisture content is too high, all the soil pores are filled with water, so that the soil becomes less compressible where the unit weight and strength characteristics decrease. The optimum energy value and optimum water content are thus of great concern. The effect of energy on soil unit weight is very large as the energy increases from 400 to 1400 KJ/m³ and after that level; the effect of energy on soil unit weight is very small. Consequently, optimal compaction energy ranges from 1200 up to value 1400 KJ/m³, where 50 to 60 blows can be applied and the optimal correlated water content is between 14% - 15%.

Keywords

Energy, Jerash Cohesive Soil, Compaction, Jordan

1. Introduction

In the construction of highways, for earth dams and other engineering structures, loose soil must be compacted to increase its unit weight, and in turn increasing soil strength characteristics, increasing soil bearing capacity, and decreasing soil settlement.

The soil compaction can be defined as "the process by which the soil grains are rearranged to decrease void space and bring them into closer contact with one another, thereby, increasing the bulk density" [1]. Compaction or densification of soil is done by removal of air which requires mechanical energy. Sheep feet rollers, rubber tired rollers and vibratory rollers may be used for soil compaction.

Knowledge of water contents in relation to the soil compaction for a particular soil can be helpful in scheduling the routine mechanical operations on that soil [2] [3].

The soil compaction is accompanied by the removal of the soil air, changes the soil structure, and macroscopically increases the soil strength [4]. The soil compaction process is highly influenced by the soil water content [5]-[7].

Vulnerability of a soil to compaction at the given soil moisture and energy level depends also on its clay content and mineralogical characteristics [8] [9]. Generally, a soil with very low moisture content is less vulnerable to compaction than a soil with high moisture content [10]. But when the moisture content is so high that all the soil pores are filled with water, the soil becomes less compressible [8].

Bulk density (dry soil mass per unit volume) is the most frequently used parameter to characterize the soil compaction [11]. In swelling-shrinking soil, it is recommendable to determine the bulk density at the standard moisture contents [12].

2. Methodology

The aim of this research is to study the effect of energy on dry unit weight of Jerash cohesive soil. Consequently; qualitative and quantitative analyses of soil compaction spot the light on soil strength characteristics and energy savings performed for soil compaction.

To perform the compaction test, samples were taken from eastern area of Jerash city in Jordan. The compaction test was conducted using "Standard Proctor Test". The number of hammer blows per each layer varied from 20 blows per layer to 70 blows per layer which in turn varied the energy per unit volume.

The soil was mixed with varying amount of water and then compacted in three equal layers by hammer that delivered 20 blows to each layer. The moisture-unit weight relationship was plotted for each sample tested to obtain the maximum dry unit weight.

The same procedure was repeated with 25, 30, 40, 50, 60, and 70 blows per layer. The optimum water content and maximum dry unit weight were obtained for each.

3. Discussion and Analysis

The degree of compaction of a soil is measured in terms of its dry unit weight. Many researchers studied the factors affect compaction characteristics of the soil; and they found that the soil type, grain size, grain distribution, and grain shape were of great influence on the maximum dry unit weight.

In this investigation, the effect of energy on the dry unit weight of Jerash cohesive soil was studied. The compaction energy per unit volume (KJ/m^3) used for Standard Proctor Test is given by the following equation:

$$\text{Energy} = \frac{(\text{Number of blows per layer})*(\text{Number of layer})*(\text{Weight of hammer})*(\text{Height of drop of hammer})}{\text{Volume of mold}}$$

To study the effect of energy on the dry unit weight of Jerash cohesive soil, samples were taken from eastern part of Jerash city. The physical properties of the soil are given in **Table 1**.

The soil was mixed with water and compacted in three equal layers by hammer that delivered 20 blows to each layer. The procedure was repeated with 25, 30, 40, 50, 60, and 70 blows per layer. The optimum water content and maximum dry unit weight were obtained for each trial as shown in **Figure 1** and **Table 2**.

While the compaction energy increases from 480 to 1680 KJ/m^3, the dry unit weight of Jerash cohesive soil increases from 14 to 16 KN/m^3 and the optimal water content (OWC) decreases from 26% to 12% (**Figure 2**, **Figure 3**). These ranges provide information about how we may manage the compaction to a wise procedure

Table 1. Physical properties of Jerash cohesive soil.

Specific Gravity	Plasticity Index	Plastic Limit	Liquid Limit	Clay	Silt	Sand
2.67	22%	36%	58%	54%	39%	7%

Table 2. The results of standard Proctor test on Jerash cohesive soil.

Number of Blows per Layer	Energy per Unit Volume (KJ/m^3)	Optimum Water Content OWC %	Unit Weight γ_d (KN/m^3)
20	480	26	14
25	600	24	14.4
30	720	21	14.7
40	960	17	15.3
50	1200	15	15.7
60	1440	14	15.9
70	1680	12	16.0

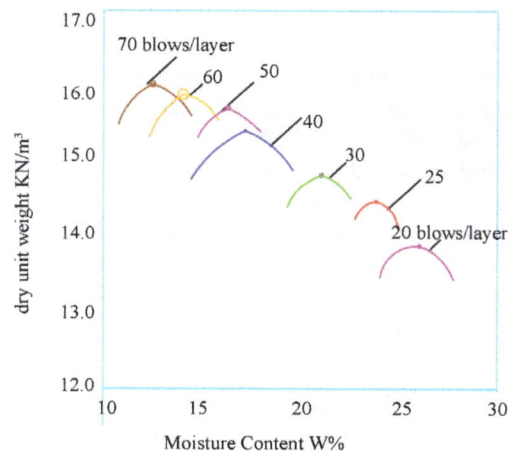

Figure 1. The values of unit weight moisture content varying within different compaction blows for Jerash soil.

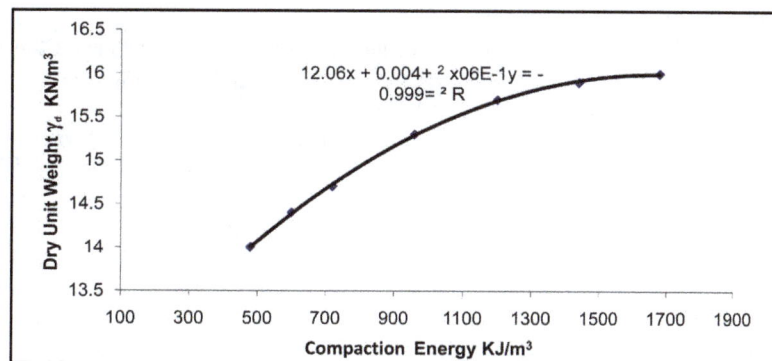

Figure 2. Relation between unit weight and compaction energy for Jerash soil.

without wasting time, effort, and energy. The effect of energy on dry unit weight is very large as the energy increase from 480 to 1400 KJ/m^3 and after that level, the effect of energy on the unit weight is very small as shown in **Figure 2**.

It is worthy to recall that for certain number of blows and while increasing water content, that the soil unit weight increases to optimal value and then decreases is because the soil with low moisture content is less susceptible to compaction than the soil with high moisture content. But when the moisture content is too high, all the soil pores are filled with water, so that the soil becomes less compressible as that is interpreted previously in **Figure 1**. Generally, while the compaction energy increases, the unit weight of the Jerash cohesive soil increases as the optimum water content decreases. Thus optimum energy value and optimum water content are of great concern about soil unit weight, soil strength characteristics, and soil energy compaction.

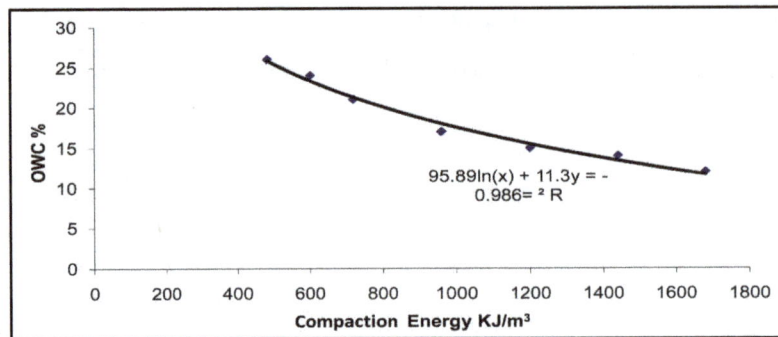

Figure 3. Relation between optimal water content and compaction energy for Jerash soil.

4. Conclusions

This investigation focuses on the effect of compaction energy on dry unit weight of Jerash cohesive soil. The optimum energy value and optimum water content are of great concern about soil unit weight, soil strength characteristics, and soil energy compaction.

Results on Jerash cohesive soil show that as the compaction energy increases, water content decreases, and the unit weight increases. The effect of compaction on unit weight of Jerash cohesive soil is very large as the energy varies from 400 to 1400 KJ/m^3, and the effect after that level is very small, which means that any further compaction of Jerash cohesive soil is not wise. In fine, optimal compaction energy ranges from 1200 to 1400 KJ/m^3, where 50 to 60 blows can be delivered, and optimal water content is between 14% - 15%.

References

[1]   SSSA (1996) Glossary of Soil Science Terms. Soil Science Society of America, Madison.

[2]   Batey, T. (2009) Soil Compaction and Soil Management—A Review. *Soil Use and Manage*, **25**, 335-345. http://dx.doi.org/10.1111/j.1475-2743.2009.00236.x

[3]   Ohu, J., Folorunso, O., Adeniji, F. and Raghavan, G. (1989) Critical Moisture Content as an Index of Compactibility of Agricultural Soils in Borno State of Nigeria. *Soil Technology*, **2**, 211-219. http://dx.doi.org/10.1016/0933-3630(89)90007-X

[4]   Taylor, H.M. (1971) Effects of Soil Strength on Seedling Emergence, Root Growth and Crop Yield. In: Barnes, K.K., Carleton, W.M., Taylor, H.M., Throckmorton, R.I. and van den Berg, G.E., Eds., *Compaction of Agricultural Soils*, American Society of Agricultural Engineers, St. Joseph, 292-305.

[5]   Hamza, M.A. and Anderson, W.K. (2005) Soil Compaction in Cropping Systems. A Review of the Nature, Causes and Possible Solutions. *Soil and Tillage Research*, **82**, 121-145. http://dx.doi.org/10.1016/j.still.2004.08.009

[6]   Horn, R., Doma, H., Sowiska-Jurkiewicz, A. and van Ouwerkerk, C. (1995) Soil Compaction Processes and Their Effects on the Structure of Arable Soils and the Environment. *Soil and Tillage Research*, **35**, 23-36. http://dx.doi.org/10.1016/0167-1987(95)00479-C

[7]   Mosaddeghi, M., Hajabbasi, M., Hemmat, A. and Afyuni, M. (2000) Soil Compactibility as Affected by Soil Moisture Content and Farmyard Manure in Central Iran. *Soil and Tillage Research*, **55**, 87-97. http://dx.doi.org/10.1016/S0167-1987(00)00102-1

[8]   Smith, C.W., Johnston, M.A. and Lorentz, S. (1997) Assessing the Compaction Susceptibility of South African Forestry Soils. I. The Effect of Soil Type, Water Content and Applied Pressure on Uni-Axial Compaction. *Soil and Tillage Research*, **41**, 53-73. http://dx.doi.org/10.1016/S0167-1987(96)01084-7

[9]   Wakindiki, I. and Ben-Hur, M. (2002) Soil Mineralogy and Texture Effects on Crust Micromorphology, Infiltration, and Erosion. *Soil Science Society of America Journal*, **66**, 897-905. http://dx.doi.org/10.2136/sssaj2002.8970

[10]   Gysi, M., Ott, A. and Flühler, H. (1999) Influence of Single Passes with High Wheel Load on a Structured, Unploughed Sandy Loam Soil. *Soil and Tillage Research*, **52**, 141-151. http://dx.doi.org/10.1016/S0167-1987(99)00066-5

[11]   Panayiotopoulos, K.P., Papadopoulou, C.P. and Hatjiioannidou, A. (1994) Compaction and Penetration Resistance of an Alfisol and Entisol and Their Influence on Root Growth of Maize Seedlings. *Soil and Tillage Research*, **31**, 323-337. http://dx.doi.org/10.1016/0167-1987(94)90039-6

[12]   Håkansson, I. and Lipiec, J. (2000) A Review of the Usefulness of Relative Bulk Density Values in Studies of Soil Structure and Compaction. *Soil and Tillage Research*, **53**, 71-85. http://dx.doi.org/10.1016/S0167-1987(99)00095-1

2

Energy Integration in South America Region and the Energy Sustainability of the Nations

Miguel Edgar Morales Udaeta, Antonio Gomes dos Reis, José Aquiles Baesso Grimoni, Antonio Celso de Abreu Junior

GEPEA/EPUSP, Energy Group of the Department of the Electrical Energy and Automation Engineering/ Polytechnic School of the University of São Paulo, São Paulo, Brazil
Email: udaeta@pea.usp.br

Abstract

The objective of this manuscript is to analyze relation involving the energy sector and socioeconomic growth and, then, contextualize the process of energy integration within the development policies in South America. The methodology considers data related to the world's economy and energy consumption and energy integration policy in countries and regions; and, South America's energy potential and the energy integration process. Results show that despite the political and institutional difficulties involving the process, energy integration can bring a lot of benefits for countries development. The process of energy integration in South America is divided in three moments, but in both periods the transnational energy projects were restricted, mostly, by a bilateral plan and the creation of physical links in a region. In the 21th century's context, it should be noted Brazil's participation which has been consolidated as a lead country in this process, and, also the IIRSA (Initiative for the Integration of Regional Infrastructure in South America, nowadays renamed as COSIPLAN) like the main initiative in energy integration in the continent, in a context where the projects are no longer limited to traditional economic blocs. Finally, we note a lack of consensus in defining a comprehensive model of integration and solving asymmetries both within countries and between them.

Keywords

Energy Integration, Energy Planning, Energy Resources, Regional Geo-Energy, South America, Energy Policy, Development

1. Introduction

Despite the economic integration process arising in Europe (including the energy integration), the related discussions soon spread worldwide, leading initiatives in other regions, including South America. From the second half of the 20th century, some economic integration mechanisms have been developed in the South-American region such as the creation of the Andean Community of Nations (CAN), the Southern Common Market (Mercosur) and the Union of South American Nations (UNASUR), plus some bilateral initiatives geared to the use of shared energy resources or trade them.

In this last, century has noticed a significant increase in the number of energy projects in South America, largely associated with the Initiative for the Integration of Regional Infrastructure in South America (IIRSA), the resulting economic growth in the region and, thus, the increase of demand energy. Indeed, studies of the International Energy Agency [1] and the World Energy Council [2], show that the energy demand of developing countries has increased significantly—due to the considerable growth of their economies—a phenomenon which also includes America South and specifically Brazil, which in 2011 occupied the 6th position in the world ranking of Gross Domestic Product (GDP) [3] and was seventh country to consume more energy in the world [4].

Energy and Development

The availability of energy is necessary for human development throughout history base, so that the use of different energy sources is the thread of history man stuff having made possible the two major changes in their relationship with nature: Neolithic Revolution and the Industrial Revolution [5].

In fact, when looking at **Figure 1**, below, one sees that the energy needs of society monitor the evolution and development of mankind. From an energy consumption of about 2000 kcal per day, which characterized the primitive man, energy consumption increased by 1 million years to 230,000 kcal per day, taking into account the consumption pattern of the so-called technological man [6].

The invention of the steam engine (a mark of the Industrial Revolution, which started in the 18th century in England) created the technical basis for the development of new forms and sources of exploitation and use of energy, replacing human labor with machines and subsequently developing means of transport [7]. By enabling the large-scale production and fast shipping of goods, the Industrial Revolution spurred the development of capitalism, specifically the economic development of industrialized nations. Moreover, this process resulted in the gradual increase in world energy demand (especially in developed countries) and, thus, for new sources and ways to harness and convert energy. This historical process shows a dialectical relationship between energy and development, in which the ability to use and power is at the same time, therefore the level of technical and eco-

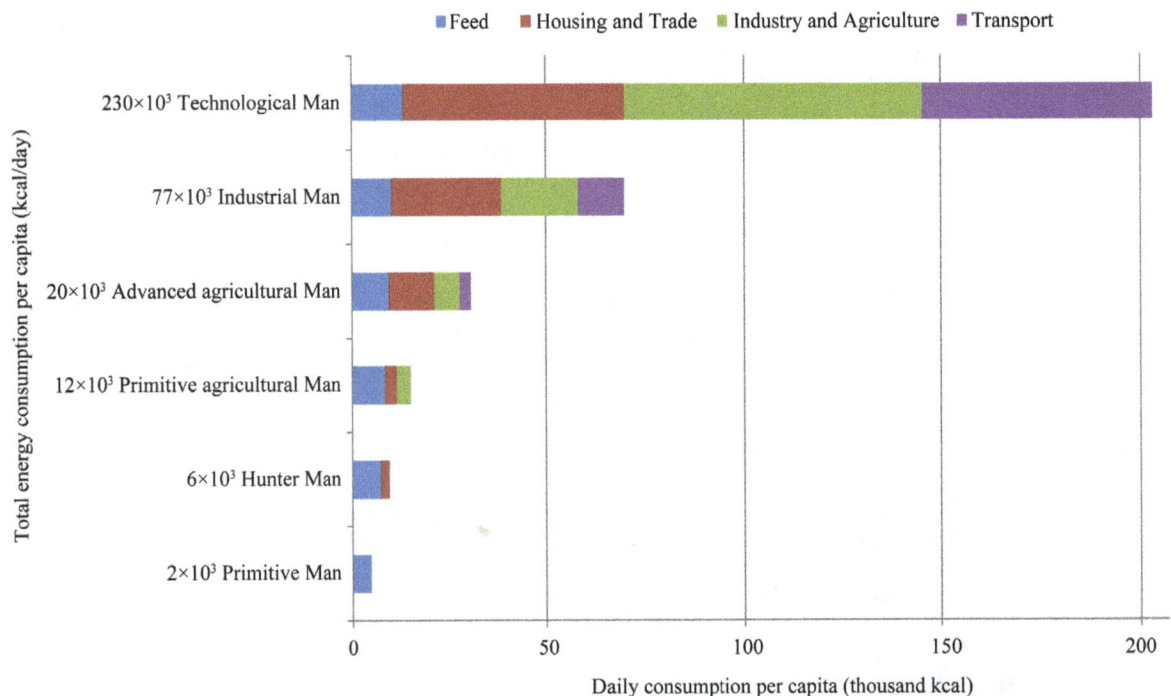

Figure 1. Evolution of human energy consumption troughout history (based on [6]).

nomic development of a society, but also the catalyst of this same development.

The relationship between energy and development is evidenced in **Table 1** and **Table 2**.

Indeed, by analyzing countries' data, related to GDP and energy consumption, we notice that countries that have higher energy consumption tend to be those with higher GDP. Of the 10 countries with the highest GDP in the world, only Italy and the UK do not appear among the 10 largest energy consumers—showing the relationship between economic development and energy consumption.

However, this relation between GDP and energy consumption, measured by calculating energy intensity—where it calculates the energy required to produce one unit of GDP (in Dollar case)—has been changing in recent times. Thus, OECD countries show a GDP growth without corresponding to a proportional increase in energy consumption, reducing their energy intensities. According to the World Energy Council (2004), this phenomenon is due to energy efficiency policies adopted in these countries, as well as changes in their economic structures or the energy matrices [2].

On the other hand, for non-OECD countries energy demand has increased considerably as a consequence of population, economic, urban and industrial growth. According to studies by the IEA (2010) [1], between the years 2008 and 2035 energy consumption in non-OECD countries is expected to increase 64%, while in OECD the growth is only 3%. Thus, it is expected that the share of developed countries in world energy consumption countries, which declined from 61% in 1973 to 44% in 2008, decrease to only 33% in 2035.

The same studies show that the OECD countries have, on average, an annual growth of primary energy demand of 0.1% (and in some cases, like Japan, rates are negative), while China and India have an annual increase of over 2.0%. These two countries are those that contribute most to the increase in global energy consumption. According to the IEA studies, the increase in primary energy demand in China in the period 2008-2035 is expected to be 75%, corresponding to 36% of global growth and resulting, therefore, in 22% of the total demand

Table 1. Ranking of the energy comsumption by countries (2011) [4].

Country	Consumption (mtoe)
1° China	2648
2° USA	2225
3° India	759
4° Russia	725
5° Japan	469
6° Germany	317
7° Brazil	268
8° Canada	266
9° South Korea	257
10° France	257

Table 2. Ranking of global GDP (2011) [3].

Country	GDP (millions USD)
1° USA	2648
2° China	2225
3° Japan	759
4° Germany	725
5° France	469
6° Brazil	317
7° United Kingdom	268
8° Italy	266
9° Russia	257
10° Canada	257

on the planet. India is the second largest contributor to the increase of energy consumption, accounting for 18% of it and having the highest annual growth rate (over 3.0%). Not coincidentally, these two countries are those whose studies point to have the highest rates of growth of GDP (forecast 5.7% and 6.4%, respectively). Brazil is expected to be the third country in the world regarding the growth rates of GDP and energy demand in the period 2008-2035 settling with one of the largest economies in the world and configured as one of the largest energy consumers in the world.

Still, besides being associated with the issue of economic development, we must not forget the social function involving energy production, since the primary purpose of energy services is to meet the needs of human beings, which means production and consumption, means to achieve it [8]. Thus, countries with higher HDI (Human Development Index) tend to be those with higher energy consumption [9], with an income factor that most influences this consumption (among the indicators that make up the HDI) [6]. Still, it must be emphasized that the low power consumption is not the only cause of poverty and development, however, a good indicator of its causes—as, for example, low levels of education and poor health systems.

Once established the relationship between energy and human development, it is important to highlight that the use of energy, by being associated with the technical development and the appropriation of space, notably affects the spatial configuration and, therefore, the environment, resulting in socio-cultural changes and significant demographic. Thus, projects involving energy production should be considered in all stages of the production chain, the social impacts of these types of development [9]. Otherwise, it will be creating a great contradiction, since above all, energy production is primarily intended to provide the satisfaction of human needs.

2. Essentials of Energy Integration

In the current context of increasing demand for natural resources, and the prospect of depletion of many of these, energy planning becomes very important and includes the research and development of alternative sources of energy production technologies that are associated with renewable resources and causing an environmental impact and minimum share.

As noted in the introduction, the energy planning involves access to energy resources. Due to the fact that they are unevenly distributed across the planet, access to them is the subject of disputes filed by various interests, being a matter of great geopolitical importance to a state.

Discussing the dependence of developed for natural resources located in other regions of the world, Hobsbawm (2007) [10] points out that one of the motivations of European imperialism over other regions of the world, occurred between the years 1875 and 1914 was the technological development that generated the need for "raw materials, due to weather or geological chance would be found exclusively or deep in remote places" (p. 96) supply. In many cases, even having plenty of resources in their territories, developed countries seek to exploit them in other regions. A clear example would be the American experience with oil after the Second World War, when the country encouraged their companies to exploit the oil fields of the Middle East, preserving located in its territory, economic, national security strategy [11]. Harvey (2011) [12] discusses the US interests in the Middle East associating with the fact that the region has the largest oil reserves in the world. Thus, by controlling the region, the United States also controls the access to this feature and with it, the global oil Market [12].

Several other authors discuss the relationship between energy security and military action in countries outside their national territories. Triola (2008) [13], an employee of the US Navy, maintains Harvey's assertions (2011) by arguing that the energy supply of the United States is a matter of national security and, therefore, also involves military interests. Nagy (2009) [14], in turn, suggests the militarization of energy security as a responsibility of the Organization of the North Atlantic Treaty (NATO), leaving her secure supply of energy resources.

These kinds of assertions show us that the access to energy resources involves different kinds of interests between countries that can be conflicting. Thus, as a reasonable alternative, it is considered that policies aimed at energy integration can meet harmoniously interests involved. The central idea of the energy integration is noted the contribution that economic and energy sectors in each country can the social and economic development process within the framework of regional integration [9] [15]. By enabling the commercialization of energy resources, or electricity itself, based on multilateral agreements, energy integration can provide a more reliable and efficient supply to large consumers of energy, also bringing economic gains for countries that sell their energy resources and its surplus electricity. In the long term, is optimized energy production, while taking advantage of the diversity resulting from connection to energy sources from neighboring countries, eliminating the dependence on a single source of energy and reducing supply costs. Also, the creation of economic blocs and energy strengthens the integrated region, leveraging the commercial, political, social and cultural relations between its members.

Moreover, despite the potential benefits related to cross-border energy integration, there are many elements that hinder their achievement, they order being political, technical, economic and environmental.

One of the main difficulties associated with the implementation of integration projects refers to the articulation of rules and congruent with the stimulus to investment and energy interdependence policies. It involves a number of agreements, targets and regulations that involve complex legal issues facing opening markets and thereby enabling the creation of rules to facilitate transactions and equity investments (state, private and national is required private multinational). This process involves the countries internal political issues—related to the approval and acceptance of laws and internal projects involving diverse interests within the nation—as well as elements associated with the foreign policy of each state and its geopolitical interests in the region.

With regard to differences of interests among countries in the South American case, one can use the question as an example of Bolivian gas, in which Bolivia nationalized refineries belonging to Petrobras, claiming that the contracts had been established the wounded interests of the Bolivian nation. Another example relates to historical differences between Chile and Bolivia, involving Bolivia access to the Pacific Ocean as a barrier to agreements between the two countries [16].

Obviously, the larger the number of agents involved in the process, the greater the difficulty in establishing policies of interest to everyone. That's why the most successful experiments were those made them bilaterally arising from projects with strong participation of national states.

From a technical standpoint, the interconnections require an infrastructure with bi-reaching goals—or multi—that includes the participation of all involved and interested. So that the integration process is done in a cohesive manner, it is essential to studies that provide adequate planning be made—with regard to the generation, transmission and distribution of energy as well as the interests and economic returns for the various agents involved in the issue. The greater the need for infrastructure and technical complexity related to the projects become more expensive the same—which implies the need for large investments of money (and, most often, in various financing). In the case of South America, for example, infrastructure integration projects to sizable proportions by both distances, as the natural difficulties imposed by the environment.

Resourcefulness along the World of Energy Integration

A) European Union

Throughout the twentieth century a number of policy initiatives and energy integration have been deployed worldwide, the most successful being developed within the European Union—that is, in a larger context of economic and political integration. Unlike what happened historically in most cases, the European experience has been guided by multilateralism and the creation of supranational regulators.

The first step in this process occurred in 1951, with the signing of the ECSC Treaty, which established the creation of the European Coal and Steel in a context in which the countries of the continent sought to economically rebuild the region after the Second World War. With this treaty, signed by France, Germany, Italy, Netherlands, Belgium and Luxembourg, we sought to integrate the Franco-German production of coal and steel—raw materials essential to industrial activity and the local economy at the time—through the creation of a common market aimed at economic development, job creation and improved quality of life.

In 1957, were signed the Treaties of Rome establishing the European Economic Community (EEC) and the European Atomic Energy Community (Euratom), establishing the creation of a common market on the continent and recognizing the importance and need to develop common energy policies the member countries in the context of regional economic and social development. To overcome the uncertainties related to traditional energy sources, the Member States of Euratom sought on nuclear energy a means to ensure energy security and independence. Thus, according to documents of the European Union itself (2013), "as the cost of investing in energy beyond the means of individual States, the founding members joined together to form the Euratom".

Throughout the following decades, the process of integration into the European continent was deepening, also encompassing the energy sector—seen as crucial to regional socioeconomic development. The main frame of this integration initiative came in 1992 with the Maastricht Treaty, which created the European Union and in which it commits to the creation and development of trans-European networks in the sectors of transport infrastructure, telecommunications and energy. Thus, it is for the authority to "promote the interconnection and interoperability of national networks as well as access to such networks" through the actions of its supranational political bodies [17].

From the descriptions above, it can be seen that the integration of the energy sector is part of European policies since the mid-twentieth century, are subordinate, in turn, to the initiatives of economic and political integration—and therefore cannot be analyzed outside this context. Similarly, we note that, over time, these policies are no longer focused on specific energy sources (coal and nuclear), for, after the Maastricht Treaty, extended to the whole European energy system and further increase use of sources clean and renewable energy—increasingly promoted by the European Union's energy policy, with the aim of reducing emissions of greenhouse gases on the continent—and also the integration into the natural gas supply.

The strategy of integration of renewable sources resulted in a less centralized and diversified system, strengthening the European integrated network. The Policy of 2009 set by the European Parliament and the Council of the European Union, concerning common rules for the internal European market and the unequal terms of trade of electricity between the member states must be overcome by the right of free choice suppliers reassured consumers.

The transmission of electricity on the continent is through the ENTSO-E network, established in July 2009, according to Policy 2009 and composed of 42 operators in 34 countries, with 305.000 km of transmission and 828 GW of generation, to supply and demand of 3400 TWh/year, serving more than 525 million citizens. The purpose of this initiative was to integrate the different operators of the system to European legislation, promoting development through reliable operation, technical and administrative support, and security in meeting the demand of the system. In this context, the intelligent transmission networks (smart grids) send electricity from points of generation to consumers using a monitoring system with digital technology, allowing the integrated use of decentralized energy sources—like solar and wind—by assimilating its entry in periods of wind and sun.

Naturally, the process of integration of the European energy sector also faces obstacles, but in the context of this work concerns us understand how is the process of multilateralism and integration with respect to interconnections and to political and economic technical aspects involved—so compared with other experiences around the world.

B) Africa

Apart from the European experience, energy integration initiatives were implemented in other continents, so that, in most cases, they gave bilaterally and were much less extensive than in the European Union.

In Sub-Saharan Africa, for example, the first cross-border electricity interconnections were deployed around large hydroelectric projects. The first interconnection was built in 1958, with the transmission line of 132 kV connecting the hydroelectric plant of Owens Falls, in Uganda, the capital of Kenya (Nairobi). Built in the Democratic Republic of Congo and interconnected with neighboring countries such as Zambia, Congo and others—later binational hydroelectric, as the Kariba North (on the border between Zambia and Zimbabwe) and the development of the Inga hydroelectric complex fall were built the southern region of the continent [18]. Parts of such linkages were incorporated into the Southern African Power Pool (SAAP), one regional system operating on the continent until 2008, when it began operating the West African Power Pool. A Power Pool is an interconnected system obtained by joining two or more interconnected electric systems—managed as if it were a system—by reallocating the demand and supply of energy and generation capacity, in order to operate more efficiently and secure.

The SAPP was created in August 1995 as an association of electric power enterprises vertically integrated, representing the 12 nations of the Commonwealth of the Southern African Development Community (SADC, for its acronym in English): Angola, Botswana, DRC, Lesotho, Malawi, Mozambique, Namibia, South Africa, Swaziland, Tanzania, Zambia and Zimbabwe [19].

C) North America

Energy integration of North America shall be considered in the context of the Free Trade Agreement (NAFTA)—an agenda of neoliberal policies designed to achieve deeper levels of regulatory, institutional and policy integration that enable the integration of markets the continent [20]. But despite NAFTA establish a "trilateral landmark" for trade in energy and electricity [21] features the experiences of energy integration in the region are, paradoxically, based on bilateral relations, in which the United States appear as large buyer power and energy resources (such as oil and natural gas) from Mexico, and especially in Canada—energy trade between Mexico and Canada is almost non-existent [22].

In fact, 99% of all Canadian energy exports are destined for the United States [23], indicating the absence of trade in energy resources and electricity from Canada to Mexico. Even Mexico is a major oil supplier to the

United States, its energy market is poorly integrated into the US (compared to Canada)—due to political and economic and internal [22] issues.

With respect to the electric integration, specifically, it is less advanced than in the natural gas and oil industries. Even so, there is a trade of electricity in the region, made possible by interconnections linking the United States to Mexico, and especially in Canada. In this context, the United States are characterized as major importers of electricity, while also exporting (to a lesser extent) to its neighbors.

Finally, it is important to note that there are differing opinions regarding the benefits of the integration process in North America. While Doucet (2007) [22] argues that this process is extremely beneficial for Canada, as the United States are the major buyers, Campbell (2007) [20] believes that these relationships undermine the energy security of the country. To justify his point of view, this author takes as argument the fact that, by the rules of NAFTA, Canada can only reduce its exports of oil and gas to the United States in the same proportion that reduce their production to market internal, with this, the country loses the right to reduce its exports to the US, even aiming to prolong the durability of their internal reserves or reduce its imports. For Campbell (2007), the fact that Mexico to impose certain restrictions on the integration model imposed by the United States represents a defense of local energy security , and is therefore beneficial for the country.

3. Energy Integration in South America

3.1. South American Energetic Potential

Because it is a region rich in energy resources, especially oil and water resources, various integration projects in terms of energy already deployed or are in process in South America is worth mentioning that many of these features occur so border (making it complex to manage and operate), highlighting the Andes and the Amazon River.

The Andes is a large mountain range spanning five countries in South America—as well as being a natural border between them—and constitutes a major mineral area in the world: gold, silver, copper, zinc, nickel, iron granite among others. The Amazon River, its waters represent much of the available freshwater in the world. This river originates in Peru, crossing the border with Colombia until you reach the tri-border involving these two countries and Brazil. Not coincidentally, these three countries are those with the potential hydroelectric May in South America, as shown in **Figure 2**.

By observing the above data, it is seen that the total hydropower potential of the American South is 590 GW, so that Brazil is by far the country with the greatest potential, exceeding 250 GW. No wonder that this country is one that has the highest hydroelectric power generation on the continent (403 TWh in 2010), corresponding to

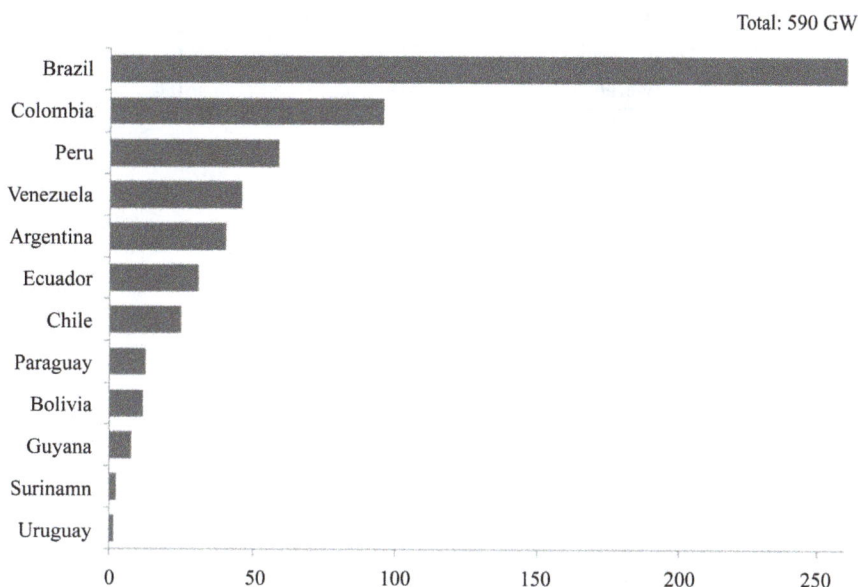

Figure 2. Hydroelectric potential in South America (2010).

11.5% of the world's hydroelectric production, according to **Table 3**. Likewise, it should be highlighted Venezuela, since this country occupies the 9th place ranking producer of hydroelectricity in the world and is the second largest South American producer.

Regarding hydrocarbons should be highlighted reserves of oil and natural gas on the continent. South America has about 4% of global natural gas reserves, so that the majority is located in Venezuela. **Figure 3** shows the distribution of proved reserves of natural gas in the region.

Oil cannot be overlooked when it comes to meeting the South American energy potential. The region has large proven reserves of this resource, highlighting, once again, to Venezuela. According to data from the US Energy Information Administration—an agency linked to the US government—in Venezuelan territory is the second largest proven oil reserves in the world (211.2, billion barrels of oil) smaller only than that of Saudi Arabia. Brazil also deserves mention, being in 15th place in the global ranking of countries with the largest oil reserves. **Figure 4** shows that, after Venezuela and Brazil, Ecuador has the third largest proven oil reserves in South America.

3.2. Regional Integration Process in South America

The process of South American energy integration starts from the mid-twentieth century, by means of natural gas, having two axes of the main action, oriented from the actions of the CAN (Andean Community of Nations)

Table 3. The biggest hydroelectricity producers in the world (IEA, 2010).

Producers	TWh	% of the global installed capacity
China	722	20.5
Brazil	403	11.5
Canada	352	10
United States	286	8.1
Russia	168	4.8
Norway	118	3.4
India	114	3.3
Japan	91	2.6
Venezuela	77	2.2
France	67	1.9
Other Countries	1118	31.7
World	**3516**	**100**

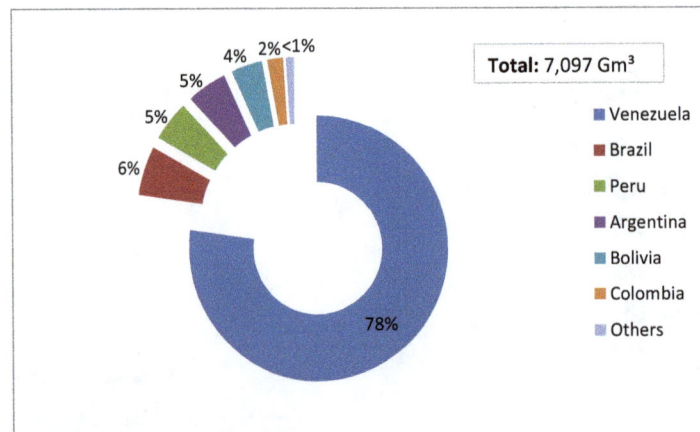

Figure 3. Regional distribution of proved reserves of natural gas in South America in 2010.

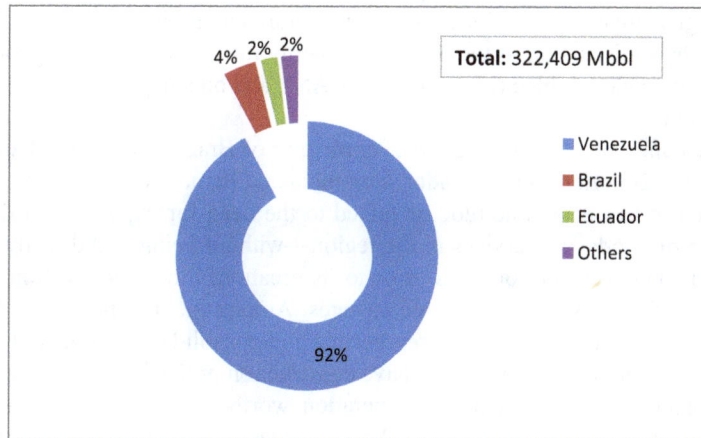

Figure 4. Distribution of proved oil reserves in South America (2010)[1].

and Mercosur [7]. It should be noted that these economic blocks have economic complementation agreements among themselves and with Chile—a country that is not a member of both.

The process of energy integration occurred in the twentieth century can be divided into two stages.

The first, between the early 1970s and the late 1980s, was marked by a great performance of National States in building binational projects. These must be highlighted, the Bolivia - Argentina, hydroelectric plants Salto Grande (Argentina and Uruguay), Itaipu (Brazil/Paraguay) and Yacyretá (Paraguay/Argentina), and transmission lines associated with these binational plants.

The period that occurred after the end of the 1980s was marked by economic and political reforms in neoliberal, resulting in decreased performance of states and the increased private participation in the South American country's economy. Thus, has begun the implementation of projects with varying degrees of participation of private, mixed and public enterprises—involving mainly the hydrocarbons sector. In this period many binational pipelines were constructed, demonstrating the importance of this resource within the energy integration projects in South America[2].

Despite the differences involving the form of state action, it is noticed that, in both periods mentioned above, projects were restricted to the bilateral framework, demonstrating the absence of a regional integration policy indeed.

The political and economic changes on the continent, in the 21th century turned, once again, the landscape of energy integration in the region. First, the election of heads of state of the Left parties made to gather strength in the region's anti-neoliberal and anti-imperialist discourse, changing the logic of financing projects in the region and strengthening the participation of States in the economies of the countries again. Meanwhile, economic growth achieved by regional countries, especially Brazil, resulted in increased energy demand of us. With this, the first decade of this century, there was a tendency to turn that vision bilaterally in order to give a more regionalist and multilateral integration projects to character. It was in this context that was created in 2000, the Initiative for the Integration of Regional Infrastructure in South America (IIRSA), in order to promote the interconnection of transport, telecommunications , energy, oil and gas pipelines.

With the construction of infrastructure related to IIRSA project, and with funding from financial institutions such as the Inter-American Development Bank (IDB), the World Bank, the Andean Development Corporation (CAF) and the Financial Fund for the Development of PrataBasin (FONPLATA), was intended to establish conditions conducive to the development of trade agreements and energy integration. Thus was deposited much confidence in the success of IIRSA, despite the logical difficulties inherent to their success—such as: history of binational conflicts; disparity markets; privilege certain actors, and difficulty of regulatory consistency between countries [9]. But in 2009, IIRSA has been discontinued and its projects were linked to the South American Council of Infrastructure and Planning (COSIPLAN)—linked to the Union of South American Nations (UNASUR). Also, COSIPLAN along with UNASUR, retains the idea of strengthening multilateral relations in South

---

[1]Source: Based on OLADE.

[2]In fact, natural gas is one of the vectors of energy integration on the continent, representing most of the energy matrix of some South American countries such as Argentina, Bolivia and Chile.

America, in order to give greater political support infrastructure integration projects.

The following is a brief discussion of the process of energy integration in South America, with a focus on initiatives both within the economic bloc (Mercosur and CAN), and outside them.

A) Mercosur Region

Mercosur is an economic bloc, created in 1991, consisting of Brazil, Argentina, Paraguay, Uruguay and Venezuela, whose members Bolivia, Chile, Ecuador, Colombia and Peru.

Despite the creation of this economic bloc be linked to the need for expansion of domestic markets and stimulate the circulation of goods and services in the region—without being tied directly to the energy issue—the binational relations among member countries prior to its creation. Since the creation of Mercosur, the energy sector of member countries showed considerable changes. Among them stand out the reform of the role of the state (acting more as a regulator than as an entrepreneur) and consolidation of natural gas as an integrating feature of the region—as all countries in the region have construction projects in the pipeline [9].

Among the transnational projects of power generation worth mentioning the construction of Itaipu, Salto Grande and Yacyretá (already mentioned above) plants, besides the Central Salta—a combined cycle power station, built by a Chilean company to generate electricity from gas natural from Argentina, but without providing energy for this country. We cannot forget that Chile is considered a poor country in terms of ownership of energy resources.

In the case of Itaipu (binational venture involving Brazil and Paraguay, aiming harness the hydroelectric potential of the Paraná River), Brazil was responsible for the setup project and investments for the construction of the plant—in addition to financing the part that would fit Paraguay [7]. In 2008, the plant secured supply 87.3% of all electricity consumed in Paraguay and 19.3% of the demand of the Brazilian Interconnected System. Importantly, Paraguay does not consume all of your energy generated, since it takes only 20% of what is produced to meet its domestic demand for energy, causing the remaining 80% are sold to Brazil.

B) Andean Community of Nations Region

The Andean Community of Nations (CAN) is formed by Peru, Bolivia, Ecuador and Colombia. In the past, Chile and Venezuela also integrated organization but abandoned (at different times) for political and economic mismatches. The region covered by the CAN has a great potential energy—both in terms of hydrocarbons (oil and gas) as regards the hydro, among others. With regard to the oil market, Venezuela, Colombia and Ecuador are configured as suppliers to countries like Brazil, Chile and Peru—since they have oil consumption that exceeds their production, in contrast with Bolivia and Venezuela that have resources that exceed their local demands. You need to highlight the situation of Chile, due to its lack of energy resources, is much interested in integration projects that enable the supply of its domestic energy demand.

According to Castro, Dassie and Delgado (2009) [24], the process of energy integration in the Andean region began in 1969 with the construction of the transmission line Zulia - La Fria, connecting Colombia and Venezuela. Although the energy exchange was not significant due to what they call security problems in the power supply, the authors argue that this project was the first step for the energy integration occurred in the region. The same authors state that the evolution of the process of electrical interconnection between the Andean countries has enabled advances, such as the prediction of building an interconnection between Bolivia and SIEPAC. Udaeta, Burani, Fagá and Oliva (2006) [9] also highlight the role of interconnections in the integration process in CAN, noting that Bolivia appears as a "hinge", because of its possibilities of interconnection with Brazil, Chile, Argentina and Peru.

Besides the physical integration, one cannot ignore the advances that have occurred in political and legal terms to enable the marketing and access to transmission networks between countries. Thanks to that, better expectations around integration projects in the region. Even so, it is of great political, economic and technical complexity to implementation of transnational projects—so that the more countries involved, and the larger the area covered, the greater the difficulty. Antunes (2007) [16] cites the proposal for energy integration taken by Chile in 2007, and that would involve this country along with Bolivia, Ecuador, Peru and Colombia, based on an Andean multilateral agreement aimed at "prevent interference geopolitical discussions on exit to the sea for Bolivia lost"(p. 5).

C) Other Energetic Integration Projects in South America

Obviously, one cannot reduce the process energy integration in South America only to economic blocs in the region. In this sense, the creation of IIRSA, and later the COSIPLAN makes integration projects that go beyond the limits of economic blocs, giving greater geopolitical cohesion to the region are encouraged. In the case of

IIRSA, a portfolio of projects divided into eight sub-regions was taken, clearly transcending the boundaries of economic blocs: the Hub; Capricorn Hub, Amazon Hub; Guianese Shield Hub, Southern Hub, Central Inter-oceanic Hub; MERCOSUR-Chile, Brazil and Peru-Bolivia.

Throughout its existence, the IIRSA delivered 524 projects, of which 451 (about 86%) belonging to the transport sector and whose investments were received from more than 55 billion dollars. The energy sector received 64 designs, but although they represent 12% of the total, his works surpassed the cost of 44 billion dollars (about 42% of the investment portfolio of IIRSA), due to the complexity and magnitude of the projects. Have the communications sector received only 9 projects, which together cost less than $50 million [25]. By analyzing these data, it is clear that the IIRSA projects favored the transport sector, to the detriment of integration of energy and communication sectors. Thus, we can say that IIRSA fostered a process that is configured much like physical interconnection, than actual integration [26]. This phenomenon is mainly related to building a road network connecting the Pacific and Atlantic interests, in order to facilitate the flow of goods across the continent.

IIRSA-related projects are targets for some criticism. Despite advances in regard to strengthening networks of transport, energy and communication sectors were left in the background. Gudynas [26] questions the interests that were behind the IIRSA project by privileging the physical interconnections (such as roads, waterways and pipelines), and seek not intrinsic to strengthen other aspects integration process—as political, productive and cultural ties. Other limitations are related to IIRSA the lack of progress in the harmonization of sectorial policies and relevant regulations and little consideration given to social and environmental aspects.

As regards the pursuit of multilateralism in the South American integration process, it should be noted the role of Bolivia, which has a role of coordination between the countries of the Southern Cone and the Andean Community. Being a country with large reserves of natural gas and has low power consumption; Bolivia is shaping up as a major supplier of this resource to countries that need to import it—like Chile and Brazil. You can use the example of the Bolivia-Brazil pipeline, named by the authors as a major milestone of the energy integration in South America, the importance of consolidating gas sector in the Brazilian energy matrix.

Energy integration also consolidates Brazil as a major buyer of energy, since, despite having large reserves of energy resources, the country has shown a significant increase in its demand. Therefore, the Brazilian government has increasingly sought to stimulate energy ventures outside its territory. However, in addition to seeking to satisfy its energy demand, there is a clear interest in encouraging the participation of Brazilian capital on projects through funding from its National Bank for Economic and Social Development (BNDES) and the participation of Brazilian construction companies civil.

In this context, the process of energy integration between Brazil and Peru should be highlighted. In 2009, was signed by both countries an agreement for the construction of six hydroelectric plants geared to supply the energy markets of both countries and located in the Amazon basin of Peru's territory. Due to the location where it will be deployed, covering areas of high biodiversity and where local communities, projects of this nature are carried live by controversy due to contradictions involving their economic gains and environmental impacts they generate. In fact, despite these initiatives have targeted to meet economic interests in the region, their social and environmental impacts are left in the background.

4. Some Conclusions Related Endogenous Energy-Integration

Despite the difficulties involving energy integration (as questions of sovereignty of countries and divergence of interests among stakeholders—be they citizens, companies, nation states or other agents), the process of energy integration can bring a lot of benefits for developing countries, since the production and consumption of energy are factors that directly influence the economic development process. Likewise, energy resources are not distributed evenly around the planet, so that trade in energy resources and electricity can benefit both the importing country and the exporting—depending on the way the procedure is conducted and interests involved.

Second, one cannot forget that the energy integration policies should be analyzed in a broader context of economic integration. Without economic integration in fact, the process of energy integration weakens or worse it become unviable, since it depends on complex and coherent political, legal and economic foundations established by supranational institutions respected by all. Often, historical political conflicts between countries are an obstacle to the process of economic and political integration, undermining thereby the energy integration. The South American case shows that, a context in which the process of economic integration is a little-advanced stage makes energy integration slow down.

By analyzing the process of historical energy integration in South America, it appears that this has gone through several changes over the past decades—following the political and economic changes occurring in the continent. In the 21th century, for example, strengthening the anti-imperialist and anti-neoliberal discourse has resulted in increased participation of states and their companies in making decisions and profits relating to the exploitation of energy resources, and reducing reliance on funding coming from the IDB and the World Bank—that guided the economic policies in the region. Also, the economic growth of countries resulted in an increase in their energy demands, requiring new developments in the energy sector. It should be noted Brazil, which has been consolidated as a lead in this process—stimulating new projects and the participation of the Brazilian capital, through BNDES financing and the involvement of construction companies—due to their economic power—and thereby politician in the region—as well as their energy requirements, which have been increasing in recent times.

Comment is also likely the fact that transnational energy projects are restricted, mostly, to bilaterally. This shows that the process of energy integration in South America was not an end in itself, but a means to meet the energy and economic needs of certain countries and agents in certain recent historical periods. In other words, projects like Itaipu and Bolivia Brazil Gas Pipeline had main objective to serve the economic interests of the parties involved—and not energy integration itself. Even in more recent initiatives, such as IIRSA, the integration of the energy sectors of the South American countries occurs in a larger plan: the integration of infrastructure (which also involves networks of movement and communication)—is therefore a means to achieve greater integration project. Even so, IIRSA—and now COSIPLAN—is the main initiative of energy integration in South America in the beginning of the 21th century, in a context where projects happen beyond the traditional economic blocs.

Moreover, the integration process is not limited to the creation of physical links in a region. It requires a series of policies and regulations to facilitate harmonic different types of flows inherent in this process as well as the narrowing of political, productive and cultural ties. In this sense, the results were still unsatisfactory, since we focused on the transportation industry related projects, at the expense of others—that cater to specific economic interests, but they do not contribute to reducing social and economic inequalities present within the southern countries Americans, and the continent in general. So far it needs to advance in South America—if we want a restricted integration only to facilitating the flow of goods across the continent

Finally, the various changes occurring in the energy integration process in the region show a lack of consensus in defining a comprehensive model of integration and satisfying interests of all States and involved actors in the region. Something that is directly related to existing internal asymmetries within countries, but also is between them. Thus, for this to be overcome, it is necessary that these asymmetries are resolved both within countries and between them.

References

[1]   Agência Internacional de Energia (IEA) (2010) World Energy Outlook 2010. Paris.

[2]   World Energy Council (WEC) - Comitê Brasileiro (2004) Eficiência Energética: Uma análise mundial. Rio de Janeiro. Disponível no www.worldenergy.org

[3]   Fundo Monetário Internacional (FMI) (2012) Dados disponíveis no www.imf.org

[4]   ENERDATA (2012) Dados disponíveis no www.enerdata.net

[5]   Cipolla, C.M. (1961) Sources d'énergieethistoire de l'humanité. In: *Annales. Économies, Sociétés, Civilisations.* 16e année, N. 3, 521-534.

[6]   Goldemberg, J. and Lucon, O. (2008) Energia, Meio Ambiente e Desenvolvimento. São Paulo, Edusp.

[7]   D'ÁVALOS, Victorio Enrique Oxília. *Raízes Socioeconômicas da Integração Energética na América do Sul: análise dos projetos Itaipu Binacional, Gasbol e Gasandes.* Tese de Doutorado. PPGE-USP. São Paulo, 2009.

[8]   Udaeta, M.E.M. (1997) Planejamento Integrado de Recursos Energéticos para o Setor Elétrico—PIR (Pensando o Desenvolvimento Sustentável). Epusp, tese de doutorado, São Paulo.

[9]   Udaeta, M.E.M., Burani, G.F., Fagá, M.T.W. and Oliva, C.R.R. (2006) Ponderação analítica para da integração energética na América do Sul. *Revista Brasileira de Energia,* **12**.

[10] Hobsbawm, E. (2007) A era dos impérios 1875-1914. Ed. Paz e Terra, São Paulo.

[11] Scarlato, F.C. (2008) O Espaço Industrial Brasileiro: Sociedade, Industrialização e Regionalização do Brasil. In: Ross, J.L.S., Org., *Geografia do Brasil*, 5th edição, Edusp, São Paulo.

[12] Harvey, D. (2011) O novo imperialismo. Edições Loyola, São Paulo.

[13] Triola, L.C. (2008) Energy and National Security: An Exploration of Threats, Solutions and Alternative Futures. *IEEE Energy* 2030, Atlanta, 17-18 November 2008, 13-35.

[14] Nagy, K. (2009) The Additional Benefits of Setting up an Energy Security Centre. *Energy*, **34**, 1715-1720.

[15] Suárez, L.P.L. (2006) O Papel das Petrolíferas para o Desenvolvimento da Integração Energética: A formação do Mercado de Gás Natural na América do Sul. Unicamp, dissertação de mestrado, Campinas.

[16] Antunes, J.C.A. (2007) Infraestrutura na América do Sul: Situação atual, necessidades e complementaridades possíveis com o Brasil. Comissão Econômica para a América Latina e o Caribe (CEPAL).

[17] Tratado De Maastricht. União Europeia, 1992.

[18] WEC—World Energy Council (2005) Regional Energy Integration in Africa. Londres.

[19] Southern African Power Pool (SAAP), 2013. http://www.sapp.co.zw/

[20] Campbell, B. (2007) Una perspectiva nacional de la integración continental del sector canadiense del petróleo y el gás. In: Vargas, R. and Ugalde, J.L.V., Org., *Dos modelos de Integración Energética—América del Norte/América del Sur*, CISAN, Ciudad de Mexico.

[21] Márquez, D. (2007) El TLCAN plus: La homologación de estándares y sus implicaciones legales para México. In: Vargas, R. and Ugalde, J.L.V., Org., *Dos modelos de Integración Energética—América del Norte/América del Sur*, CISAN, Ciudad de Mexico.

[22] Doucet, J. (2007) La integración energética norteamericana. Una perspectiva canadiense. In: Vargas, R. and Ugalde, J.L.V., Org., *Dos modelos de Integración Energética—América del Norte/América del Sur*, CISAN, Ciudad de Mexico.

[23] Us Energy Information Administration, 2012. http://www.eia.gov/countries/analysisbriefs/Canada/canada.pdf

[24] Castro, N.J., Dassie, A.M. and Delgado, D. (2009) Indicadores Mundiais do Setor Elétrico—As Experiências Latino-Americanas de Integração Energética. GESEL/UFRJ, Rio de Janeiro.

[25] IIRSA (2010) IIRSA diez años después: Sus logros y desafíos. BID-INTAL, Buenos Aires.

[26] Gudynas, E. (2008) As instituições financeiras e a integração na América do Sul. In: Verdum, R., Org., *Financiamento e megaprojetos*: *Uma interpretação da dinâmica regional Sul-Americana*, Inesc, Brasília.

3

Solar Thermal Energy Generation Potential in Gujarat and Tamil Nadu States, India

C. Nagarjuna Reddy*, T. Harinarayana

Gujarat Energy Research and Management Institute, Gandhinagar, India
Email: *naga2980@outlook.in

Abstract

Government of India has come out with an ambitious target of 100 GW of using solar energy alone by the year 2022. To reach this target, innovative ideas are required to use the solar energy more effectively. For solar electricity generation, mainly two types of technologies are presently in use, namely, solar PV and solar thermal. Being a tropical country, India has large solar PV and solar thermal energy. More research is required on economic aspects to make the solar thermal competitive to solar PV. Towards this direction, in our present study we have simulated a solar thermal power plant using Parabolic Trough Collector (PTC) technology and normalized with 1 MW solar thermal power plant at Gurgaon near New Delhi. Through simulation, we have extended our study and computed the electricity generation possible at different locations of India. For this purpose with $1° \times 1°$ spacing, computations have been carried out at 296 locations. The work is further extended for more detailed study at two representative states, namely, Gujarat and Tamil Nadu. In these two states, closer data points with $0.25° \times 0.25°$ spacing have been considered at 273 locations for Gujarat and 197 locations for Tamil Nadu. Our results indicate a large potential of electricity generation using solar thermal energy in southern states of India, namely, Tamil Nadu, Karnataka, Kerala, southern and western part of Andhra Pradesh and eastern part of Maharashtra. Good potential has also been observed in eastern parts of Gujarat and parts of Madhya Pradesh and eastern part of Rajasthan. The annual potential ranges from 1800 MWh to as much as 2600 MWh. Major parts of northern states, for example Uttar Pradesh, Bihar, West Bengal, Punjab, Jammu and Kashmir have medium range potential. Here, the annual potential ranges from 1000 to 1500 MWh. Poor range of potential is observed towards eastern parts of India and north eastern states. Here, the electricity generation potential ranges from 600 to 1200 MWh. Our results are useful to solar thermal developer and decision managers.

Keywords

Solar Thermal, Parabolic Trough, Energy, India, Gujarat, Tamil Nadu, Modelling

---

*Corresponding author.

1. Introduction

Water and energy can be considered as two eyes of any nation. If the Government provides or facilitates these two items adequately, the nation grows. It helps the economy to grow. It facilitates the people to live comfortably and happily. Many workers have provided innovative ways and suggested new methodologies for energy security [1]-[5]. Clean water and clean energy make all the people in this world healthy and help to live longer with fewer problems. However, many nations in this world are starving for both clean water and the energy. For example, majority of the islands like Papua New Guinea, Fiji islands, parts of Andaman, Nicobar and Lakshadweep in India are suffering from acute shortage of electricity. At these locations, diesel generators are being used for electricity generation. For this purpose, diesel oil is transported from nearby main land to these islands [6]. In view of this, the cost per unit electricity generation is nearly double as compared to the persons living on the main land.

It is well known that India is facing acute power shortage. The present electrical power generation is close to 250 GW. In the estimation, India is expected to generate about 1300 GW by the year 2034 [7]. This is nearly more than 5 times the present generation value. In order to reach this target, all types of fuels need to be used. In present day scenario, coal is being used as a major source for electricity generation in India, as compared to other fuels. Due to increase in demand for more power, sufficient quantity of coal could not be supplied to the thermal plants. For this purpose, we are now importing the coal to reach the power demand. It is well known that coal is not the preferred source of energy due to various issues related to pollution as it omits large CO_2 in the atmosphere and creates climate change problem. For all these above reasons, renewable energy is being projected and promoted by majority of the nations around the world.

In this direction, India is ambitiously planning to generate 100 GW of electricity from solar energy alone by the year 2022. We need to have clear roadmap to reach this target from the present generation of nearly 2 GW of installed capacity. There are mainly two types of technologies presently known to generate electricity using solar energy, namely, solar PV and solar thermal. Due to economic considerations, solar PV has become more popular and widely being used as compared to Solar thermal. However if the number of solar developers using Solar thermal technology increases, the cost of Solar thermal technology is likely to be competitive to solar PV. It is reported that among the available solar thermal technology of Parabolic Trough Collector (PTC), Linear Fresnel Reflector (LFR), Solar tower and Dish sterling, Parabolic Trough Collector is more popular and is being used at many locations around the world [8] [9].

Being a tropical country with 300 sunny days, apart from solar PV solar thermal technology, India also needs to be promoted. With more innovations in solar thermal technology and with "Make in India" campaign, cost may be likely to come down and affordable. Ministry of New and Renewable Energy (MNRE), government of India is carefully experimenting the solar thermal technology with pilot scale projects. With its full support, IIT-Bombay has installed 1 MW power plant in Gurgaon. Government of India also supports private players to initiate the electrical generation using solar thermal technology in Rajasthan, Gujarat and also in Andhra Pradesh. In such a scenario, it is of utmost importance to have an estimation of solar thermal potential for the whole country. Presently majority of the solar developers are using the solar irradiation maps. While such maps are necessary and helpful, they are not sufficient. This fact is realized that solar PV potential for India on a regional scale and more detailed maps for Gujarat, Andhra Pradesh and Telangana states has been prepared recently [7]. In a similar manner, it is of great help for the solar developers, if a solar thermal energy generation potential map for India and different states is made available. Towards this direction, an attempt has been made in our present study to estimate the solar thermal potential by simulating the PTC technology similar to the installed solar thermal plant near Delhi. For this purpose, simulation studies have been initiated and results have been compared with the existing power plant of Delhi. With the same initial parameters, computations have been extended to different locations of India. Additionally for Gujarat and Tamil Nadu states are chosen with closer station interval to prepare solar thermal energy generation potential maps for different months and also on annual basis. The methodology followed and results obtained are discussed in the following section.

2. Methodology

2.1. Design of Solar Thermal Power Plant

The solar thermal power plant installed at Gurgaon near Delhi by IIT Bombay is 1 MW capacity (**Figure 1**). In

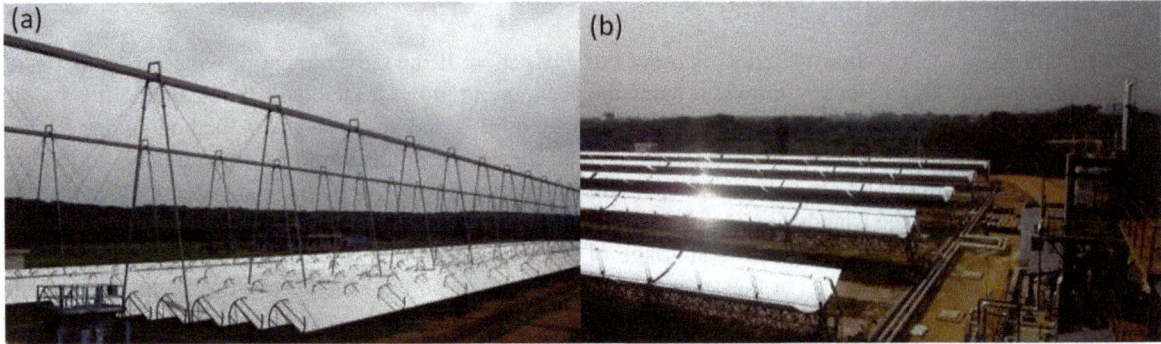

Figure 1. Photo showing the 1 MW solar thermal power plant installed at Gurgaon, New Delhi. The plant uses two types of technologies. The linear Fresnel reflector technology is shown in (a) and the parabolic trough collector technology is present in (b).

this power plant mainly two types of technologies have been used, namely, Linear Fresnel Reflector and Parabolic Trough Collector with therminol VP1 oil as fluid. The plant at Gurgaon has been studied in detail by Desai *et al.* [10]. In their study, they have simulated both Linear Fresnel field and Parabolic Trough field. The power output for different months except January, July, August and December are provided. In these months, it is reported that due to low DNI values of radiation, the plant remain non-operational. The simplified flow diagram of Gurgaon 1 MW solar thermal power plant can be seen in **Figure 1** of Desai *et al.* (2013).

The Linear Fresnel Reflector (LFR) technology (**Figure 1(a)**) contains a fixed receiver pipe while mirrors track the solar radiation. These mirrors usually consist of multiple small mirror facets. Each mirror is thin, low weight and low volume lens with short focal length. In parabolic trough collector technology, the mirror shape acts as a parabolic trough to concentrate the direct solar radiation on to a tubular receiver. This technology constitute large share in installed solar thermal power technology. Its share crosses more than 90% of the capacity as compared to different technologies available in solar thermal power technology that is in operation today. For example the installed capacity of solar power as on 2010 is 821.9 MW. Among this, 93.6% of the power comes from parabolic trough, while solar tower technology is about 5.1%, dish sterling technology is 0.2% and Fresnel technology is 1.1% [11]. Accordingly in our present study, we have simulated using PTC technology. In our simulation, we have used 3 different softwares, namely, Meteonorm, Surfer and TRNSYS. They are briefly described in the following.

2.2. Meteonorm Software

This software gives solar irradiation data for any place on the Earth for a given coordinates. The software systematically organizes the data internally and provides the data in different formats as per the requirement. Based on our requirement, data can be used for solar thermal, solar photo voltaic or building simulation [12]. The meteonorm data are based on 8325 number of weather stations all over the world. For our present requirement of solar thermal plant, we have taken TMY2 data format. This format is compatible to another software—TRNSYS. Firstly, we have defined a location by specifying latitude and longitude. From these two parameters altitude and time zone are updated. We have taken open situation mode for all the locations. From these values, the average climate data of temperature from 2000-2009 and radiation from 1991-2010 are considered. By proceeding further, we can get the weather data for a given location. They are radiation, irradiance, cloud cover fraction, air temperature, humidity, wind direction, wind speed, pressure. These climatic parameters are fed to TRNSYS software to compute solar thermal power generation for a given design of a plant.

2.3. TRNSYS Software

TRaNsient SYstem Simulation (TRNSYS) is mainly used to model and simulate energy systems [13]. It is used by engineers and researchers around the world to validate new energy concepts from simple domestic hot water systems to the design and simulation of buildings and their equipment, including control strategies, alternative energy systems-wind, solar photovoltaic, solar thermal, hydrogen systems etc. The library in the software includes many components commonly found in thermal and electrical energy systems. It is developed with many thermal and energy equations that re internally computed for a given design of the power plant. The main window is simulation studio. The input parameters are component connections of a solar thermal plant and climatic data. Firstly, we have prepared all the components in the simulation studio of TRNSYS. By using the control card the required number of hours (from 0 hour to 8760 hour-365 days) are given. From this data energy generation for each month is obtained. The values for each month for different locations have been gridded, contoured to generate maps using surfer software.

2.4. Surfer Software

Surfer is windows based software and can be used for gridding, blanking, contouring, 3D surface mapping etc, [14]. By taking XYZ data from the user maps can be generated based on the requirement. It is developed mainly with statistical equations with more user friendly as compared to the well known Excel package of MS-Office. Here we have taken X value as longitude, Y value as latitude and Z as the power output from the simulated plant for a specified location. Using these three values the data is subjected to gridding using Krigging method and color filled contouring. Firstly, a base map is prepared and color contour map has been added with uniform grid spacing maintaining in both X and Y directions. In our present study, surfer version 11 has been used.

2.5. Design of Solar Thermal Plant Using TRNSYS

Based on TRNSYS software, we have designed a simplified power plant (**Figure 2**) using Parabolic Trough Collector (PTC) technology. This is mainly due to non-availability of LFR technology in the software. For this purpose TRNSYS software Version 17.2.4 [13] has been used and simulated. Another limitation encountered is the magnitude of the power plant. Using TRNSYS software large thermal plants (>1 MW) can easily be simulated as compared to small plants. Accordingly, we have first computed the power output for a plant with initial input parameters presented in **Table 1**, using PTC for different months. For this purpose, DNI data obtained from the Meteonorm software [12] for a given location, presently the Gurgaon has been used. In our design PTC, the solar radiation falls on the parabolic trough gets reflected and concentrated on a tube like structure placed near the center of the parabolic trough at a distance. In this tube, the therminol fluid flow in a closed loop. The fluid after getting heated up enters into a high temperature tank. From there, it flows through three chambers, namely super heater, steam generator and preheater. At these three chambers, heat exchanges between the therminol and water. From the preheater, the temperature comes down to about 200°C. This fluid enters into low temperature tank, from which it enters the tube located in the parabolic trough, thus completing the closed circular movement. As can be seen from the **Figure 2**, there is another closed loop movement on the right hand side. Water gets pre heated enters into a steam generator and later on to the super heater. In super heater, the temperature of the steam is about 350°C and enters into turbine which in turn generates the electricity. From the turbine, the steam enters into condenser and reaches the deaerator chamber. At deaerator, the unwanted gases are removed. The condensed water again reaches preheated chamber, thus completing the loop. As described before, the various input parameters used in our modelling study are presented in **Table 1**.

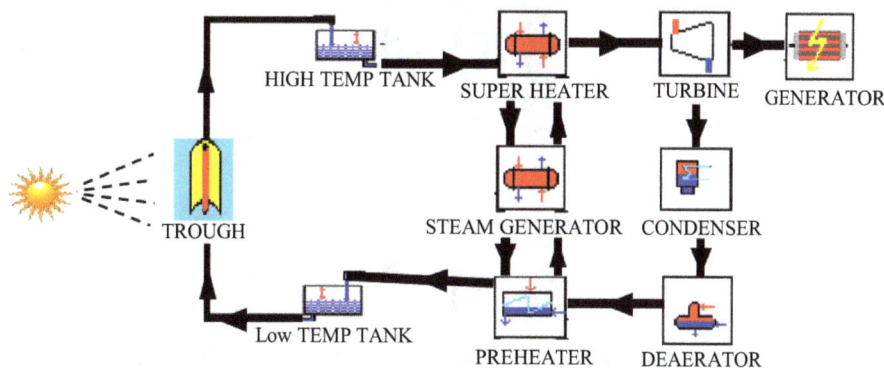

Figure 2. Schematic diagram showing the different components used in modelling studies of our study. In this diagram, the two types of fluids circulated in a closed loop. On the left hand part of the figure, therminol fluid passes through the trough, high temperature tank super heater preheat followed by low temperature tank and finally reaches to PTC again; on right hand side of figure, the water gets preheated enters into steam generator super heater and then the turbine condenser deaerator, from which it enters the preheater again.

3. Results

3.1. Comparison of 1 MW Solar Thermal Plant

In **Figure 3**, the results derived from modelling studies have been presented. The energy generation showed large variation in different months. Additionally, there is zero potential during the months of January, November and December. The annual generation is 3897 MWh. If we compare our results with the installed plant near Delhi the annual electricity generation is nearly 2.85 times the actual generation (1365 MWh/annum). Additionally it is reported that the plant remain non-operational during the months of January, July, August, December [10]. Thus our simulation results are calibrated with the installed plant near Delhi.

3.2. Solar Thermal Power Map of India

Now, it is interesting to study the energy generation if Gurgaon 1 MW capacity plant has been constructed at various locations in different parts of India. For this purpose, 1° × 1° grid spacing with a total of 296 locations has been identified (**Figure 4(a)**). Similarly detailed maps for two states, namely Gujarat in the western part of India and Tamil Nadu in the southern part of India have also been considered. In these two states, we have

Table 1. Input and initial parameters of solar thermal power plant.

S. No	Equipment	Input and initial parameters
1	PTC Field	For example, length of Solar Collector Assembly (SCA): 120 m, row spacing = 12.5 m, inlet temperature = 227°C, DNI = 2160 KJ/hr·m², Ambient Temperature = 23°C, loss coefficient parameter: A = 73.6, B = 0.0042, C = 7.40
2	Turbine	Mechanical efficiency = 0.99
3	High temperature tank	Volume = 20 m³, maximum level = 18 m³, minimum level = 5 m³, dry loss co-efficient: 4 KJ/hr·m²·K
4	Low temperature tank	Volume = 20 m³, maximum level = 18 m³, minimum level = 5 m³, dry loss co-efficient: 4 KJ/hr·m²·K
5	Super heater	Heat exchanger type: rankine cycle, input temperature = 390°C
6	Steam generator	Heat exchanger type: rankine cycle, input temperature = 260°C
7	Preheater	Heat exchanger type: rankine cycle, input temperature = 105°C

Simulated annual power:1736 MWh
Actual generation of the Gurgaon plant:1365 MWh

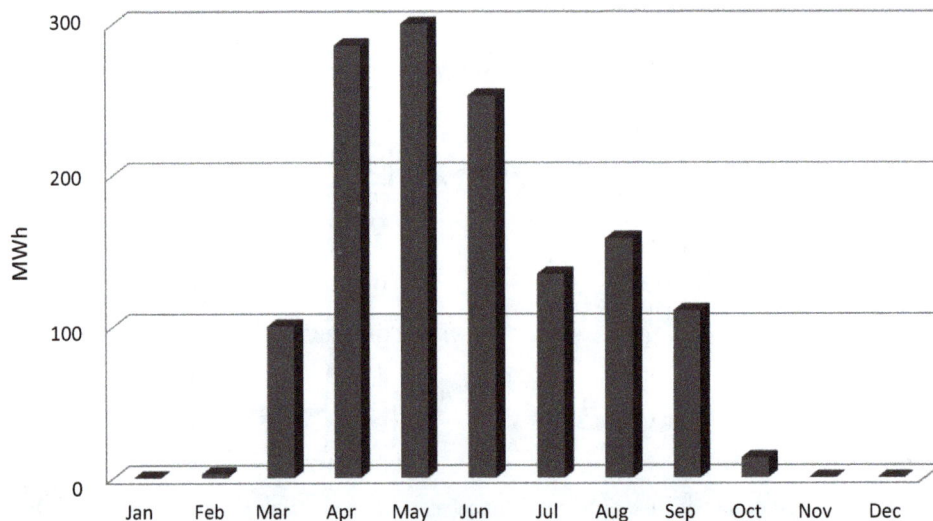

Figure 3. Our present modelling results normalized to 1 MW power plant and the power generation of the installed solar thermal power plant at Gurgaon, Delhi. Less than 5% match is achieved for most of the monthly generation data.

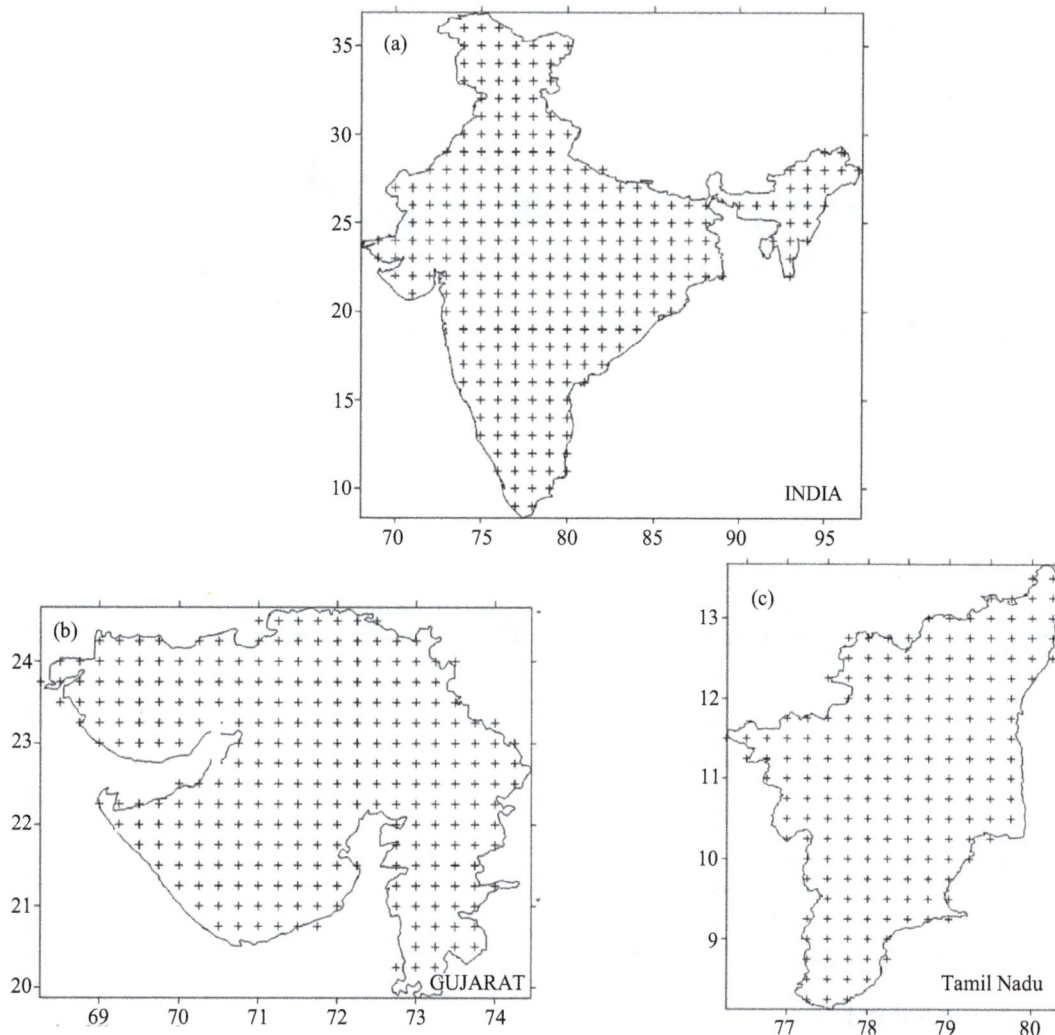

Figure 4. Different locations considered for our modelling studies. $1° \times 1°$ locations are considered for India. A total of 296 number of locations are shown with + symbol in (a). In (b) and (c) location map for Gujarat and Tamil Nadu states are shown for both the states. $0.25° \times 0.25°$ spacing is considered as can be seen in the figure.

considered $0.25° \times 0.25°$ (**Figure 4(b)** and **Figure 4(c)**). A total of 273 number of locations for Gujarat state and 197 number of locations for Tamil Nadu state have been considered. The closer station spacing helped to generate more detailed power generation maps for the two states. The simulation studies are continued at all these locations. At each location, latitude and longitude values are fed to the Meteonorm software and DNI, average temperature, wind speed, wind direction etc, parameters have been obtained. These parameters are fed to the TRNSYS software and the solar thermal energy generation averaged month wise are computed.

After getting the solar thermal energy generation for all the locations contouring have been carried out by gridding using Surfer software version 11 [14]. After getting the generation values at all the locations for all months, total value for the whole year has been computed and presented in **Figure 5**. As can be seen from the figure a large variation of 600 MWH to nearly 2600 MWH per annum is possible in India for 1 MW power plant using parabolic trough technology. It may be noted that we have first simulated power generation and calibrated with the plant near Delhi. As can be seen southern states like Tamil Nadu, Kerala, Karnataka, southern part of Andhra Pradesh and eastern part of Maharashtra and parts of Madhya Pradesh, Gujarat and southern parts of Rajasthan seems to be having higher potential as compared to other states. All the states of north eastern region have less than 1400 MWh/annum potential as compared to 1800 to 2600 MWh/annum for the southern states. While such is the case, majority of the northern states like Jammu and Kashmir, Punjab, Uttar Pradesh, Bihar,

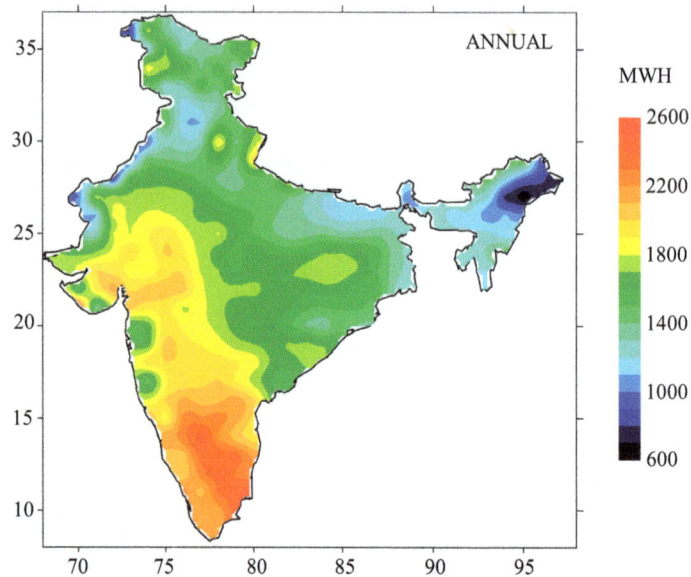

Figure 5. Solar thermal energy generation potential map for India using PTC technology. Considering 1 MW power plant at different locations. The solar thermal potential ranges from 600 MWh/annum in parts of north eastern region to about 2600 MWh/annum in parts of south India.

West Bengal eastern part of Orissa etc, are having medium range of potential with an annual energy potential of about 1000 to 1800 MWh/annum as shown in **Figure 5**.

The solar thermal energy generation potential for different months from January to December has been presented in **Figure 6**. As can be seen from the figure, during the months of January, November and December the solar thermal energy potential is near zero for most of the northern part, eastern part and north eastern parts of India. Southern states always have nearly good potential on all the months. For example, in Tamil Nadu, Kerala, Karnataka and Andhra Pradesh during June and July, the potential varies from 30 to 100 MWh as compared to other months as shown in **Figure 10**.

3.3. Gujarat Solar Thermal Power Map

Figure 7 shows the annual energy generation potential for Gujarat state. The energy potential varies from 1600 to 2300 MWh/annum. Interestingly, major part of Saurashtra region and middle part of southern Gujarat have shown higher potential as compared to other regions.

The energy generation month wise are presented except for the month of December (**Figure 8**). Due to low temperature in December the energy generation potential is near zero. As can be seen, during the months of January, February, March, April, May, October, November and December, Kutch region has exhibited low thermal potential as compared to other regions.

From **Figure 9**, we have further analyzed the weather parameters for the month of April for Gujarat state. The parameters considered are the DNI values in Watts/Sq. m, wind speed in m/s and temperature in °C. It can be seen from these parameters that the solar radiation for the month of April in Kutch region is the lowest as compared to the rest of Gujarat as shown in **Figure 9(a)**. The wind speed is maximum towards northern part of Saurashtra as compared to rest of the state as shown in **Figure 9(b)**. The average ambient temperature in the northern part of Gujarat covering Ahmedabad Gandhinagar, Patan, etc. districts have shown more than 31°C temperature as shown in **Figure 9(c)**. Accordingly, radiation (DNI) parameter plays a dominant role in solar thermal as compared to other parameters namely, the wind speed and ambient temperature.

3.4. Tamil Nadu Solar Thermal Power Map

We can see the annual energy generation for Tamil Nadu state in **Figure 10**. The annual energy ranges from 1850 to 2650 MWh. In Tamil Nadu region, western half of Tamil Nadu has better potential as compared to eastern half of the state.

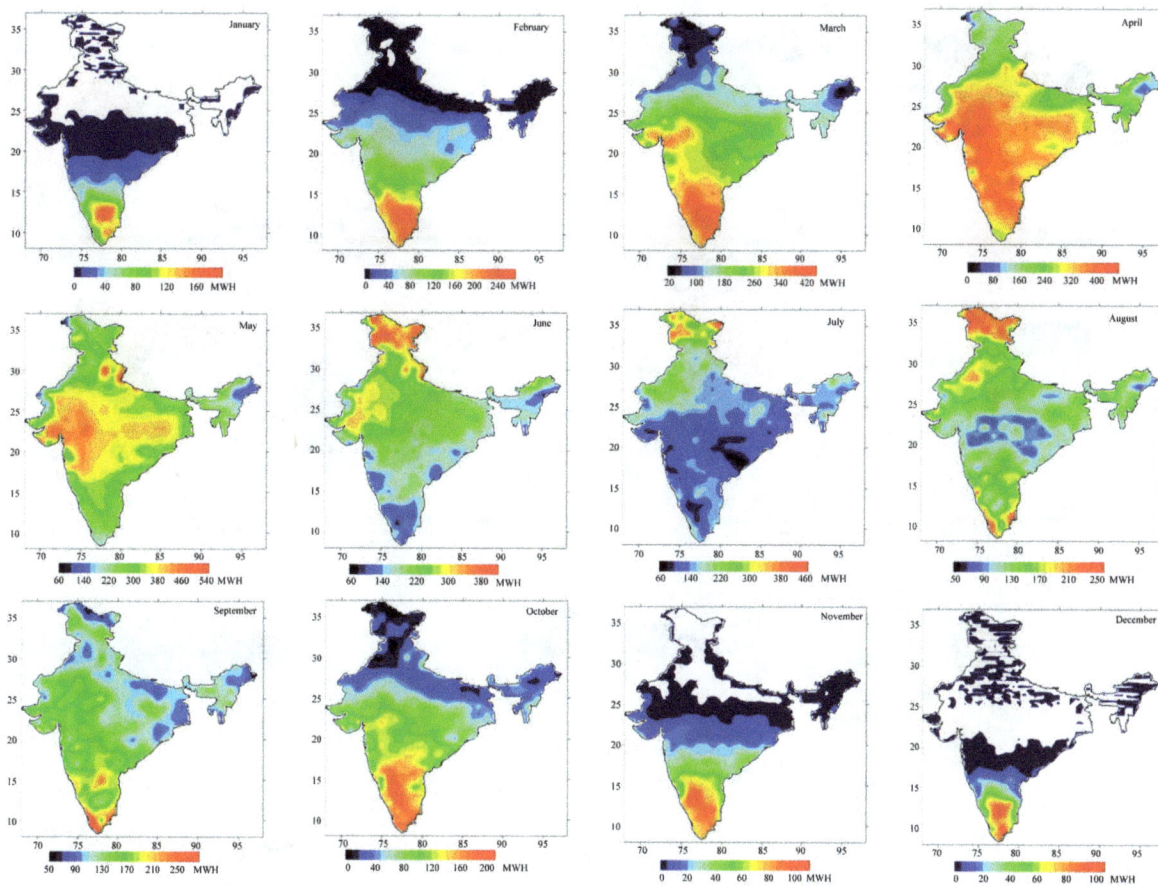

Figure 6. Solar thermal energy potential for different months in India for 1 MW power plant near zero potential for the months of January, November, December in most of the north Indian states may be observed on the other hand more potential has been observed in Sothern states namely Tamil Nadu, Kerala, Karnataka and southern part of Andhra Pradesh.

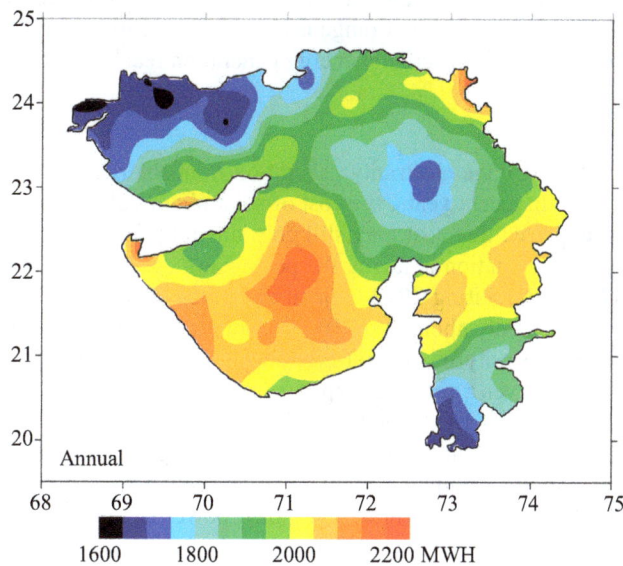

Figure 7. Solar thermal energy potential for 1 MW for Gujarat state poor potential near Kutch and southern part of south Gujarat and north eastern part of Gujarat may be observed. On the other hand, large part of Saurashtra region and middle part of south Gujarat have shown greater potential. The annual potential ranges from 1600 to nearly 2300 MWh.

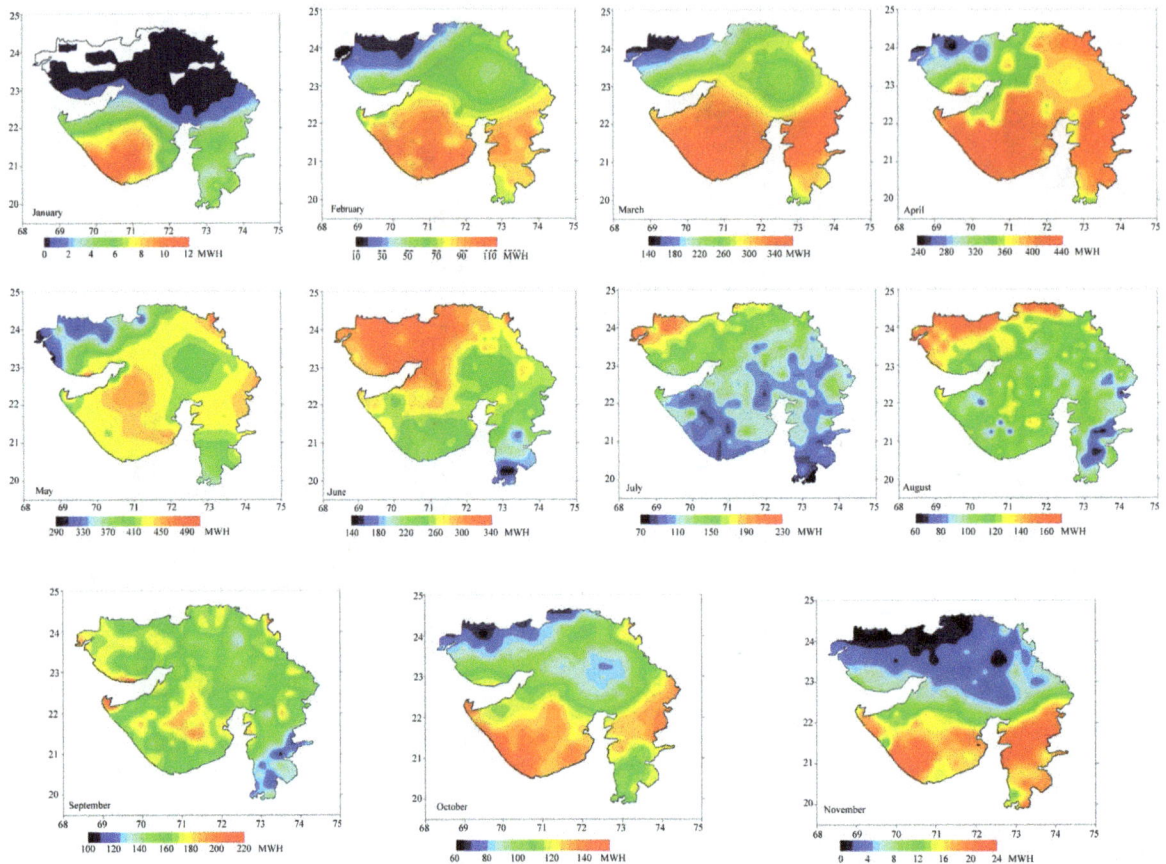

Figure 8. Solar thermal energy potential for different months of Gujarat state near zero potential during the months of January, October and November months near the Kutch and surrounding regions can be seen. Due to near zero potential in all parts of Gujarat, the 1 MW power plant may become non-operational in Gujarat.

The energy generation for all the 12 months for Tamil Nadu state can be seen in **Figure 11**. During January and December months, the maximum potential range is about 125 to 190 MWh. Maximum potential is observed during the months of March and April, where the energy generation reaches to more than 400 MWh.

4. Discussion

It is estimated that by the year 2030 the electricity generation from solar energy will be dominating as compared to other source of energy [15]. In recent years, the cost of solar photo voltaic has fallen to 50% less as compared to the prices of 2000. It is expected that cost may still goes down by 20% to 25% due to various innovations in solar technology. While the entire nation is facing the power shortage, Gujarat state has 3000 MW more power than its requirement. Additionally, during 2012 Gujarat alone has crossed 600 MW power through solar while rest of the country could produce only about 200 MW. Gujarat is going ahead using solar power with innovative concepts. For example the capital city of Gujarat-Gandhinagar, the rooftop solar has been initiated. Similarly solar panels have been installed over a canal on experimental basis with 1 MW capacity. Realizing the success story of canal top project, it has further extended to 10 MW near Baroda. In another innovation through modelling studies, it is proposed that solar panels can be installed over agricultural lands also by putting up the solar panels in a chess board like pattern. The main advantage of this concept is the dual usage of the land. The farmer can continue farming with different crops, while on top of it power can be generated for his and his village use [16]. Recognizing the importance of this concept, government of Gujarat has initiated a major project to put up solar panels at 4 different agricultural farms and to study the shadow effect on different crops. The project is presently being worked out by government of Gujarat in close association with different agricultural universities. Due to recent realization of climate change problem more stress is being given to renewable energy production [17] [18].

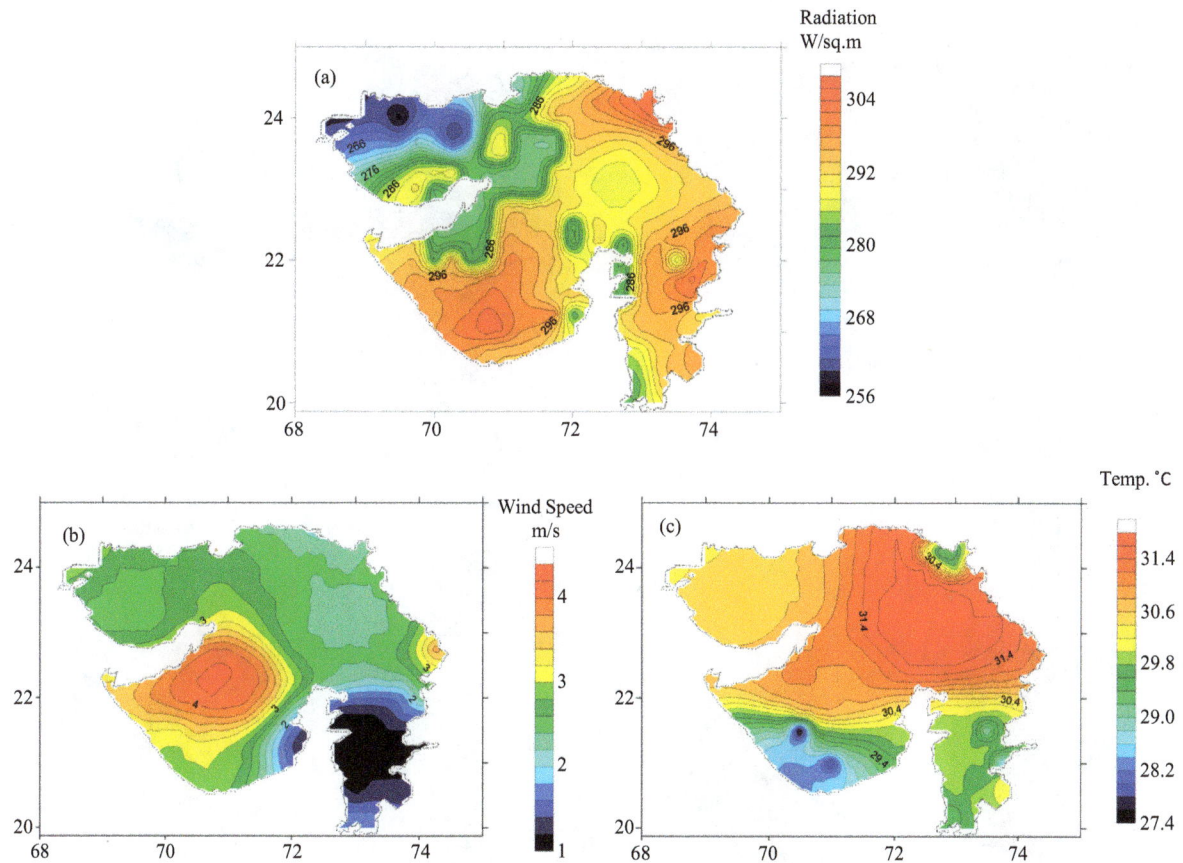

Figure 9. The radiation, the wind speed and the temperature distribution for Gujarat state during the month of April, can be seen that a poor radiation in Kutch region compared to other parts of the state.

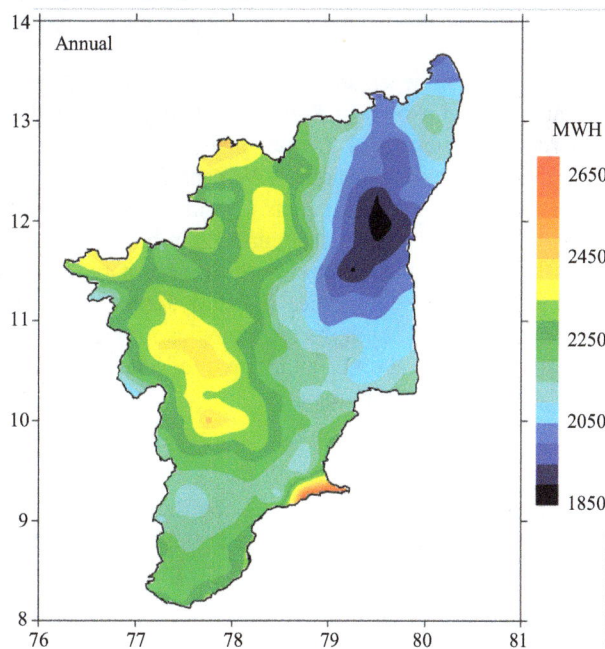

Figure 10. Solar thermal energy potential for Tamil Nadu state for 1 MW power plant the minimum annual energy potential is about 1850 MWh and reaches 2650 MWh. Western part of Tamil Nadu has greater potential as compared to eastern and southern part of the region.

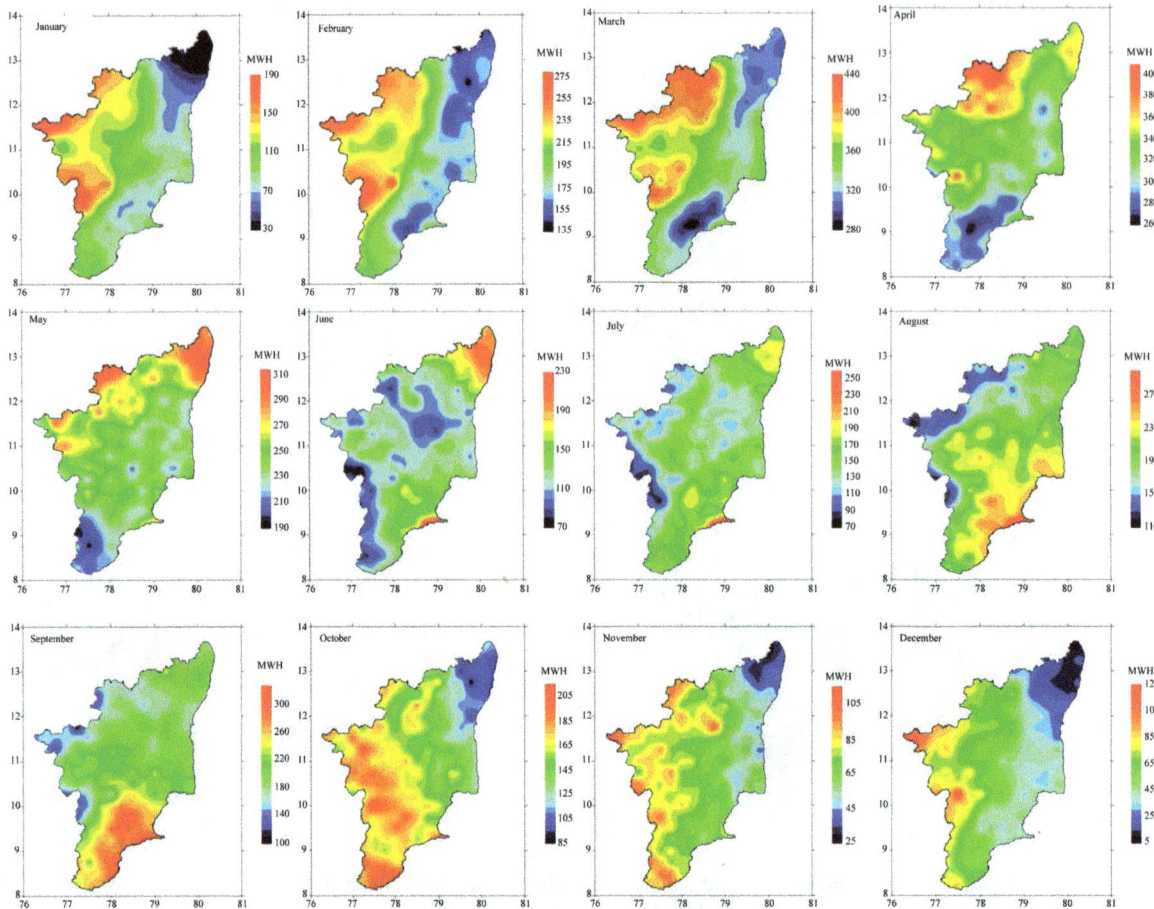

Figure 11. Solar thermal energy potential for 1 MW power plant in different months. During all the months, the potential reaches to about 100 MWh.

A comparative study has been made between solar thermal energy generation potential and solar PV potential for India and also for Gujarat state and shown in **Figure 12**. It may be noted that both of them are based on modelling studies using different strategies. Solar thermal energy potential has been estimated by simulating the technology used in the solar thermal power plant of 1 MW capacity at Gurgaon near Delhi, whereas solar PV potential has been estimated considering the solar panels kept in 1 acre of land. However, the two maps can be compared by considering the low and high potential. Interestingly both the solar PV and solar thermal technology have shown poor estimation towards eastern and north eastern parts of India. The southern part of India seems to have large potential for both the technologies. Similarly, from **Figure 12(c)** and **Figure 12(d)** one can compare the annual energy generation using both the technologies for Gujarat state. Northern and western parts of the Kutch region has shown poor potential and southern part of Saurashtra and middle part of southern Gujarat have shown good potential for both the technologies.

5. Conclusion

The present research study is based on the preparation of Solar thermal energy potential map for India on a regional scale with 1° × 1° spacing and more detailed spacing of 0.25° × 0.25° for Gujarat and Tamil Nadu states. We have simulated a solar thermal plant and compared with the installed solar thermal plant of 1 MW capacity at Gurgaon near New Delhi. This plant has helped to calibrate our results. After calibration, the solar thermal energy generation using PTC technology has been computed for the designed plant at different location points of India in a gridded fashion. Large potential is observed at most of the southern states and medium to poor potential is observed for most of the northern and north eastern states of India. Our map will be useful to all the solar energy developing industries and also to decision makers in choosing the location for possible installation of

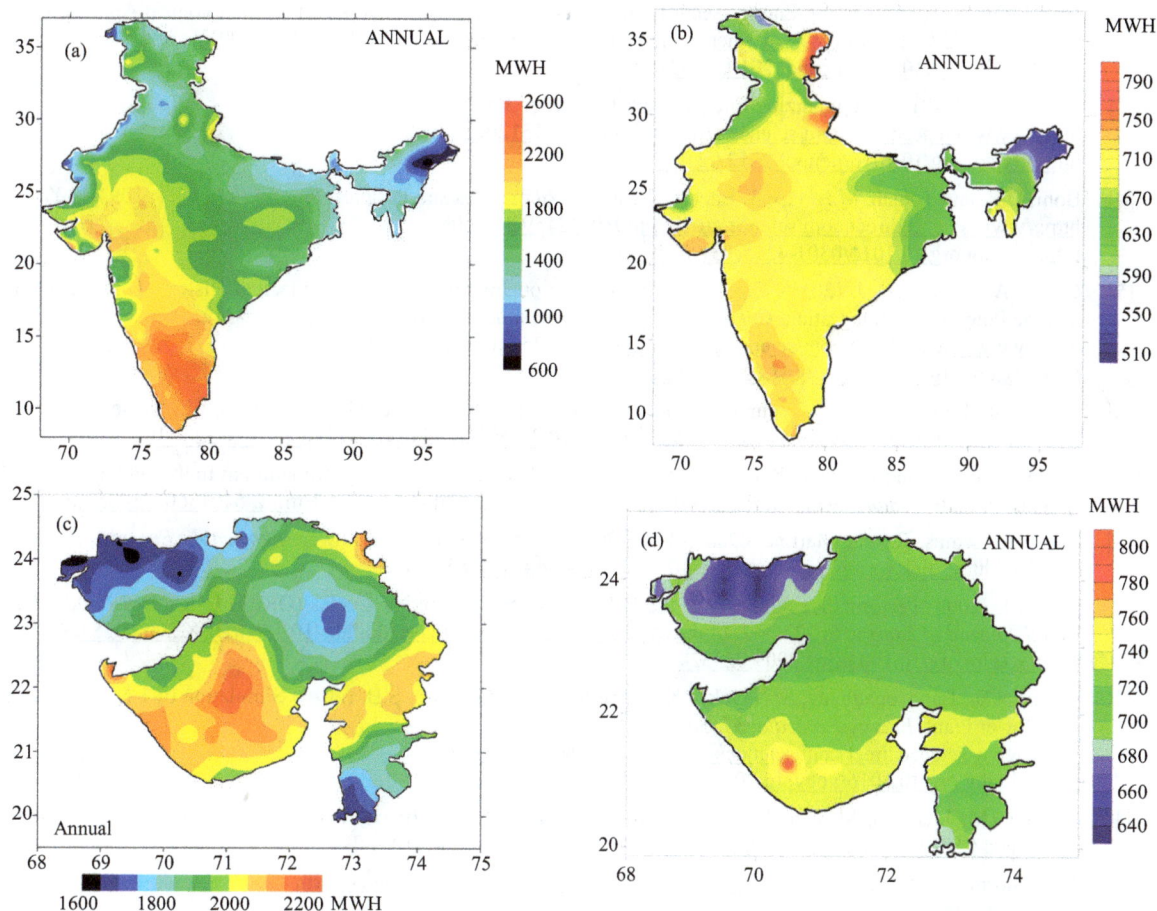

Figure 12. Comparison of solar thermal energy for 1 MW power plant (a) and solar PV potential in 1 acre of land (b). Western and southern parts of India have exhibited large potential for both solar PV and solar thermal energy generation. Medium and low potential can be seen in both the maps for eastern and north eastern parts of India ((a) and (b)). Similarly northern part of Kutch region seems to be having poor potential for both solar PV and solar thermal generation. As can be seen in solar thermal map (c) and solar PV map (d). Relatively large potential in southern part of Saurashtra and middle part of south Gujarat.

large scale solar thermal parks. This greatly helps in reducing the cost. For example, if Rs. 8 to 10 crores are spent to install 1 MW solar thermal power plant in large potential region of Tamil Nadu, it may cost nearly double to establish a solar plant if one aims to produce same amount of annual generation. This is because, annual energy generation in western part of Tamil Nadu state is more than 1800 MWh for 1 MW plant and for the same plant in NE states it is less than half. Thus, our study directly helps the decision makers and planners to choose the optimal location such that one can generate more energy with less cost.

Acknowledgements

We are thankful to VCMT, Sri DJ Pandian, IAS and MD, GSPC Sri Atanu Chakraborthy, IAS for all the encouragement. We wish to thank all the GERMI staff for giving full support. Help rendered by Dr. Sagar Kumar Agaravat, Scientist, GERMI and Sri Jaymin Gajjar, Research Associate, GERMI is gratefully acknowledged. Special thanks goes to Prasanth Gopiyani, the summer internship coordinator during execution of the project.

References

[1]   Löschel, A. Moslener, U. and Rübbelke, D.T. (2010) Indicators of Energy Security in Industrialised Countries. *Energy Policy*, **38**, 1665-1671. http://www.sciencedirect.com/science/article/pii/S0301421509002262 http://dx.doi.org/10.1016/j.enpol.2009.03.061

[2]   Vivoda, V. (2010) Evaluating Energy Security in the Asia-Pacific Region: A Novel Methodological Approach. *Energy Policy*, **38**, 5258-5263. http://www.sciencedirect.com/science/article/pii/S030142151000399X http://dx.doi.org/10.1016/j.enpol.2010.05.028

[3]   Winzer, C. (2012) Conceptualizing Energy Security. *Energy Policy*, **46**, 36-48. http://www.sciencedirect.com/science/article/pii/S0301421512002029 http://dx.doi.org/10.1016/j.enpol.2012.02.067

[4]   Bohi, D.R. and Toman, M.A. (1993) Energy Security: Externalities and Policies. *Energy Policy*, **21**, 1093-1109. http://www.sciencedirect.com/science/article/pii/030142159390260M http://dx.doi.org/10.1016/0301-4215(93)90260-M

[5]   Cherp, A. and Jewell, J. (2011) The Three Perspectives on Energy Security: Intellectual History, Disciplinary Roots and the Potential for Integration. *Current Opinion in Environmental Sustainability*, **3**, 202-212. http://www.sciencedirect.com/science/article/pii/S1877343511000583 http://dx.doi.org/10.1016/j.cosust.2011.07.001

[6]   Lal, S. and Raturi, A. (2012) Techno-Economic Analysis of a Hybrid Mini-Grid System for Fiji Islands. *International Journal of Energy and Environmental Engineering*, **3**, 10. http://dx.doi.org/10.1186/2251-6832-3-10

[7]   Harinarayana, T. and Kashyap, K.J. (2014) Solar Energy Generation Potential Estimation in India and Gujarat, Andhra, Telangana States. *Smart Grid and Renewable Energy*, **5**, 275. http://file.scirp.org/Html/2-6401360_51455.htm

[8]   Guerrero-Lemus, R. and Martínez-Duart, J.M. (2013) Concentrated Solar Power. *Renewable Energies and CO$_2$*, **3**, 135-151. http://link.springer.com/chapter/10.1007/978-1-4471-4385-7_7

[9]   Yari, M. (2009) Performance Analysis of the Different Organic Rankine Cycles (ORCs) Using Dry Fluids. *International Journal of Exergy*, **6**, 323-342. http://www.inderscienceonline.com/doi/abs/10.1504/IJEX.2009.025324 http://dx.doi.org/10.1504/ijex.2009.025324

[10] Desai, N.B., Bandyopadhyay, S. Nayak, J.K., Banerjee, R. and Kedare, S.B. (2014) Simulation of 1 MWe Solar Thermal Power Plant. *Energy Procedia*, **57**, 507-516. http://www.sciencedirect.com/science/article/pii/S1876610214015719 http://dx.doi.org/10.1016/j.egypro.2014.10.204

[11] Günther, M. Joemann, M. Csambor, S. Guizani, A. Krüger, D. and Hirsch, T. (2011) Parabolic Trough Technology, Advanced CST Teaching Materials. *enerMENA*, Chapter 5. http://goo.gl/VWBwUB

[12] Meteonorm v7.1.5. http://www.meteotest.ch/en/footernavi/solar_energy/meteonorm http://meteonorm.com

[13] TRNSYS v17.2.4. http://sel.me.wisc.edu/trnsys http://www.trnsys.com

[14] Surfer v11. http://downloads.goldensoftware.com/guides/Surfer11TrainingGuide.pdf http://www.goldensoftware.com/products/surfer

[15] European Photovoltaic Industry Association (2014) Global Market Outlook for Photovoltaics. http://www.epia.org/news/publications/

[16] Harinarayana, T. and Vasavi, K.S.V. (2014) Solar Energy Generation Using Agriculture Cultivated Lands. Smart Grid and Renewable Energy. http://file.scirp.org/Html/2-6401313_42763.htm

[17] Dincer, I. (2000) Renewable Energy and Sustainable Development: A Crucial Review. *Renewable and Sustainable Energy Reviews*, **4**, 157-175. http://www.sciencedirect.com/science/article/pii/S1364032199000118 http://dx.doi.org/10.1016/s1364-0321(99)00011-8

[18] Boyle, G. (2004) Renewable Energy, Oxford University Press, Oxford.

4

Energy Analysis and Exergy Utilization in the Residential Sector of Cameroon

Talla Konchou Franck Armel[1,2]*, Aloyem Kazé Claude Vidal[2,3], Tchinda René[2]

[1]Environmental Energy Technologies Laboratory (EETL), University of Yaounde I, Yaounde, Cameroon
[2]LISIE, University Institute of Technology Fotso-Victor, University of Dschang, Bandjoun, Cameroon
[3]HTTTC, Department of Electrical and Power Engineering/FSC, University of Bamenda, Bamenda, Cameroon
Email: *tkfarmel@yahoo.fr

Abstract

In this paper, we present an analysis of energy and exergy utilization in the residential sector of Cameroon by considering the sectoral energy and exergy flows for the years of 2001-2010. Exergy analysis of Cameroon residential sector utilisation indicates a less efficient picture than that obtained by the energy analysis. Cooking stands out as the most inefficient end use in the Cameroon's residential sector. In 2010, the energy and exergy efficiency are determined and were respectively 58.74% and 22.63%. Energy and exergy flows diagrams for the overall efficiencies of Cameroon residential sector are illustrated and a comparison with the residential sector of other countries is also done. To carry out this study, a survey of 250 households was conducted and the sharing of the end uses of energy was done and data were gathered.

Keywords

Exergy, Efficiency, Energy Flow, Residential Sector

1. Introduction

The search for better living conditions has pushed the country to develop new areas of research including that of energy management, and the optimization of energy use. Thus today the industrialization of a country is determined by the rational and efficient use of energy. It has been observed that in Cameroon, the use of energy has taken on a new dimension this can be explained by the growing population (which increases energy requirements) for the past decade, and also by the government policy that wants to raise the rank of Cameroon an emerging country in 2035. Thus the consumption of wood energy, fuel and electricity continue to rise, so it will be wise to reconcile high population growth, government projects and the efficient and rational use of energy. In

---
*Corresponding author.

2010, energy consumption in Cameroon was estimated to 5722.22 kilo Ton Oil Equivalent (TOE) [1]. In the same year, the residential sector represented approximately 70% of the total Cameroon energy consumption [1]. World-wide residential sector consumes about 17% - 70% of the total energy as can be seen in **Table 1** [2].

From the first law of thermodynamics, it is clear that energy is conserved. The first law tells us only about the amount of energy that is put into play, and it takes into account only the internal parameters in the system, thus neglecting external parameters and the external environment to the system. Following these imperfections, it was introduced a new concept that will not only study the quality of energy used, but will take into account external parameters to the system: it is exergy. The analysis based on exergy was based on the two principles of thermodynamics hence its complexity and efficiency. Nowadays, exergy analysis is the best tool in the analysis of the efficiency of a system [3].

This concept has been used for the first time by Reistad (1975) [4]. He applied it in the economic analysis of US in 1970. Several studies have been done in the same vein and have been applied indifferent countries, including Italy [5], Turkey [6], the UK [7], Norway [8] China [9] and Saudi Arabia [10]-[12]. Other researches in the residential sector were made in Malaysia [2], Turkey [13] Jordan [14].

Energy and exergy analysis is an important approach to identify the various losses and determine the specific sector with low energy efficiency, which is necessary to find new ways to improve the efficiency of energy.

The primary objective of this work is to determine the overall energy and exergy efficiency of the residential sector in Cameroon from 2001 to 2010, then we will discuss the residential sector Cameroonians compare to other countries in this literature.

2. Methods

The methodology adopted in this study was first used by Dincer *et al.* [9], who applied Resitad's approach [3].

2.1. Exergy Calculation

In this section, we will discuss some basic quantities and mathematical relations to exergy analysis.

2.1.1. Chemical Exergy
One of the most common mass flows is hydrocarbon fuels at near-ambient conditions, for which the specific exergy reduces to chemical exergy and can be written as follows [2]:

$$\varepsilon_{ff} = \gamma_{ff} H_{ff} \tag{1}$$

where γ_{ff} denotes the fuel exergy grade function, defined as the ratio of fuel chemical exergy. **Table 2** ([15]-[17]) shows typical values of H_{ff}, ε_{ff} and γ_{ff} for the fuels encountered in the present study. As shown in this table, all the values of the exergy grade function are very close to unity.

Ertesvag and Mielnik [7] and Erstesvag [13] reported that the major parts of the input to the system are fuel and other energy carriers. For mechanical and electrical energy, the exergy content is equal to the energy content. Exergy factors for different energy carriers are shown in **Table 3**.

Table 1. Worldwide residential energy consumption.

	Sweden	Italy	USA	UK	Canada	Mexico	Japan	Malaysia	Jordan	Cameroon
Percentage	19%	17%	25%	31%	24%	23%	22.60%	19%	29%	70%

Source: Saidur *et al.*, 2007.

Table 2. Properties of selected fuels.

Fuel	H_f (kj/kg)	ε_f (kj/kg)	γ_f
Diesel	44.800	42.265	0.943
Butane	49.463	48.272	0.976

Table 3. Exergy factors of energy carriers.

Energy carriers	Exergy factors
Waterfall anergy	1
Electrical energy	1
Oil, petroleum products	1.06
LPG	1.06
Coal	1.06
Diesel oil	1.07
Fuel wood (20% humidity)	1.11

2.1.2. Energy and Exergy Efficiencies

In this study, the expression of the energy (η) and the exergy (ψ) efficiencies for the principle type of processes are based on the following definitions [2]:

$$\eta = \left(\frac{\text{energy in product}}{\text{total energy input}} \right) \times 100\% \tag{2}$$

$$\psi = \left(\frac{\text{exergy in product}}{\text{total exergy input}} \right) \times 100\% \tag{3}$$

2.1.3. Heating and Cooling Process

Electric process is taken to generate product heat Q_p at the constant temperature T_p either from electrical energy W_e. The energy and exergy efficiencies for electrical heating are [9].

$$\eta_{h,e} = \frac{Q_p}{W_e} \tag{4}$$

$$\psi_{h,e} = \frac{E^{Q_p}}{E^{W_e}} = \frac{\left[1 - (T_0/T_p) Q_p \right]}{W_e} \tag{5}$$

$$\psi_{h,e} = \left(1 - \frac{T_0}{T_p} \right) \eta_{h,e} \tag{6}$$

2.2. Methodology and Data Sources

In order to determine the energetic and exergetic output of Cameroon's residential sector, it is essential to know how this energy is used in this sector. Therefore, we have used the data of two institutions namely the National Statistical Institute (NSI) and the Ministry of Energy and Water Resource. Then, a further survey was carried out on 250 households of different locations and different living standards in Cameroon. The aim of this survey was to collect necessary data for the realization of this work. It was carried out by the students of the Analysis Laboratory of Technologies of Energy and Environment (ALTEE). The questionnaire used covered the following aspects:

- Types of appliances functioning with fuel.
- Information detailed over the electrical appliance, their powers, numbers, and the utilization time.
- Types of appliances used in cooking.
- Types of lighting and its utilization time and power rating.

In all questions, the level of ownership was being asked for each type of equipment and appliance. To ensure the good quality of such a survey, the students were given a tutorial lecture on how to collect and record the required data correctly and efficiently. At the end of the survey, the sharing of the end uses of electrical energy was determined.

2.3. Energy and Exergy Efficiency of Electrical Appliances

Each appliance operates in a range of efficiencies. The efficiency of an appliance is usually described in terms of energy only. But there we add another term called exergy efficiency that correlates the second law of thermodynamics [14]. The reference temperature T_0, appliances energy efficiency η_e, and its product temperature T_p, are needed to get the exergy efficiency and these values are presented in **Table 4**.

If electrical or mechanical energy is output, the following equation is valid since the quality factor of both energy forms is equal to one based on Utlu and Hepbasli [18]

Exergy efficiency

$$\psi = q_{elec} \times W_{elec(Mech)} \big/ (Mech) \big/ q_{elec} \times W_{elec} = \text{energy efficiency, } \eta \tag{7}$$

The exergy efficiency of some of refrigerator-freezer has been calculated with the aid of methodologies described by Dincer *et al.* [9] and Utlu and Hepbasli [19].

2.4. Energy and Exergy Efficiency of a Cooking Appliance

Using the methodology describe in Section 2.1, the exergy efficiency for this appliance can be calculated using following known values:

- Energy efficiency, $\eta_f = 65\%$.
- Reference temperature, $T_o = 298\,\text{K}$.
- Product temperature, $T_p = 374\,\text{K}$.

These values are needed in calculating the overall energy and exergy efficiencies.

2.5. Weighted Mean Energy and Exergy Efficiencies

Weighted mean energy and exergy efficiencies are calculated for the residential sector using a three-step process.

2.5.1. Weighted Mean Electrical Energy an Exergy Efficiencies (First Step)

First, weighted means are obtained for the electrical energy and exergy efficiencies for the device categories listed in **Table 4**, where the weighting factor is the ratio of electrical energy input to the device category to the total electrical energy input to all device categories in the sector [2].

The calculation of the overall weighted mean energy and exergy can be illustrated as follows:

$$R_{\eta_e} = \frac{\left(e_{\text{Fluorescent light}} \times \eta_{\text{Fluorescent light}}\right) + \left(e_{TV} \times \eta_{TV}\right) + \cdots + \left(e_{\text{Radio player}} \times \eta_{\text{Radio player}}\right)}{e_{\text{Fluorescent light}} + e_{TV} + \cdots + e_{\text{Radio player}}} \tag{8}$$

where R_{η_e} is the weighted mean energy efficiency, $e_{\text{appliance}}$ is the appliance's energy consumption, $\eta_{\text{appliance}}$ the appliance's energy efficiency.

Table 4. Energy efficiency, product and environment temperatures, exergy efficiency and shared of used energy.

Appliances	η_e (%)	T_p (K)	T_o (K)	ψ_e (%)	Share of used energy (%)
Fluorescent light	12.50	-	-	11.50	35.92
TV	80	-	-	80	11.37
Fan	80	-	-	80	7.18
Refrigerator-freezer	60	298	265	7.47	23.95
Mobile phone charger (MPC)	70	-	-	70	2.40
Personal computer (PC)	70	-	-	70	11.94
Radio player	80	-	-	80	7.18
Cooking appliances (LPG and diesel oil)	65	298	374	13.20	

$$R_{\psi_e} = \frac{\left(e_{\text{Fluorescent light}} \times \psi_{\text{Fluorescent light}}\right) + \left(e_{TV} \times \psi_{TV}\right) + \cdots + \left(e_{\text{Radio player}} \times \psi_{\text{Radio player}}\right)}{e_{\text{Fluorescent light}} + e_{TV} + \cdots + e_{\text{Radio player}}} \tag{9}$$

where R_{ψ_e} is the weighted mean exergy efficiency, $e_{\text{appliance}}$ is the appliance's energy consumption, $\psi_{\text{appliance}}$ the appliance's exergy efficiency.

2.5.2. Overall Weighted Means Energy and Exergy Efficiencies (Third Step)

Third, overall weighted means are obtained for the energy and exergy efficiencies for the electrical and fossil fuel processes, where the weighting factor is the ratio of total fossil fuel or electrical energy input to the residential sector to the total energy input to the sector [2].

Weighting factor for electrical energy can be calculated using following equation:

$$WF_e = \frac{\sum e}{\sum e + \sum f} \tag{10}$$

WF_e is the overall weighting factor for electrical energy, $\sum e$ the overall electrical energy consumption, and $\sum f$ the overall fossil fuel energy consumption.

Similarly weighting factor for fossil fuel can be calculated using following equation:

$$WF_f = \frac{\sum f}{\sum e + \sum f} \tag{11}$$

Finally, using the following equation, the overall weighted energy and exergy can be calculated as:

$$R_{\eta_o} = \left(R_{\eta_e} \times WF_e\right) + \left(R_{\eta_f} \times WF_f\right) \tag{12}$$

where R_{η_o} is the overall energy efficiency, R_{η_e} is the weighted mean electrical energy efficiency, and R_{η_f} the weighted mean fossil fuel energy efficiency.

$$R_{\psi_o} = \left(R_{\psi_e} \times WF_e\right) + \left(R_{\psi_f} \times WF_f\right) \tag{13}$$

where R_{ψ_o} is the overall exergy efficiency, R_{ψ_e} is the weighted mean electrical exergy efficiency, and R_{ψ_f} the weighted mean fossil fuel exergy efficiency.

3. Results and Discussion

3.1. Electrical Energy and Exergy Calculation

Using the sharing of energy in the residential sector, the energy consumption in the residential sector is given in **Table 5**.

Table 5 shows a significant growth in the consumption of electrical energy. That is simply due to the growing demand of the population. It is important to note that much of the Cameroonian population is still without access to electricity. Indeed according to the National Institute of Statistics in 2005, 49.7% of the population has access to electricity; this figure rose to 60% in 2010 due to the expansion and intensification of the power grid efforts. Between 2001 and 2010, access to electricity rate increased nationally and in urban areas. But there was a slight decline in rural areas. This may be due to the fact that the number of households increases without that new subscribers will have the same pace because households are more likely to subscribe to electricity from a neighbour than from AES Sonel.

3.2. Fossil Fuel Energy and Exergy Use

In Cameroon, the residential sector mainly uses Liquefied Petroleum Gas (LPG) and diesel oil as their fuel for cooking, while electrical energy is rarely utilized. Their energetic and exergetic consumptions are given in **Table 6**. It is important to mention that this energy is essentially used by cooking appliances.

It should be noted that the domestic gas consumption is still very low in Cameroon as evidenced by recurrent shortages and limited access especially in rural areas and in the savannah zone of Northern Cameroon. The offer

Table 5. Total energy and exergy consumption (assuming energy consumption = exergy consumption) in TJ.

Appliances	2001	2002	2003	2004	2005	2006	2007	2008	2009	2010
Fluorescent light	689.66	732.76	786.64	898.00	977.02	1052.45	1074.00	1156.62	1167.40	1275.16
TV	218.30	231.98	249.00	284.25	309.26	333.14	339.96	366.11	369.52	403.63
Fan	137.85	146.67	157.24	179.50	195.29	210.37	214.68	231.19	233.35	254.89
Refrigerator-freezer	459.84	488.58	524.50	598.75	651.44	701.73	716.10	771.19	778.37	850.22
MPC	46.08	48.96	52.56	60.00	65.28	70.32	71.76	77.28	78.00	85.20
PC	229.82	244.18	262.14	299.25	325.58	350.72	357.90	385.43	389.02	424.93
Radio player	137.85	146.67	157.24	179.50	195.29	210.37	214.68	231.19	233.35	254.89
Total	1920	2040	2190	2500	2720	2930	2990	3220	3250	3550

Table 6. Total fossil fuel energy consumption (TJ).

	2001	2002	2003	2004	2005	2006	2007	2008	2009	2010
Butane energy	1920	2040	2190	2500	2720	2930	2990	3220	3250	3550
Oil energy	5600	5780	6710	6350	5600	5660	5340	5180	5180	5240
Total energy	7520	7820	8900	8850	8320	8590	8330	8400	8430	8790
Total exergy	7898.30	8210.40	9355.10	9279.90	8698	8965.70	8683.90	8741.10	8770.60	9132.50

is limited today by subsidized and administered system in one hand and by structural problems (lack of storage capacity, lack of security stocks, high cost of equipment for primary consumption, low interchangeability packaging system, etc.). According to the HSPC, such provision of LPG increased by 9% between 2010 and 2011.

The consumption of oil has increased from 5600 TJ in 2001 to 5240 TJ in 2010, a decrease of approximately 34%. This decline has been observed since 2003. This may be explained by the increase in prices of petroleum products and the increase in electrified rural areas. This continuous decline in consumption of kerosene causes a change in the structure of consumption of petroleum products by households.

3.3. Energy and Exergy Efficiencies

3.3.1. Energy and Exergy Efficiencies of Electrical Appliances

Energy and exergy efficiencies of electrical appliances are determined using the methodology stated in Section 2.3. The exergy efficiency depends on the energy efficiency, the environmental temperature and the product temperature. Using the electrical energy efficiencies as inputs, exergy efficiencies (ψ_e) have been determined and presented in **Table 4**. For example, the exergy efficiency for the refrigerator-freezer appliance is calculated using Equation (6) as follows:

Assuming:

Energy efficiency, $\eta_e = 60\%$.

Reference Temperature, $T_0 = (273 + 25)\,\text{K} = 298\,\text{K}$.

Product temperature, $T_p = (273 - 8)\,\text{K} = 265\,\text{K}$.

Then, exergy efficiency is calculated using the following formula

$$\psi_e = \left(1 - \frac{T_0}{T_p}\right)\eta_e = \left(1 - \frac{265}{298}\right) \times 60 = 7.47\%$$

The calculation of the overall energy and exergy efficiencies, η_0 and ψ_0, respectively, of all appliances can be done using the share or percentage use of each appliance in the total used energy as a weighting factor.

$$\eta_0 = \sum_i \eta_{ei} \times F_{ei}/100 \qquad\qquad (14)$$

$$\psi_0 = \sum_i \psi_{ei} \times F_{ei}/100 \tag{15}$$

where η_{ei} and ψ_{ei} are the energy and exergy efficiencies of appliance i (%), F_{ei} i the fractional energy usage for appliance i (%). Applying these equations, the overall weighted energy and exergy efficiencies for all appliances are 70.24% and 50.61%, respectively.

3.3.2. Energy and Exergy Efficiencies of a Cooking Appliance
Energy and exergy efficiencies of cooking appliances have been calculated using the input data shown in Section 2.4 and Equations (10) and (11) and presented in **Table 4**.

3.3.3. Energy and Exergy Efficiencies of Lighting
According to the survey, lighting electrical consumption in Cameroon is found to be divided approximately between fluorescent and bulb types with their energy and exergy efficiencies are 20% and 18.5%, and 5% and 4.5%, respectively [20] [21]. Using these two values, the energy and exergy efficiencies for lighting are calculated and were 12.5% and 11.5%, respectively.

3.4. Weighted Mean Energy and Exergy Efficiencies

The weighted mean energy and exergy efficiencies are calculated using Equations (8) and (9) and shown in **Table 7**. A sample example for the year 2010 is shown below:

$$R_{\eta_e} = (1275.16 \times 12.5 + 403.63 \times 80 + 254.89 \times 80 + 850.22 \times 60 + 85.20 \times 70 + 424.93 \times 70$$
$$+ 254.89 \times 80)/(1275.19 + 403.63 + 254.89 + 850.22 + 85.20 + 424.93 + 254.89)$$
$$= 49.50\%$$

$$R_{\psi_e} = (275.16 \times 11.5 + 403.63 \times 80 + 254.89 \times 80 + 850.22 \times 7.47 + 85.20 \times 70 + 424.93 \times 70$$
$$+ 254.89 \times 80)/(1275.19 + 403.63 + 254.89 + 850.22 + 85.20 + 424.93 + 254.89)$$
$$= 36.56\%$$

3.4.1. Weighting Factor
The weighting factors for fossil fuel and electricity is determined by applying Equations (12) and (13) and presented in **Table 8**. An example for the year 2010 is calculates as follows:

Table 7. Weighted mean electrical and fossil fuel energy and exergy efficiency (%).

Year	Electrical		Fossil fuel	
	Energy efficiency R_{η_e}	Exergy efficiency R_{ψ_e}	Energy efficiency R_{η_f}	Exergy efficiency R_{ψ_f}
2001	49.50	36.56	65	13.20
2002	49.50	36.56	65	13.20
2003	49.50	36.56	65	13.20
2004	49.50	36.56	65	13.20
2005	49.50	36.56	65	13.20
2006	49.50	36.56	65	13.20
2007	49.50	36.56	65	13.20
2008	49.50	36.56	65	13.20
2009	49.50	36.56	65	13.20
2010	49.50	36.56	65	13.20

Table 8. Overall weighting factors for electrical and fossil fuel energy.

Year	WF_e	WF_f
2001	0.25	0.74
2002	0.26	0.74
2003	0.25	075
2004	0.28	0.72
2005	0.33	0.67
2006	0.34	0.66
2007	0.36	0.64
2008	0.38	0.62
2009	0.38	0.61
2010	0.40	0.59

$$WF_e = \frac{3550}{5240+3550} = 0.40 \text{ for the electricity}$$

$$WF_f = \frac{5240}{5240+3550} = 0.59 \text{ for the fossil fuel}$$

3.4.2. Overall Energy and Exergy Efficiencies

Using the Equations (14) and (15), the overall energy and exergy efficiencies are determined and presented in the **Figure 1**. The example for the year 2010 is illustrated

$$\text{Overall energy efficiency} = (49.50 \times 0.40) + (65 \times 0.59) = 58.15\%$$

$$\text{Overall exergy efficiency} = (36.56 \times 0.40) + (13.2 \times 0.59) = 22.41\%$$

We note that, exergy efficiency was lower than its corresponding energy efficiency for the years 2001-2010. The most significant differences between energy and exergy efficiencies are attributed to cooking and Refrigerator-freezer.

In **Table 9** we classified the end use of energy into three great families (electrical appliances, lighting and cooking). It can be seen that, electrical appliances are the most efficient end uses for energy and exergy utilization. Owing to the fact that most electrical appliances have identical energy and exergy efficiency values since their output is either electrical or mechanical energy (**Figure 2**).

For the proper utilization of exergy, it is desirable to have a value for η as close to unity as practical and also a good match between the supply energy quality and the end use process. **Figure 2** shows the energy efficiency, exergy efficiency as well as the ratio between them for different end uses of the residential sector in Cameroon. A proper match is made whenever the value of the energy efficiency/exergy efficiency ratio approaches unity. The inefficient utilization of energy resource is characterized by the higher value [14]. Cooking stands out as the most inefficient end use in the Cameroon's residential sector, due to the large gap between reference temperature (298 K) and product temperature (374 K).

Cameroon is not the only country where the exergy efficiency of his residential sector is lower than the energy efficiency many other researches where made and the same conclusion was done. According to the **Table 10**, it is observed that the exergy efficiency of the residential sector in Cameroon is higher than in many countries. This can be explained by the standard of living of the Cameroonian population and climate. Indeed because of their standard of living, Cameroonians can not afford some appliances with high exergy loss as rice cooker washing machine and air conditioner. The favorable climate also avoids certain need such as water heating and

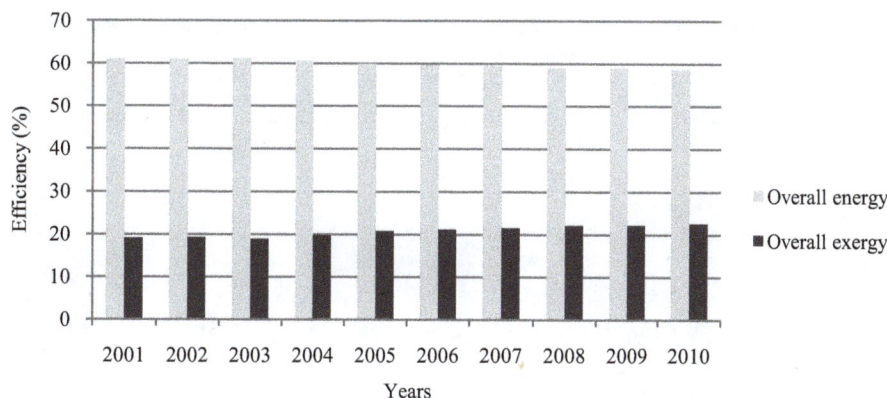

Figure 1. Overall energy and exergy efficiency of residential sector in Cameroon.

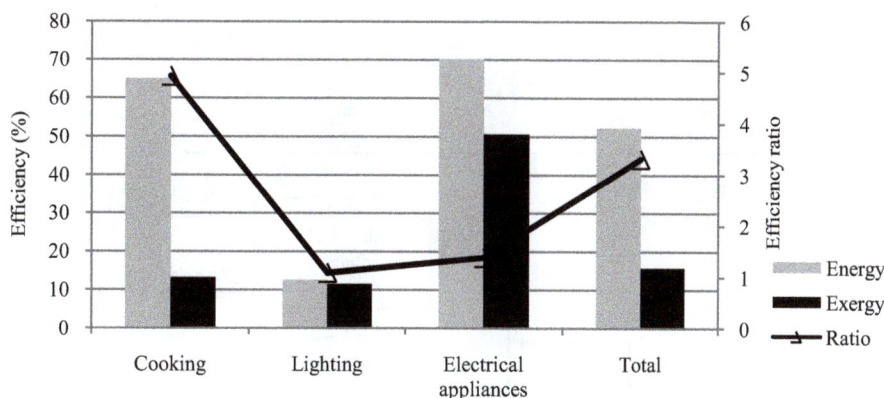

Figure 2. Energy efficiency, exergy efficiency for different end uses.

Table 9. Calculated average energy efficiency, exergy efficiency of Cameroon's residential sector in 2010.

End use	η (%)	ψ (%)	Share (%)
Cooking	65	13.20	59.61
Lighting	12.50	11.50	14.50
Electrical appliances	70.24	50.61	25.87
Total	58.15	22.41	100

space heating. These two needs are the main causes of low exergy efficiency in the residential sector in Jordan.

3.5. Energy and Exergy Products and Losses

To identify the energy and exergy products and losses, it is relevant to utilize energy and exergy flow. These flow diagrams are also helpful in distinguishing the high or low energy and exergy efficient devices. Based on the result of the analyses presented above, we established the energy and exergy flow diagrams of Cameroon's residential sector for the year 2010. These diagrams are presented in **Figure 3** and **Figure 4** which show energy and exergy inputs, productions and losses for each end use of the sector.

To raise the exergy efficiency in the residential sector in Cameroon, more actions are undertaken. One of them is the creation in the Ministry of Water and Energy a Sub-Directorate of Energy Management. One of its mission is to promote energy efficiency in the Cameroon territory and thus in his actions, several awareness campaigns are carried out in this direction.

Table 10. Overall energy and exergy efficiencies at the residential sector of few countries.

Countries	Year	Overall energy efficiency (%)	Overall exergy efficiency (%)	Reference
China	2005	-	10	[8]
Canada	1986	50	15	[7]
USA	1970	50	14	[7]
Brazil	2001	35	23	[7]
Italy	1990	-	2	[7]
Malaysia	1997-2004	69.32	29.98	[12]
Jordan	2006	66.60	15.40	[14]
Turkey	2004-2005	80.98	22.17	[15]
Norway	2000	-	12	[7]
Saudi Arabia	2004	76	9	[9]
Cameroon	2010	58.15	22.41	-

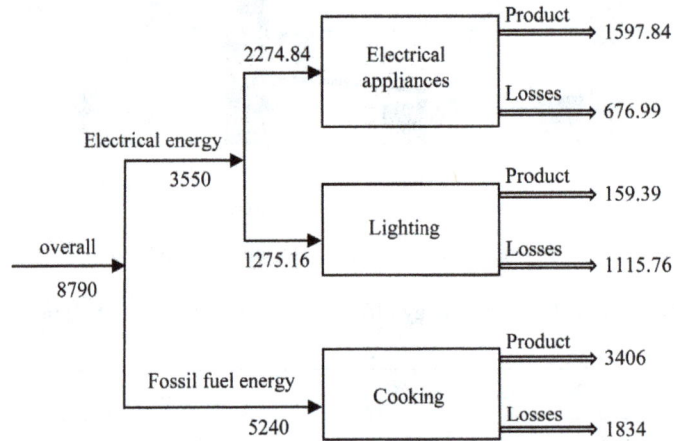

Figure 3. Energy flow (TJ) diagram for the residential sector in Cameroon for the year 2010.

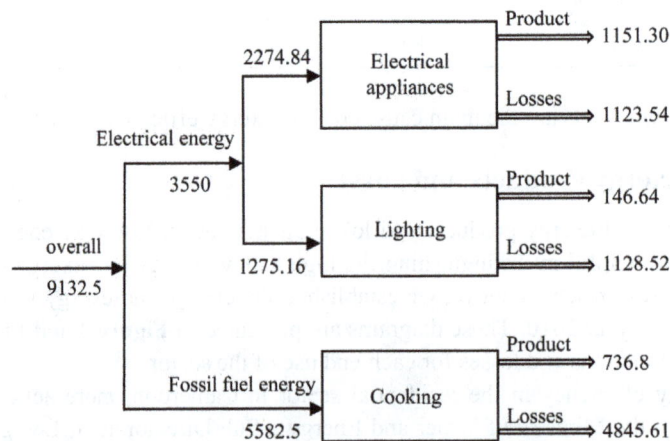

Figure 4. Exergy flow (TJ) diagram for the residential sector in Cameroon for the year 2010.

4. Conclusion and Policy Implications

In this paper, the overall energy and exergy efficiencies of the residential sector in Cameroon for a period of 10 years (2001-2010) have been determined by applying exergy analysis techniques used by Dincer *et al.* [9]. In 2010, the overall energy efficiency was 58.74% whereas the overall exergy was 22.63%. The exergy efficiency appears to be much lower than its corresponding energy efficiency due to the large amount of losses taking place in cooking end use. Analyses in this study have shown that the highest energy and exergy efficient end uses are the electrical appliances while lighting and cooking end uses are the least respectively.

The obtained efficiency values are comparable to those reported in China, Canada, USA, Brazil, Italy, Malaysia, Jordan, Turkey, Norway and Saudi Arabia. The exergy efficiencies of the Cameroon residential sector appear to be higher than those of Saudi Arabian and Norwegian but less than those of Malaysian and Turkish.

The data obtained from the study can be used as a basis for calculating cost benefit analysis for the implementation of new renewable sources for electricity generation and developing an emission abatement program in Cameroon.

References

[1]   (2010) Energy Information System.

[2]   Saidur, R., Masjuki, H.H. and Jamaluddin, M.Y. (2007) An Application of Energy and Exergy Analysis in Residential Sector of Malaysia. *Energy Policy*, **35**, 1050-1063. http://dx.doi.org/10.1016/j.enpol.2006.02.006

[3]   Moran, M. (1982) Availability Analysis: A Guide to Efficient Energy Use. Prentice-Hall, Upper Saddle River.

[4]   Reistad, G. (1975) Available Energy Conversion and Utilization in the United States. *Journal of Energy Power*, **97**, 429-434.

[5]   Wall, G., Sciubba, E. and Naso, V. (1994) Exergy Use in the Italian Society. *Energy—The International Journal*, **19**, 1267-1274.

[6]   Ozdogan, O. and Arikol, M. (1995) Energy and Exergy Analyses of Selected Turkish Industries. *Energy*, **20**, 73-80. http://dx.doi.org/10.1016/0360-5442(94)00054-7

[7]   Hammond, G.P. and Stapleton, A.J. (2001) Exergy Analysis of the United Kingdom Energy System. *IMech-Journal of Power and Energy*, **215**, 141-162. http://dx.doi.org/10.1243/0957650011538424

[8]   Ertesvag, I.S. and Mielnik, M. (2000) Exergy Analysis of the Norwegian Society. *Energy*, **25**, 957-973. http://dx.doi.org/10.1016/S0360-5442(00)00025-6

[9]   Xi, J. and Chen, G.Q. (2005) Exergy Analysis of Energy Utilization in the Transportation Sector in China. *Energy Policy*, **34**, 1709-1719.

[10]   Dincer, I., Hussain, M. and Al-Zaharnah, I. (2004) Analysis of Sectoral Energy and Exergy Use of Saudi Arabia. *International Journal of Energy Research*, **28**, 205-243. http://dx.doi.org/10.1002/er.962

[11]   Dincer, I., Hussain, M.M. and Al-Zaharnah, I. (2004) Energy and Exergy Utilization in Transportation Sector of Saudi Arabia. *Applied Thermal Engineering*, **24**, 525-538. http://dx.doi.org/10.1016/j.applthermaleng.2003.10.011

[12]   Dincer, I., Hussain, M.M. and Al-Zaharnah, I. (2004) Energy and Exergy Use in Public and Private Sector of Saudi Arabia. *Energy Policy*, **32**, 1615-1624. http://dx.doi.org/10.1016/S0301-4215(03)00132-0

[13]   Ertesvag, I.S. (2005) Energy, Exergy, and Extended-Exergy Analysis of the Norwegian Society 2000. *Energy*, **30**, 649-675. http://dx.doi.org/10.1016/j.energy.2004.05.025

[14]   Al-Ghandoor, A., Al-Hinti, I., Akash, B. and Abu-Nada, E. (2008) Analysis of Energy and Exergy Use in the Jordanian Urban Residential Sector. *International Journal of Exergy*, **5**, 413-428.

[15]   Szargut, J., Morris, D. and Steward, R. (1988) Exergy Analysis of Thermal, Chemical, and Metallurgical Processes. Hemisphere Publishing Corporation, New York.

[16]   Petchers, N. (2003) Combined Heating, Cooling and Power Handbook: Technologies and Application. The Fairmont Press, Lilburn.

[17]   Utlu, Z. and Hepbasli, A. (2007) A Review on Analyzing and Evaluating the Energy Utilization Efficiency of Countries. *Renewable and Sustainable Energy Reviews*, **11**, 1-29. http://dx.doi.org/10.1016/j.rser.2004.12.005

[18]   Utlu, Z. and Hepbasli, A. (2005) Analysis of Energy and Exergy Use of the Turkish Residential-Commercial Sector. *Building and Environment*, **40**, 641-655. http://dx.doi.org/10.1016/j.buildenv.2004.08.006

[19]   Utlu, Z. and Hepbasli, A. (2003) A Study on the Evaluation of Energy Utilization Efficiency in the Turkish Residential-Commercial Sector Using Energy and Exergy Analyses. *Energy and Buildings*, **35**, 1145-1153.

http://dx.doi.org/10.1016/j.enbuild.2003.09.003

[20]   Rosen, M.A. and Dincer, I. (1997) Sectoral Energy and Exergy Modeling of Turkey. *Transaction ASME*, **119**, 200-204.

[21]   Ileri, A. and Gurer, T. (1998) Energy and Exergy Utilization in Turkey during 1995. *Energy*, **23**, 1099-1106.
       http://dx.doi.org/10.1016/S0360-5442(98)00063-2

5

Net-Zero Energy Buildings and Communities: Potential and the Role of Energy Storage

Marc A. Rosen

Faculty of Engineering and Applied Science, University of Ontario Institute of Technology, Oshawa, Canada
Email: Marc.Rosen@uoit.ca

Abstract

Net-zero energy buildings and communities, which are receiving increasing interest, and the role of energy storage in them, are described. A net-zero energy building or community is defined as one that, in an average year, produces as much energy from renewable energy as it consumes. Net-zero energy buildings and communities and the manner in which energy sustainability is facilitated by them are described and examples are given. Also, energy storage is discussed and the role and importance of energy storage as part of net-zero buildings and communities are explained. The NSERC Smart Net-zero Energy Buildings Research Network, a major Canadian research effort in smart net-zero energy buildings and communities, is described.

Keywords

Net-Zero Energy Building, Net-Zero Energy Community, Energy Storage

1. Introduction

Net-zero energy buildings and communities, which in an average year produce as much energy from renewable energy sources as they consume, are becoming increasingly applied and gaining interest. By reducing energy use to net-zero and expanding the use of renewable energy, such buildings and communities facilitate sustainability. Energy storage is a key component of most net-zero energy buildings and communities and a key to their achieving net-zero energy use. This is primarily due to the intermittency of most renewable energy, which necessitates an energy storage capability in order for them to be used to a large extent. In this article, we examine and review the status of net-zero energy buildings and communities and energy storage, as well as needs for better linkages between them. The objective is to improve understanding of these technologies and their needs, and thereby to facilitate increased applications of them and more sustainable use of energy.

2. Net-Zero Energy Buildings and Communities

A net-zero energy building is defined as one that, in an average year, produces as much energy (electrical plus thermal) from renewable energy sources as it consumes. A net-zero energy community is similarly defined, but extended from an individual building to a community made up of a group of buildings as well as other components. Various technologies are required in smart net-zero energy buildings and communities (e.g., building-integrated solar systems, high performance windows with active control of solar gains, heat pumps, combined heat and power technologies, energy storage technologies ranging from short- to long-term, and smart controls. Much research is ongoing and has been reported in this area [1]-[5].

An example of the growing importance of net-zero energy buildings and communities can be observed through the NSERC Smart Net-zero Energy Buildings Research Network (SNEBRN) (http://www.solarbuildings.ca). SNEBRN is currently the major Canadian research effort in smart net-zero energy buildings and communities, bringing together 29 Canadian researchers from 15 universities, as well as experts from Natural Resources Canada, Hydro-Québec, and other industrial partners. International researchers are also participating in the SNEBRN. The aim is to develop the smart net-zero energy homes, commercial buildings and communities of the future. The vision of SNEBRN is to facilitate widespread adoption in key regions of Canada, by 2030, of optimized net-zero energy building design and operation concepts suited to Canadian climatic conditions and construction practices. The various technologies required in smart net-zero energy buildings and communities, mentioned in the previous paragraph, are all undergoing research in SNEBRN.

The Drake Landing Solar Community (DLSC) [6], located in Okotoks, Alberta, Canada and completed in 2006, is an example of a community energy system with several sustainable technologies, including solar energy, thermal energy storage, district heating and heat pumps. DLSC includes 52 low-rise detached homes (138 - 151 m^2 in gross floor area), which are located within a 835-home subdivision. DLSC is not a net-zero energy community, as only 90% of heating and 60% of hot-water needs are designed to be met using solar energy, but it could be expanded to net-zero energy use. The DLSC project demonstrates the feasibility of replacing substantial residential conventional fuel energy use with solar energy, collected during the summer and utilized for space heating during the following winter, in conjunction with seasonal thermal energy storage.

3. Energy Storage

Energy storage is an often hidden—but nonetheless important—energy technology. It can enhance the performance of energy systems, and improve their efficiency, economics, reliability and environmental impact. It is also a key to facilitating the widespread use of many renewable energy resources. Consequently, energy storage, although utilized quietly at present, is likely to be increasingly and more overtly used in the future as pressures from energy costs, security of energy supply and environmental damage expand.

This growth in interest in energy storage was recently emphasized in Ontario, via that province's recent Long-Term Energy Plan, which calls for the procurement of 50 MW of energy storage capacity by the end of 2014. Consequently, an energy storage procurement framework was jointly submitted by the Independent Electricity System Operator (IESO) and Ontario Power Authority (OPA), and supported by the Minister of Energy in the province. The procurement framework allows for a diverse portfolio of energy storage technologies, so as to foster improved understanding of and experience with 1) the services energy storages can provide, 2) the benefits they bring to operations and 3) how storage can best be integrated into electricity markets. In Phase I of the procurement framework, released March 12, 2014, the IESO issued a request for proposals for up to 35 MW of grid energy storage capacity from various storage technologies that can provide ancillary and other grid services (e.g., energy time-shifting, transmission congestion relief).

Some of the benefits of energy storage, which often are quite significant, follow:

- When energy demand varies significantly temporally, energy storage allows peak demands to be reduced or shifted to periods of lower demand, often with significant economic advantages.
- Energy storage is especially helpful for bridging periods between energy supply and energy demand, and reducing the mismatches between periods of energy supply and demand.
- The performance of energy systems that incorporate intermittent energy sources is made much more effective through the use of energy storage.
- Energy storage facilitates the utilization of many renewable energy sources which, due to their intermittent nature, are often not available when energy services are required. Wind and solar energy technologies, for instance, often benefit from storages that allow the energy products to be held until needed, if they are not required when the wind is blowing or the sun is shining.
- Increased operational performance flexibility can be provided through the use of energy storage.
- Energy storage can reduce use of fossil fuels and their associated environmental impacts.

A broad range of types of energy storage technologies and systems exists (see **Figure 1**). These can be separated into as chemical, electrical, thermal, thermochemical, mechanical and other classifications. Specific types of energy storage include battery storage, hydrogen energy storage, flywheel energy storage, compressed-gas energy storage, pumped storage, magnetic storage, capacitor storage, chemical storage, thermal energy storage (both sensible and latent), thermochemical energy storage, organic and biological energy storage, and others. Although much energy storage is mature and commercially available, new storage technologies are being actively investigated and improvements to existing ones continually sought. Some technologies are both commercial at present but also undergoing extensive research, e.g., thermal energy storage [6].

Energy storage can be utilized in a wide range of applications (see **Figure 2**). The main types of applications include utility and other electrical power systems, conventional and renewable power generation, renewable energy sources, transportation, heat pumps, building heating and cooling and district energy systems. The ability of energy storage to facilitate the efficient, effective and economic operation of renewable energy systems is particularly noteworthy, e.g., energy storage to improve solar power plants, as is the ability of energy storage to facilitate the design and operation of net-zero energy buildings and communities.

A recent book by the author describes energy storage from several perspectives, providing a broad understanding and appreciation of the technology and its importance and benefits [7]. Included in the book are descriptions of energy storage technology and systems, updates on the status of technological development of energy storage technology, and recent advances in energy storage technologies, systems and applications. The book is organized into several parts, mainly based on category of energy storage, which highlight the diversity of storage types and applications:

- General energy storage concepts, technologies and methods, including an overview of the technology and methods for evaluating efficiencies of energy storage systems.
- Hydrogen energy storage, including descriptions of hydrogen storage on such materials as magnesium hydride, porous materials and hydrogen-based compounds.
- Electrical energy storage, including emerging energy storage technologies in utility power systems, water storage for storing electricity to facilitate renewable energy use, double layer capacitors for electrical storage, and energy storage technologies for future power systems.
- Thermal energy storage, including descriptions of thermal storage media for solar thermal power plants, the performance of phase change materials and geothermal storage, and applications of thermal energy storage in building heating and cooling and district energy.

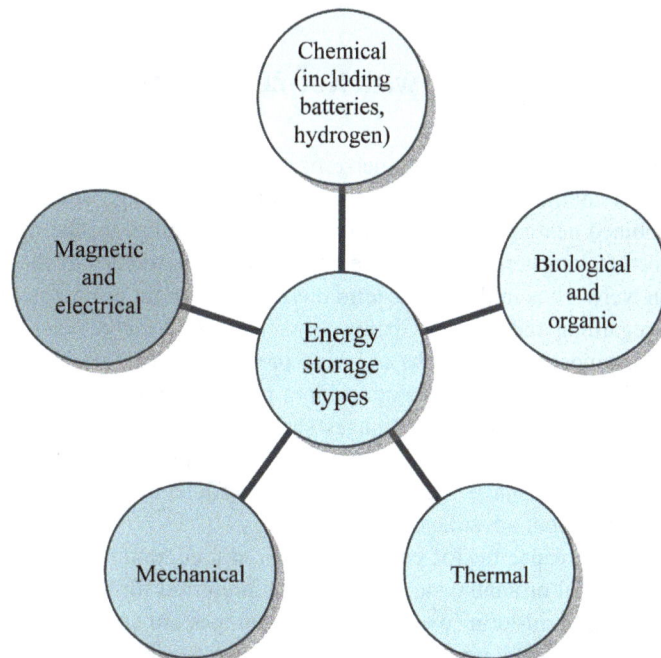

Figure 1. Energy storage types classified by type of stored energy.

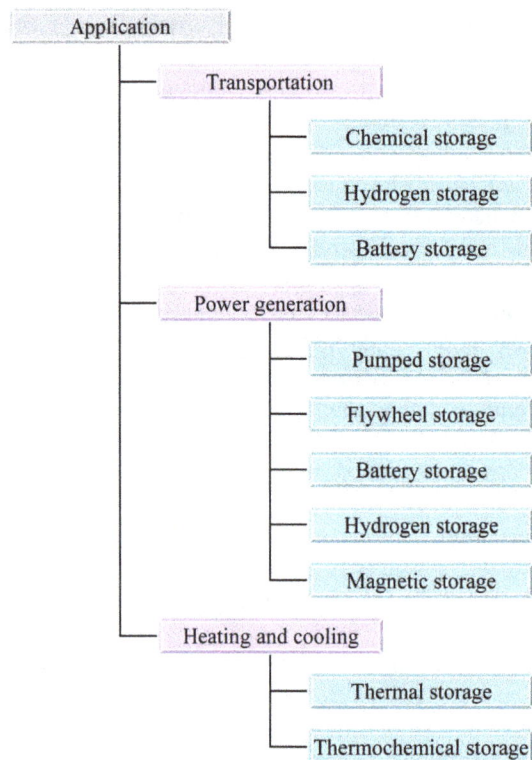

Figure 2. Selected energy storage application, based on area of utilization.

- Thermochemical energy storage, including a description of the technology and applications.

The benefits of and need for energy storage make it of interest wherever energy is used throughout the world. Global interest in energy storage technologies and applications is evident from the chapters in the above mentioned book, which originate from several countries: Canada, United States, Argentina, France, Italy, Denmark, Croatia, Iran, Japan and Thailand.

4. Needs for Linking Energy Storage with Net-Zero Energy Buildings and Communities

Research is required on linking energy storage with net-zero energy buildings and communities. Although smart net-zero energy buildings and communities require various technologies (e.g., solar energy, high performance windows, heat pumps, combined heat and power, smart controls), they hinge on the use of energy storage technologies, ranging from short- to long-term. In fact, one of the five main themes of the NSERC Smart Net-zero Energy Buildings Research Network is mid- to long-term thermal storage for buildings and communities.

Numerous needs exist regarding the linkages between energy storage and net-zero energy buildings and communities, so as to facilitate more advantageous and widespread applications:

- The appropriate scale and number of energy storages in a net-zero energy community need to be better assessed. Some efforts have focused on single storages scaled to a size appropriate for a given community, while others have focused on multiple smaller storages appropriately located throughout the community. Although smaller storages tend to have higher losses, other efficiency advantages of multiple storages in community settings can make them advantageous.
- The appropriate time duration capacities of energy storages in a community need to be determined. Long-term storage (based on seasonal or annual storage cycles) is preferred for some community energy systems, while short-term (diurnal) and mid-term (weekly) thermal storages are appropriate for others. Sometimes, a combination of short-, medium- and long-term storage is required to yield the most benefits from community energy systems. For example, the Drake Landing Solar Community described earlier uses a combination of

seasonal ground-based thermal storage with short-term liquid storage tanks.

- The appropriate type of energy storage(s) in a community energy system need to be better analyzed and identified. For instance, thermal energy storage is often employed, based on sensible, latent, thermochemical and combined modes.
- Developments are needed in energy storage technology and systems for net-zero buildings and communities, in terms of factors such as efficiency, reliability, economics, environmental impact and others, so as to achieve optimal performance.

5. Conclusion

Net-zero energy buildings and communities are being applied and interest in them is growing. Energy storage plays a key role in net-zero buildings and communities. Net-zero energy buildings and communities are being investigated actively in many countries, e.g., the NSERC Smart Net-zero Energy Buildings Research Network is expanding knowledge in this field and bringing together researchers and practitioners. Net-zero energy buildings and communities and energy storage technologies are becoming more effective and advantageous, making applications likely to expand and permitting better and more sustainable energy systems.

Acknowledgements

Support was provided by the Natural Sciences and Engineering Research Council of Canada.

References

[1]　Bucking, S., Athienitis, A. and Zmeureanu, R. (2013) An Information Driven Hybrid Evolutionary Algorithm for Optimal Design of a Net Zero Energy House. *Solar Energy*, **96**, 128-139. http://dx.doi.org/10.1016/j.solener.2013.07.011

[2]　Berggren, B., Hall, M. and Wall, M. (2013) LCE Analysis of Buildings: Taking the Step towards Net Zero Energy Buildings. *Energy and Buildings*, **62**, 381-391. http://dx.doi.org/10.1016/j.enbuild.2013.02.063

[3]　Gaiser, K. and Stroeve, P. (2014) The Impact of Scheduling Appliances and Rate Structure on Bill Savings for Net-Zero Energy Communities: Application to West Village. *Applied Energy*, **113**, 1586-1595. http://dx.doi.org/10.1016/j.apenergy.2013.08.075

[4]　Cellura, M., Guarino, F., Longo, S. and Mistretta, M. (2014) Energy Life-Cycle Approach in Net Zero Energy Buildings Balance: Operation and Embodied Energy of an Italian Case Study. *Energy and Buildings*, **72**, 371-381. http://dx.doi.org/10.1016/j.enbuild.2013.12.046

[5]　Mohamed, A., Hasan, A. and Sirén, K. (2014) Fulfillment of Net-Zero Energy Building (NZEB) with Four Metrics in a Single Family House with Different Heating Alternatives. *Applied Energy*, **114**, 385-399. http://dx.doi.org/10.1016/j.apenergy.2013.09.065

[6]　Dincer, I. and Rosen, M.A. (2011) Thermal Energy Storage. 2nd Edition, Wiley, London.

[7]　Rosen, M.A. (2012) Energy Storage. Nova Science Publishers, Hauppauge.

Disk Bimorph-Type Piezoelectric Energy Harvester

V. Tsaplev[1,2], R. Konovalov[1], K. Abbakumov[1]

[1]Department of Electroacoustics and Ultrasonic Engineering, Saint-Petersburg State Electrotechnical University, Saint-Petersburg, Russia
[2]Department of Physics, North-West Open Technical University, Saint-Petersburg, Russia
Email: valery@convergences-fr.ru

Abstract

The study of the experimental investigation of a disk-type piezoelectric energy harvester presented. The harvester contains disk bimorph piezoceramic element of the umbrella form and contains two disk PZT plates. The element is excited at the base point at its center. The element is supplied by a loading ring mass to decrease its resonance frequency. The dependences of the vibration displacement along the radii of the bimorph and the ring mass from the frequency of excitation are presented and the output voltage frequency response is also presented as well. The idle mode and the load duty are investigated. The value of the internal resistance of the harvester is obtained using the load characteristic. The piezoelectric specific power is estimated experimentally.

Keywords

Piezoelectric Energy Harvester, Umbrella Bimorph Plate, Amplitude-Frequency Characteristic, Flexural Vibration, Specific Power, Output Voltage

1. Introduction

Design, study and creation of microminiaturize piezoelectric energy harvesters now is a subject of interest of a great number of researches. The well known reason of this interest depends on the possibility of creation of small, independent and practically inexhaustible sources for a great number of different autonomous electronic devices. These sources transform the waste vibration energy, that exists practically everywhere, to the electrical energy. They do not need external energy sources and do not need expenses to periodic recharges of batteries or their chemical treatment.

One of the most early works on the usage of the energy vibrations for the needs of the power feeding of microelectronic devices [1] substantiates the advantages of the piezoelectric harvesting method compared with electromagnetic or electrostatic. Very detailed reviews [2] [3] show, that the main advantage of the piezoelectric method of harvesting by comparison with other methods of conversion is the greater output power density, and the simplicity of construction. The range of obtained values of voltage is more than obtained with the help of other types of harvesters. The main shortcoming of the piezoelectric method of harvesting is the great output resistance, but it is easily possible to eliminate this difficulty by proper choice of piezoelements in the harvester and by proper coupling them together in parallel or in series mode of connection. Moreover, this disadvantage one can turn into an advantage-by proper combination of elements and their connection one can obtain any values of output voltage and output impedance desired.

Piezoelectric harvesters generate less power density, than solar cell batteries, but our estimations show, that this parameter of piezoharvesters may also be competitive with that of solar ones, if use the nonlinear mode of piezoelements. Therewith, the efficiency of solar cell batteries strongly depends on the solar activity at the place.

Various types of piezoelectric harvesters were in detail reviewed in [4]. It was shown, that it is possible to use both longitudinal and transversal modes of oscillations, but mostly piezoelectric harvesters are made in the form of cantilever beam, one end of it being clamped and the other is free, or may be loaded by a passive mass. One or two piezoelectric layers are pasted to the beam thus making a unimorph or bimorph construction. Three or more piezoelectric layers are also possible. The beam is excited at the base point. The passive load mass is used to reduce the main resonance frequency of the beam thus lowering the frequency range of the whole construction.

Thus, we can conclude, that the piezoelectric method of energy harvesting provides much more freedom in the choice of specific construction. In this paper we describe disk bimorph-type piezoelectric transducer for energy harvesting. Unlike to the cantilever beam this construction seems to be more rational in relation to the arrangement.

2. Construction

Bimorph piezoelement for the piezoelectric energy harvester is shown in diagram form in **Figure 1**. The theory of a disk bimorph piezoelement, supported at the periphery and vibrating at the bending mode, is described in [5] [6]. Here we consider a bimorph umbrella-type piezoceramic element, free at the periphery and excited at the center, the vibration amplitude being small in comparison with its thickness.

The initial equations of piezoeffect are written as following:

$$u_{rr} = s_{11}^E \sigma_{rr} + s_{12}^E \sigma_{\theta\theta} + d_{31} E_z,$$
$$u_{\theta\theta} = s_{12}^E \sigma_{rr} + s_{11}^E \sigma_{\theta\theta} + d_{31} E_z, \tag{1}$$
$$D_z = d_{31} \sigma_{rr} + d_{31} \sigma_{\theta\theta} + \varepsilon_{33}^\sigma E_z,$$

where u_{rr}, $u_{\theta\theta}$-strain tensor components, σ_{rr}, $\sigma_{\theta\theta}$-bending stress tensor components, E_z-electric field strength, D_z-electric induction, s_{11}^E, s_{12}^E-elastic compliance tensor components, electric field strength being constant, d_{31}-piezoelectric modulus tensor components, ε_{33}^σ-dielectric permittivity tensor components of piezoceramics, under constant value of elastic stress. Bending elastic strain tensor components are defined as following:

$$u_{rr} = -z \frac{d^2\eta}{dr^2}, \quad u_{\theta\theta} = -\frac{z}{r} \frac{d\eta}{dr} \tag{2}$$

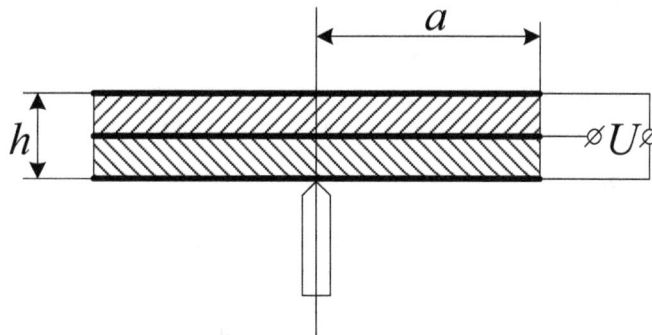

Figure 1. Bimorph transducer with central excitation.

where z-axial coordinate, r-radial coordinate, θ-angle, η-bending deflection. One must mind, that in the absence of volume charges $\text{div}\boldsymbol{D} = 0$, and without the magnetic field $\text{rot}\boldsymbol{E} = 0$. Then the voltage across the bimorph plates is

$$U(z > 0) = \int_0^{h/2} E_z \text{d}z, \quad U(z < 0) = -\int_{-h/2}^0 E_z \text{d}z \; , \tag{3}$$

And the transducer current

$$I = j\omega \int_0^a D_z 2\pi r \text{d}r \; , \tag{4}$$

where h-the whole thickness of the plate, a-disk radius, ω-circular frequency. Equation of motion for bending vibrations under harmonic excitation at the centre of bimorph plate we can write as

$$\frac{\text{d}M_t}{\text{d}r} - \frac{\text{d}^2(rM_r)}{\text{d}r^2} - \omega^2 rh\rho\eta = F\cos\omega t \; , \tag{5}$$

where M_t and M_r-bending torque intensity per length unit:

$$M_t = \int_{-h/2}^{h/2} \sigma_{\theta\theta} z\text{d}z, \quad M_r = \int_{-h/2}^{h/2} \sigma_{rr} z\text{d}z \; , \tag{6}$$

And ρ-piezoceramic density.
Boundary conditions at the centre of the disk:

$$\eta\big|_{r=0} = 0, \quad M_r\big|_{r=0} = 0 \tag{7}$$

But mind, that "pure bimorph" construction is well suitable for usage as the receiving transducer, but is not very suitable for energy harvesting. Mechanical stress in the layers of piezoceramics, adjacent to the neutral plane, is very small, and the electrical energy in these layers is also very small. Such mode of operation is good to receive weak acoustic signals that are subject to subsequent amplification. The weak signal provides low nonlinear distortions and clipping the signal. Nonlinear distortions have no importance in the case of energy harvesting. On the contrary, piezoceramics in nonlinear mode has much greater piezoelectric modulus, than at low levels of excitation. It was shown experimentally [7], that if the level of mechanical stress in piezoceramics is about 60 - 80 MPa, the transverse piezoelectric modulus d_{31} of the soft piezoceramics (as, for example, PZT-19) has the value about three times as much as the ordinary value (about 300×10^{-12} C/N, instead of specified value 100×10^{-12} K/H). The electric power density is directly proportional to the second power of piezoelectric modulus, that is why, the usage of the nonlinear vibration mode provides gain in energy by order of magnitude greater than in linear mode.

It means, that the transducer must have rather thick base metal plate with two bimorph piesoceramic plates, having oppositely directed polar vectors. If the base is rather thick, the piezoceramics works in the range of significant mechanical stresses.

But in this case the thickness of the whole construction also increases, and the piezoelectric transducer becomes high-frequency and comes out of the necessary range of vibration frequencies. The main frequency of the umbrella resonance also increases. To reduce this main frequency it was used the metal circumferential passive mass. The whole transducer is shown in **Figure 2**.

The nonlinear elastic properties of PZT-19 piezoceramics are in detail described in [7] [8], and in this paper we do not touch these problems. Here we just concern the study of experimental model at low-level excitation.

2. Experimental Results

The model of the piezoelectric energy harvester for experimental investigation is shown in **Figure 3**. The model consists of piezoelectric bimorph element on the bronze base plate pasted into the massive bronze ring. The cross-section of the piezoelement is shown in **Figure 3(a)**. Bimorph piezoelement consists of two piezoceramic PZT-19 plates, their dimensions are $\varnothing 32.0 \times 0.2$ mm. The piezoelement is pasted into the bronze ring (the internal diameter is 32 mm, the external one is 70 mm, and the thickness is 8 mm). The cylindrical base is pasted

Figure 2. Bimorph transducer with passive circular mass.

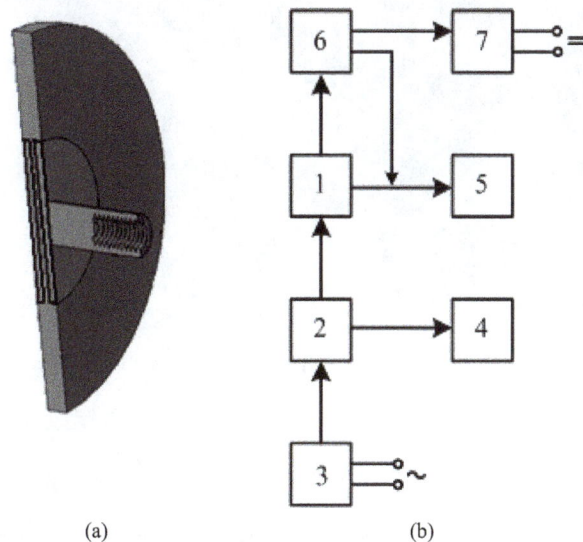

(a)   (b)

Figure 3. Cross-section of the bimorph transducer and the experimental setup.

to the centre, thus making an "umbrella" construction. This base serves to fasten the transducer to the shaker.

The piezoceramic plates were pasted to the base bronze plate in such a way, that their polarization vectors looked towards each other. The outside silver electrodes were connected electrically together, as it is showed in **Figure 1**, and the internal silver surfaces were soldered on to the base bronze plate.

The experimental setup is shown in **Figure 3(b)**. The piezoelectric transducer 1 was drived by the electrodynamic shaker 2, that in turn was excited by oscillator 3. The frequency was measured by the frequency meter 4. The output signal was measured by the electronic oscilloscope 5 (LeCroy WaveAce101). The amplitude of the vibrational displacement was measured by small vibration pick-up 6 (KD 91 PCB Piezotronics Inc.) with flat amplitude response from 1 Hz up to 15,000 Hz and 7 is the digital amplifier and registrator.

The values of amplitudes of vibration displacements across the surface of piezoelectric transducer were measured at three discrete points: at the centre and at two points −13 and 27 mm away from the centre. The sensor was small enough and its mass did not affect essentially the results of measurements.

Figure 4 is the overview of the experimental setup.

Figure 5 shows the amplitude-frequency response of the open-circuit, piezoelectric energy harvester. Maximal amplitude values of vibration displacements on the surface of transducer versus frequency f are plotted for three points on the surface. Curve 1 is for the point at the centre of the bimorph element, curve 2 corresponds to the point 13 mm away from the centre, and curve 3 is for the point 27 mm from the centre. One can observe that the frequency, corresponding to the maximal amplitude displacement decreases as the point of measurement moves off the centre. This may be explained by the influence of the mass of sensor, though other reasons may be possible as well. Следует отметить, что наблюдается смещение максимумов кривых в сторону более низких частот. This may be a subject of further study. The maximal vibration displacement on the surface of the bimorph plate is at the centre, and minimal at the periphery.

Figure 6 shows the amplitude-frequency response of the open-circuit harvester. The maximal value of the 0pen-circuit output voltage was 32 V and was restricted only by possibilities of the shaker.

The internal impedance characteristic was measured at the main resonance frequency by output load voltage characterization. It was about 17 kOhm. The output electric power at the matched load was about 15 mWt, and the output power density about 50 mWt/sm^3.

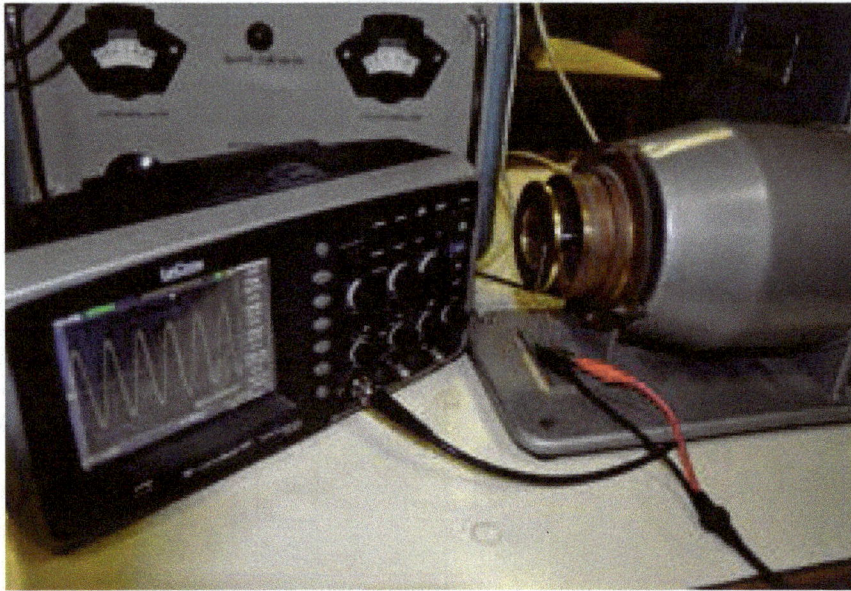

Figure 4. The overview of the experimental setup.

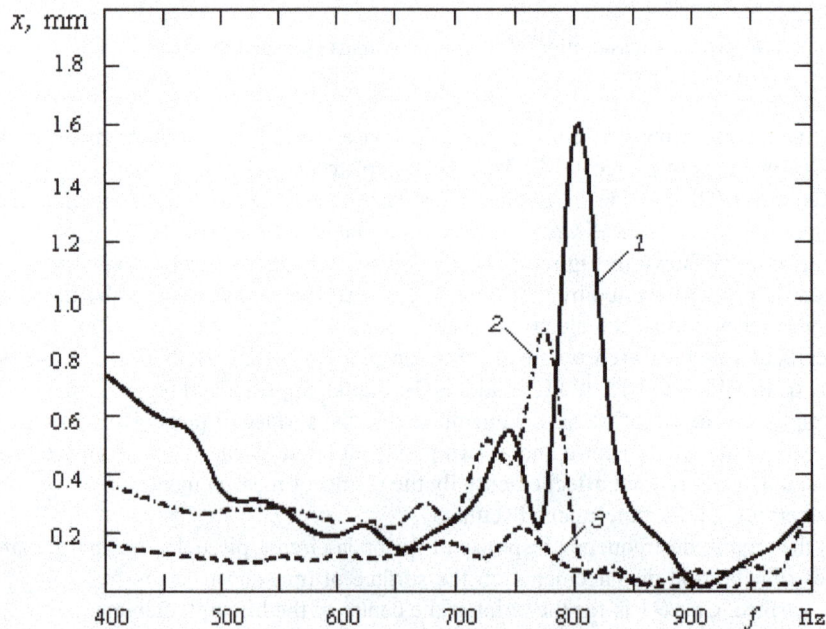

Figure 5. The amplitude-frequency response of the open-circuit transducer.

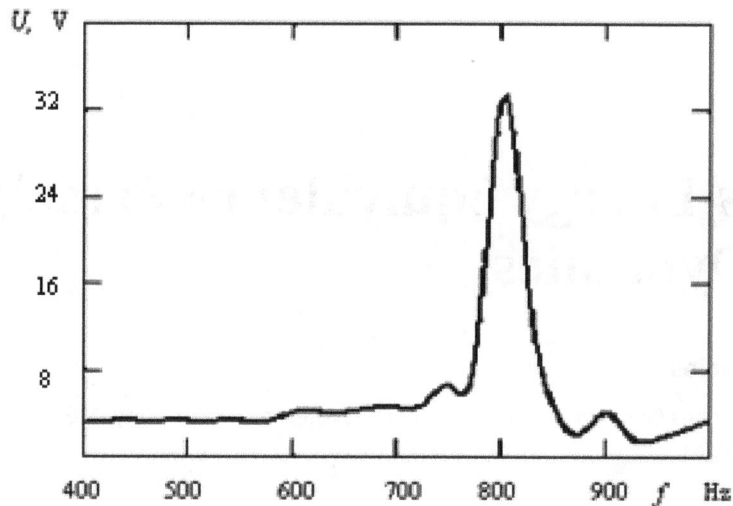

Figure 6. The amplitude-frequency response of the open-circuit transducer.

3. Conclusion

The results of the experiment show, that the design of this type of piezoelectric harvester provides some advantages as compared with cantilever beam-type harvesters. The main advantage is the possibility of increasing the total number of piezoelements, their total volume, and creation of multilayer constructions. This in turn makes possible the usage of nonlinear mode of vibration, thus increasing the output power.

Acknowledgements

This work has been accomplished as a part of the Ministry of Education and Science of the Russian Federation research assignment 'Realization of scientific research (fundamental studies, applied research and advanced developments)'. Project code: 2548.

References

[1]   Williams, C.B. and Yates, R.B. (1996) Analysis of a Micro-Electric Generator for Microsystems. *Sensors and Actuators A*: *Physical*, **52**, 8-11. http://dx.doi.org/10.1016/0924-4247(96)80118-X

[2]   Sodano, H., Park, G. and Inman, D.J. (2004) A Review of Power Harvesting from Vibration Using Piezoelectric Materials. *The Shock and Vibration Digest*, **36**, 197-205. http://dx.doi.org/10.1177/0583102404043275

[3]   Anton, S.R. and Sodano, H.A. (2007) A Review of Power Harvesting Using Piezoelectric Materials (2003-2006). *Smart materials and Structures*, **16**, R1-R21. http://dx.doi.org/10.1088/0964-1726/16/3/R01

[4]   Erturk, A. and Inman, D.J. (2011) Piezoelectric Energy Harvesting. John Wiley & Sons Ltd., New York.

[5]   Brailov, E.S. and Vassergisser, M.E. (1980) Evaluation of Flexural Bimorph Disk Piezoelement Characteristics. *Akusticheskij Zhurnal*, **26**, 590-595.

[6]   Antonyak, Yu.T. and Vassergiser, M.E. (1982) Calculated Characteristics of Flexural Membrane Type Piezoelectric Transducer. *Akusticheskij Zhurnal*, **28**, 294-302.

[7]   Tsaplev, V.M. (2013) Nonlinear Acoustoelastisity of Piezoceramic Materials: p. I. Physical Acoustics of Piezoceramics. Saint-Petersburg State Electrotechnical University, Saint-Petersburg.

[8]   Tsaplev, V.M. (2003) Nonlinear Properties and Creep in Piezoceramics. North-West State Technical University, Saint-Petersburg.

7

The Mass-Energy Equivalence Principle in Fluid Dynamics

Angel Fierros Palacios

Instituto de Investigaciones Eléctricas, División de Energías Alternas, Mexico City, México
Email: afierros@iie.org.mx

Abstract

From Lagrangian formalism as in Classical Field Theory and within the theoretical scheme of the Hamilton-Type Variational Principle, the mass-energy equivalence principle for any fluid is obtained.

Keywords

The Mass-Energy Equivalence Principle, Fluid Dynamics

1. Introduction

In the relativistic formulation of particle mechanics, it is demonstrated that the energy of a free particle does not vanish when its speed goes to zero. Instead, it reaches a finite value called the energy at rest of the particle. This is one of the best known, spectacular and important results of the Special Theory of Relativity of A. Einstein. The $E = mc^2$ Equation expresses the fact that mass and energy are equivalent, that is, they conform a single invariant denominated the mass-energy relation. It has been confirmed through multiple practical applications and according to Einstein its validity extends to the whole Universe. As a consequence, it is natural to state that this result can also be derived from the Hamilton-Type Principle of Fluid Mechanics within the theoretical scheme of Lagrange's Analytical Mechanics.

2. The Mass-Energy Relation

Consider any fluid confined within an arbitrary region R of the three-dimensional Euclidian space. It is desired to obtain the mass-energy equivalence principle for this continuous system with the help of the Lagrange formalism. In the analytical treatment of fluid dynamics [1], an action functional which is a space-time integral

$$W = \int\limits_{t_1}^{t_2} \int\limits_{R} \ell \, \mathrm{d}V \mathrm{d}t \tag{1}$$

of a Lagrangian density

$$\ell = \ell\left(J; \overline{\mathrm{grad}} J\right) \tag{2}$$

is commonly used. In this case the Lagrangian density is a function of the Jacobian J and its first gradient [1] [2]; which implies that it is a geometric Lagrangian density. According to the Hamilton-Type Principle, the action integral remains invariant for continuous infinitesimal geometric variations with respect to a set of continuous time independent geometric parameters $\{\alpha\}$, from which the field variables and coordinates depend; that is [1]

$$\delta W = \frac{\partial W}{\partial \alpha} \delta \alpha = 0. \tag{3}$$

Furthermore, the following boundary condition of general character is imposed on the coordinates

$$\delta x^i (t_1) = \delta x^i (t_2) = 0. \tag{4}$$

It can be shown that the local variation of the action integral (1) and the application of the Hamilton-Type Principle provide as a direct consequence the following result [1]

$$\int\limits_{t_1}^{t_2} \int\limits_{R} \rho \delta \lambda \, \mathrm{d}V \mathrm{d}t = 0 \; ; \tag{5}$$

where

$$\lambda = \lambda\left(J; \mathrm{grad} J\right) \tag{6}$$

is the specific Lagrangian. The Lagrangians (2) and (6) are related as follows [3] [4]

$$\ell = \rho \lambda \; ; \tag{7}$$

where $\rho(x,t)$ is the mass density. According to the functionality of λ given in (6) is it evident that

$$\rho \delta \lambda = \rho \frac{\partial \lambda}{\partial J} \delta J + \rho \frac{\partial \lambda}{\partial \nabla J} \delta \nabla J . \tag{8}$$

It can be demonstrated that [1]

$$\delta J = J \mathrm{div}(\delta x) \tag{9}$$

and

$$\delta \nabla J = \mathrm{grad} J \frac{\partial}{\partial x^i}\left(\delta x^i\right).$$

In that case, the following result can be obtained from (8):

$$\rho \delta \lambda = \frac{\partial}{\partial x^i}\left[\rho\left(J\frac{\partial \lambda}{\partial J} + \nabla J \frac{\partial \lambda}{\partial \nabla J}\right)\delta x^i\right] - \frac{\partial}{\partial x^i}\left[\rho\left(J\frac{\partial \lambda}{\partial J} + \nabla J \frac{\partial \lambda}{\partial \nabla J}\right)\right]\delta x^i \; ; \tag{10}$$

where integration by parts has been performed. If the first term of the right hand side of (10) is used in (5) and the Green's theorem is applied, the following result is obtained

$$\int\limits_{t_1}^{t_2} \int\limits_{S} \rho\left[J\frac{\partial \lambda}{\partial J} + \nabla J \frac{\partial \lambda}{\partial \nabla J}\right]\delta x^i \, \mathrm{d}a_i \mathrm{d}t \; ; \tag{11}$$

where S is the surrounding area of region R and $\mathrm{d}a$ is the differential of area. The surface integral is null due to the following. Consider a continuous medium contained within a region R which is not deformed at infinity and the integration surface extended to infinity, where $J = 1$, in such a way that $\nabla J = 0$ and the integral vanishes. The substitution of the second term of the right hand side of (10) into (5) provides the following result

$$\int_{t_1}^{t_2}\int_R \left\{ \frac{\partial}{\partial x^i}\left[\rho\left(J\frac{\partial \lambda}{\partial J} + \nabla J\frac{\partial \lambda}{\partial \nabla J} \right) \right] \right\} \delta x^i \mathrm{d}V \mathrm{d}t = 0 \,. \tag{12}$$

As the local variations of x are arbitrary and linearly independent among them, and $\mathrm{d}V$ as well as $\mathrm{d}t$ are completely arbitrary increments and therefore different from zero, the previous equation is satisfied only if the integrand vanishes; that is

$$\rho\,\mathrm{grad}\left[J\frac{\partial \lambda}{\partial J} + \nabla J\frac{\partial \lambda}{\partial \nabla J} \right] + \left[J\frac{\partial \lambda}{\partial J} + \nabla J\frac{\partial \lambda}{\partial \nabla J} \right]\mathrm{grad}\,\rho = 0 \tag{13}$$

where the derivative has been calculated. Nevertheless, it can be seen in the first term of (13), that

$$\rho\left[\frac{\partial J}{\partial x^i}\frac{\partial \lambda}{\partial J} + \frac{\partial \nabla J}{\partial x^i}\frac{\partial \lambda}{\partial \nabla J} \right] = 2\rho\frac{\partial \lambda}{\partial x^i} = 0 \,; \tag{14}$$

where the chain rule has been used to obtain the result. The term is zero because λ, $\partial \lambda / \partial J$ and $\partial \lambda / \partial \nabla J$ are not explicit functions of x. As $\mathrm{grad}\,\rho \neq 0$, it follows from (13) that

$$\frac{\partial \lambda}{\partial J} + \frac{1}{J}\frac{\partial \lambda}{\partial \nabla J}\mathrm{grad}\,J = 0 \,. \tag{15}$$

This is the field differential equation for the mass density in terms of the specific lagrangian. Let

$$\lambda = \varepsilon_o\left[J\Gamma_{ik}^i u_k + u_k\frac{\partial J}{\partial x^k} \right] \tag{16}$$

be the explicit form of the specific Lagrangian [1] [2]. Where, ε_0 is the constant equilibrium value of the internal specific energy, $\Gamma_{ik}^i = \Gamma_{ki}^i$ is the contracted Christoffel symbol and $u(x)$ is the displacement vector [1] [5] [6]. Hence, using (16) in (15) produces

$$\rho(x,t) = \frac{\rho_o}{J}u\cdot\mathrm{grad}\,J \tag{17}$$

as by definition [1] [2]

$$\Gamma_{ik}^i u_k = -\frac{\rho(x,t)}{\rho_o}\,. \tag{18}$$

Equation (17) is the scalar equation for the mass density [1] [2]. On the other hand, within the theoretical scheme of the Hamilton-Type Variational Principle, it is demonstrated that the action integral (1) remains invariant to an infinitesimal continuous transformation with respect to time; thus,

$$\delta^+ W = 0 \tag{19}$$

with

$$\delta^+(\ ) \equiv \frac{\mathrm{d}}{\mathrm{d}t}(\ )\delta^+ t \tag{20}$$

being the time variation definition [1]. Besides, the temporary variations are subject to the following general boundary condition [1]

$$\delta^+ t_1 = \delta^+ t_2 = 0 \,. \tag{21}$$

Hence, the invariance condition (19) applied to the action integral (1) provides the following as a general result

[1] [4]

$$\int_{t_1}^{t_2} \int_R \left(\delta^+ \ell - \frac{d\ell}{dt} \delta^+ t \right) dV dt = 0 .$$ (22)

According to the functionality of ℓ given in (2)

$$\delta^+ \ell = \frac{\partial \ell}{\partial J} \delta^+ J + \frac{\partial \ell}{\partial \nabla J} \delta^+ \nabla J .$$ (23)

Besides and by the definition in (20), it follows that

$$\delta^+ J = \frac{dJ}{dt} \delta^+ t = J \operatorname{div} v \delta^+ t ;$$ (24)

where the Euler relation [1]

$$\frac{dJ}{dt} = J \operatorname{div} v ,$$ (25)

has been considered. Moreover, it is clear that [1]

$$\delta^+ (\operatorname{grad} J) = (\operatorname{grad} J) \operatorname{div} v \delta^+ t ;$$ (26)

and hence

$$\delta^+ \ell = \left[\frac{\partial \lambda}{\partial J} + \frac{1}{J} \frac{\partial \lambda}{\partial \nabla J} \operatorname{grad} J \right] \rho J \operatorname{div} v \delta^+ t .$$

In that case $\delta^+ \ell = 0$, because the field Equation (15) is contained within the square bracket, and that term is zero. Therefore, (22) becomes

$$\int_{t_1}^{t_2} \int_R \frac{d\ell}{dt} \delta^+ t \, dV dt = 0 ,$$ (27)

which is satisfied only if the integrand vanishes [1]. Indeed, as the temporary variations $\delta^+ t$ are arbitrary and linearly independent among them, and dV as well as dt are arbitrary increments and therefore different from zero, it follows that

$$\frac{d\ell}{dt} = 0 ;$$ (28)

which is the mass balance equation. Effectively, it can be demonstrated (see Appendix) that the relativistic lagrangian density for any fluid free of forces is

$$\ell = -\rho c^2 \sqrt{1 - \frac{v^2}{c^2}} .$$

In the case of a fluid at rest, $\ell = -\rho c^2$, and according to (7), $\lambda = -c^2$; in such a way that neither ℓ, and of course nor λ are functions of the velocity field; so that

$$\frac{d\ell}{dt} = \rho c^2 \operatorname{div} v .$$ (29)

From the hydrodynamics derivative definition [7] and taking in to account the relationship (7) it is easy to obtain that

$$\frac{d\ell}{dt} = \lambda \left[\frac{\partial \rho}{\partial t} + \operatorname{div}(\rho v) \right] + \rho \frac{\partial \lambda}{\partial t} - \lambda \rho \operatorname{div} v ;$$ (30)

where an integration by parts was made, and it was considered that λ is not an explicit function of x. Finally, from (29) and (30), the following result is obtained

$$\lambda\left[\frac{\partial\rho}{\partial t}+\operatorname{div}(\rho v)\right]+\rho\frac{\partial\lambda}{\partial t}=0 . \tag{31}$$

Time uniformity has as a consequence that the specific Lagrangian does not become an explicit function of time; so that $\partial\lambda/\partial t=0$, and then in (31) we have that

$$\frac{\partial\rho}{\partial t}+\operatorname{div}(\rho v)=0 ; \tag{32}$$

because $\lambda\neq 0$. This is the mass balance equation.

Furthermore, the invariance of the action under transformations with respect to the evolution parameter and due to the uniformity of time, more than the mass balance equation, the energy balance equation should be obtained from Equation (28). As a consequence from the obtained result, it seems natural to assume that in the field of fluid dynamics there is a close relationship between the densities of mass and energy of any continuous medium. To prove the previous assumption, consider the general definition of the hamiltonian density [1] [4]

$$\mathcal{H}=\frac{\partial\ell}{\partial v^{i}}v^{i}-\ell ; \tag{33}$$

here v^{i} is the ith-component of the velocity field. As for the present case the geometric lagrangian density (2) does not depend on the velocity field, it holds that $\mathcal{H}=-\ell$ and therefore from (28) the following result can be obtained

$$\mathcal{H}=\text{constant} . \tag{34}$$

Assume that

$$\mathcal{H}=\rho\beta^{2} , \tag{35}$$

with ρ again as the mass density and β^{2} as a constant with units of velocity squared that has the purpose of balancing dimensions in Equation (35). It will be shown in the appendix that $\beta^{2}=c^{2}$, being c the velocity of light in vacuum space. Furthermore

$$\rho=\frac{\rho_{o}}{\sqrt{1-v^{2}/c^{2}}} ; \tag{36}$$

where ρ_{o} is the mass density at rest and v the magnitude of the flow velocity. Using the previous definitions, it follows that

$$\mathcal{H}=\frac{\rho_{o}c^{2}}{\sqrt{1-v^{2}/c^{2}}} ; \tag{37}$$

and the condition imposed in (34) is satisfied because the flow rate is constant. The previous equation shows that in the domain of relativistic dynamics, the kinetic energy of a fluid does not vanish when its flow rate vanishes, since for such case $\mathcal{H}$ has the following value as it can be seen from (37)

$$\mathcal{H}=\rho_{o}c^{2} . \tag{38}$$

This is the equation for the energy density of a continuous at rest medium. It expresses the equivalence between energy and mass densities in the field of fluid dynamics. It is clear that as $\mathcal{H}$ is the energy per unit volume and ρ_{o} the mass of fluid also per unit volume, Equation (38) reduces to the well known Einstein Equation of relativistic mechanics, $E=m_{o}c^{2}$; except that here m_{o} is not the mass of a free particle at rest, but the mass of the continuous medium at rest, free of forces.

Furthermore, for low flow rates when $v/c\ll 1$, Equation (37) can be expanded into a power series of v/c to obtain

$$\mathcal{H} \approx \rho c^2 + \frac{1}{2}\rho v^2 .$$ (39)

The form of the previous equation is due to the fact that when $v/c \to 0$, $\rho \to \rho_o$. This is the Newtonian limit of the general equation for the relativistic mass density ρ. Except for the energy at rest, the previous equation is the expression for the kinetic energy density of the continuous medium under study. It is clear from Equation (36) that the relativistic relationship between mass and velocity is satisfied [8]; that is

$$m = \frac{m_o}{\sqrt{1 - v^2/c^2}} ,$$ (40)

where m_o is the mass of the fluid at rest and v the magnitude of the flow velocity. Finally, from the total derivative of Equation (35) with respect to time and as $\rho\beta^2 \neq 0$, it can be directly obtained that

$$\mathrm{div} v = 0 .$$ (41)

This is the continuity equation for the case in which the mass density is referred to the system at rest. The previous relation is identically satisfied because the system under consideration is not in motion, and so the flow velocity is zero. In addition, as the kind of fluid contained in R has not been specified anywhere, the *mass-energy relation* is valid for any continuous medium, and so are the rest of the results obtained.

3. Conclusions

The theoretical scheme of the Hamilton-Type Variational Principle provides the methodology required to obtain the mass-energy equivalence principle for any fluid. The problem is enclosed within the Lagrange formulation of theoretical mechanics, in such a way that the required methodology is configured with the help of the action integral and a Lagrangian density depending only on purely geometric entities. The local variation of the action integral and the usage of appropriate boundary conditions within the Hamilton-Type Principle scheme produce as a result of the field equation for the mass density. The scalar equation for the mass density is obtained using the explicit form of the specific Lagrangian. As a result of the temporary variation of the action integral, the referred Hamilton-Type Principle and the corresponding boundary conditions, the mass balance equation is obtained.

Finally, from the mass balance equation and the general definition of the Hamiltonian density, the equivalence principle between the mass and energy densities in fluid dynamics is obtained. Furthermore, the relationship between the mass density and the flow velocity is provided, and the corresponding continuity equation for the case of mass density referred to a continuous medium at rest is obtained.

References

[1] Fierros Palacios, A. (2006) El Principio tipo Hamilton en la Dinámica de los Fluidos, primera y segunda ediciones. McGraw-Hill Interamericana Editores, S.A. de C.V. (1998 y 1999). México. Buenos Aires. Madrid. Nueva York. Montreal. Sidney. etc. The Hamilton-Type Principle in Fluid Dynamics. Fundamentals and Applications to Magnetohydrodynamics, Thermodynamics, and Astrophysics.Springer-Verlag. Wien.

[2] Fierros Palacios, A. (1994) La ecuación de campo para la densidad de masa. *Rev. del IMP*, **XXIV**.

[3] Viniegra, H.F., Salcido, G.A. and Fierros Palacios, A. (1979) Las ecuaciones de balance de un fluido perfecto a partir de un principio variacional tipo Hamilton. *Rev. del IMP*, **XI**.

[4] Fierros Palacios, A. (1992) Las ecuaciones de balance para un fluido viscoso a partir de un principio variacional tipo Hamilton. *Rev. Mex. Fis.*, **38**, 4518-4531.

[5] Landau, L.D. and Lifshitz, E.M. (1959) Theory of Elasticity. Addison-Wesley Publishing Co., Boston.

[6] Landau, L.D. and Lifshitz, E.M. (1962) The Classical Theory of Fields. Addison-Wesley Publishing Co., Boston.

[7] Landau, L.D. and Lifshitz, E.M. (1959) Fluid Mechanics. Addison-Wesley Publishing. Co., London, Paris, Frankfurt.

[8] Resnick, R. (1976) Conceptos de relatividad yteoría cuántica. Editorial Limusa, México.

Appendix

Let R be a region in the three-dimensional Euclidian space containing any fluid. Let ℓ be the lagrangian density such that the classic lagrangian has the following value

$$L = \int_R \ell \, dV \; ; \tag{A-1}$$

where dV is the volume element. According to relativistic dynamics, the classic lagrangian for a free particle [6] of mass m in motion with a constant velocity v is

$$L = -mc^2 \sqrt{1 - \frac{v^2}{c^2}} \; . \tag{A-2}$$

Never the less, the previous relationship also holds for an extended body; in particular, it must be valid for any continuous medium whose total mass is given by

$$m = \int_R \rho(x,t) \, dV \; ; \tag{A-3}$$

where $\rho(x, t)$ is the mass density. Substituting (A-3) in (A-2) and comparing the result with (A-1), it follows that

$$\ell = -\rho c^2 \sqrt{1 - \frac{v^2}{c^2}} \; . \tag{A-4}$$

In the newtonian limit of relativistic mechanics; that is, when $v/c \to 0$, (A-4) yields $\ell \to -\rho c^2$. In that case and since $H = -\ell$, $\beta^2 = c^2$, the assumption made after Equation (35) is satisfied. Besides, according to the *Lorentz contraction* [6] the *proper volume* of the continuous medium under consideration is

$$V = V_o \sqrt{1 - \frac{v^2}{c^2}} \; . \tag{A-5}$$

Since $V = m/\rho$ and $V_o = m/\rho_o$, Equation (36) is satisfied.

8

Research on Energy Consumption of Traditional Natural Villages in Transition: A Case Study in Zhejiang Province

Meiyan Wang[1,2]*, Shenglan Huang[2], Xinxin Lin[1], Didit Novianto[1], Liyang Fan[3], Weijun Gao[1], Zu Wang[4]

[1]Environmental Engineering, University of Kitakyushu, Kitakyushu, Japan
[2]School of Landscape and Architecture, Zhejiang A&F University, Lin'an, China
[3]Nikken Sekkei Research Institute, Osaka, Japan
[4]School of Architecture & Civil Engineering, Zhejing University, Hangzhou, China
Email: *wangmeiyan0606@163.com

Abstract

Traditional agriculture is in the direction of increasing integration of the primary industry, secondary industry, and tertiary industry in Zhejiang province. A survey was undertaken on energy consumption of traditional natural villages by taking Anji Ligeng village for an example. This paper firstly studied rural buildings, rural family structure, occupants' activity and the usage of household appliances in the form of a questionnaire. Then, the household energy resource structure and energy consumption structure were analyzed and compared with other surveys. The results show that, the electric energy consumption was 6 kWh/(m²·a), which was far less than urban residential household. In rural household energy resource structure, the proportion of non-commercial energy resource was higher than commercial energy resource. Firewood accounted for 83%, electricity for 12%, LPG for 3% and solar energy for 2%. In building energy consumption structure, cooking and hot water took up 33%, appliances 31%, lighting 20%, heating 12%, cooling 4%. In all influential factors, frequently used area, number of air conditioner per household and building function were obviously correlated with energy consumption; income, building shape factor and window to wall area ratio had no correlation with energy consumption in the low energy consumption area.

Keywords

Natural Village, Rural Building, Energy Resource, Energy Consumption Structure, Influential

---

*Corresponding author.

Factor

1. Introduction

Since the Reform and Opening-up policy was initiated in 1978, influenced by the tide of world economic globalization, China's rural areas are experiencing unprecedented rapid development, especially in the countryside along the coast areas where the economy developed rapidly. In 2010 the Chinese government launched the "New Rural Construction" campaign, in order to improve the rural living standards. Under the dual stimulation of politics and economy, rural building energy consumption increased from 80 kgce in 2000 to 246 kgce in 2012 [1].

By the radiation and drive of the China Yangtze River Delta region, the Engel coefficient of rural residents in Zhejiang province dropped from 50.4 in 1995 to 37.7 in 2012 (**Figure 1**), which indicated that the rural living standards reached the rich level. In recent years, Zhejiang government put forward the "one product for one village, one industry for one town" development plan, and then "Farmhouse", "E-commerce village" grew up. Traditional agriculture is in the direction of the integration of the primary industry, secondary industry, and tertiary industry in Zhejiang province. With the rural production and living patterns changing, rural household energy structure and energy consumption have had a huge change. The consumption of coal reduced; electricity and LPG increased fast; firewood, straw and other biomass substantially reduced [1]-[3] (**Figure 1**). The rapid development of economic and the tremendous change of rural energy structure have brought great energy pressure. The aim of this paper is to explore the features of rural household energy consumption in traditional natural rural villages by taking Anji Ligeng village for an example, and analyze its influential factors.

2. Previous Studies

Since the late 1980s, domestic and foreign scholars have done a lot of research on rural energy consumption in China [4]-[6], which were mainly based on the statistical data published by the government, or a wide range of questionnaire. Wang *et al.* analyzed the main energy sources used in rural areas and the end-used energy consumption according to the 30 provinces energy consumption statistics [7]. Guo *et al.* studied rural energy consumption level in 1996 from four sectors—lighting, cooking and hot water, heating and household appliances, by using both macro- and micro-approaches [8]. Tsinghua University has undertaken a large-scale survey in rural areas from 2006 to 2008, and compared the difference of rural energy consumption between the North and South rural areas [9].

In terms of the influences of energy consumption, the relationship between income and energy consumption has been studied. Wang *et al.* believes that income is related to electricity consumption, but less related to com-

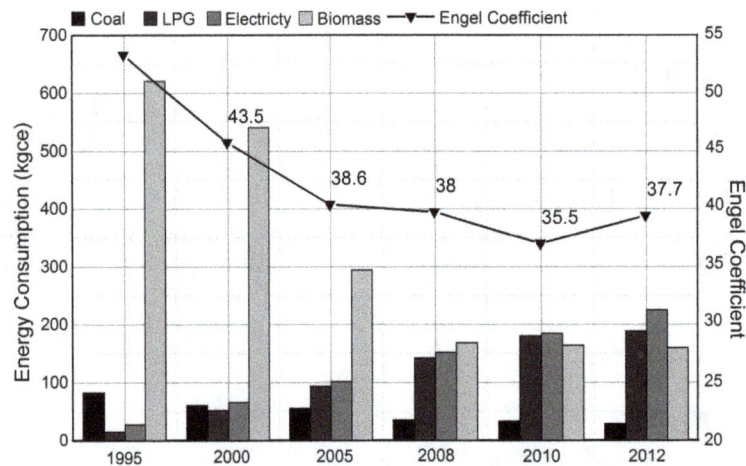

Figure 1. Rural energy consumption and Engel coefficient in Zhejiang province (1995-2012).

mercial energy consumption [10]; some other researches reveals the relationship between them [8] [11]. In other ways, such as climate, terrain, education, consuming behavior, appliances, building area are also discussed [10]-[14].

Previous researches have revealed the nationwide characteristics of energy structure in China from macro aspects, and it was very clear that the energy consumption in Northern China is higher than Southern. However, they are too general and not precise enough. Some scholars focus on the regional areas. Li *et al.* surveyed the rural energy resources in Zhangziying town near Beijing [15]. Yutaka *et al.* took a survey about the energy consumption in winter especial the usage of stoves [16]. Wu *et al.* investigated 30 households in Jinzhou area and analyzed the heating, cooking, lighting and other end-used energy consumption [17]. Besides the study of rural energy consumption, some scholars also analyzed the indoor environment, carbon dioxide emissions and other aspects problems [18]-[21]. These researches explored the concrete conditions in Northern China, but studies in Southern China are lacking [22]-[24]. And all these results will be compared with our study later on. Moreover, we found that above-mentioned researches were from the point of view of geographic position or climate, but lacking from different types in rural industrial structures.

3. Survey Methodology

3.1. Overview of the Study Village

The Engel coefficient of Anji county in 2013 is 32.8%, which belongs to affluence level [25]. Ligeng village which we studied is located in Anji country of Zhejiang province, with the longitude of E30°45'59.61", latitude of N119°29'37.47", and altitude of 263 m. The coldest monthly average temperature is −8.5°C, the hottest monthly average temperature is 38.7°C, with an annual average temperature of around 16.6°C. Ligeng village is next to shimen village, Zhongshan village, Dayinshan village and consists of three groups named Li Geng, Li Cun and Li Ming. It has 120 households, and more than 400 people. The village lies in a small valley, and is famous for cool and beautiful nature landscape that attracting lots of tourists every year to enjoy "Agritainment" in recent years.

3.2. Sample Design and Investigation Method

Sample size selection is crucial to the results of survey. If sample size is too large, it will cause the waste of manpower, material and financial resources. However, if the sample size is too small, it will affect the reliability of survey. The overall survey scale is between 100 and 1000, the appropriate sample volume of empirical judgment interval is 50% - 20% [26]. The simple random sampling method was applied to collect the data, and we finally chose 49 household as the object of our survey, accounting for 40.8% of the total households in Ligeng village. **Figure 2** shows the distribution of the samples which were uniformly distributed. The questionnaire survey was conducted from house to house. Since most farmers we surveyed were illiterate, the face-to-face interview was adopted to get more detailed data. The interviewers asked a set of questions and filled in the answers. We made a pre-survey in May 2014. In order to improve the reliability of the survey, we repeated our interview in September 2014 and July 2015.

3.3. Questionnaire Design

The types of household energy consumption include commercial energy (coal, LPG and electricity etc.) and non-commercial energy (firewood, straw, mash gas etc.) [1] [3]. The household energy utilization modes consist of five categories: cooking and hot water, heating, cooling, lighting, and household appliances [27]. The household energy consumption is related to the number of family members, rural buildings and energy usage habits. So the questionnaire we designed includes five aspects: family information, building information, energy usage, household appliances and human behavior (**Table 1**).

4. Survey Results

4.1. Features of Households

4.1.1. Family Composition

The family size we surveyed is 2 to 9, and the average size is 4.6 persons per household. Three-generation family is very common, as high as 73.5%. Five person in one family accounts for 44.9%, with the mode of 2 (old couples)-2 (young parents)-1 (child). The ratio of migrant population working outside is very large, which is

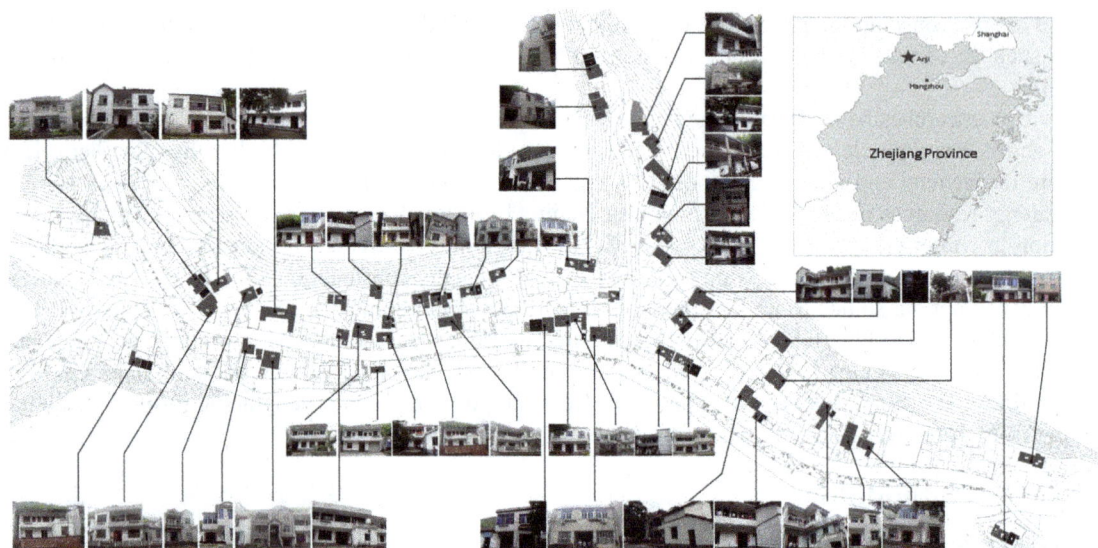

Figure 2. Plan of Ligeng village and distribution of samples.

Table 1. Questionnaire content.

Items	Main Content
Household Information	Family population, permanent population, age, cultural level, income etc.
Building Information	Built time, layers, building orientation, building area, material and construction, shape coefficient, window to wall ratio etc.
Energy Consumption	Electricity consumption, usage of LPG, firewood and solar energy etc.
Household Appliance	Number of household appliances, model, power etc.
Human Behavior	Activity schedule, usage frequency of appliances and time table etc.

about 53%. They work in big cities or towns most of the year, and only return home during holidays which is about two months. Two permanent population family accounts for 59.2%, most of them are over 50-aged elders and school aged children. In permanent population, the old people over the age of 50 years old is up to 73%, and over 60 years old is 48%. According to international standard, the ratio of people aged over 60 reaches to 10%, it will be considered as an aging society. It shows that Ligeng village has a significant population aging characteristic (**Figure 3**, **Figure 4**).

4.1.2. Household Income

Most household income in Ligeng village is made up of two parts: one is from traditional agricultural production and processing, mainly includes high mountain tea, bamboo and bamboo products processing. The other part is from migrant labor working in cities, which has become the main source of household income. In recent years, with the development of rural tourism, some families began operating agritainment. **Table 2** shows that the household annual income from $5000 to $10,000 accounts for 49%, from $15,000 to $20,000 takes up 30%.

4.2. Rural Building Status

4.2.1. Built Time Building Form Layers

Figure 5 shows the typical rural buildings built in different periods. The buildings which were built before the 1980s had been demolished or rebuilt. The existing buildings, built in 1980s and 1990s, account for 69%, and 53% were built in 1990s. Lots of buildings were rebuilt in different periods, and we can see different styles in one building.

Most rural buildings have two or three stories, and two-story buildings account for 88%. Living room, dining

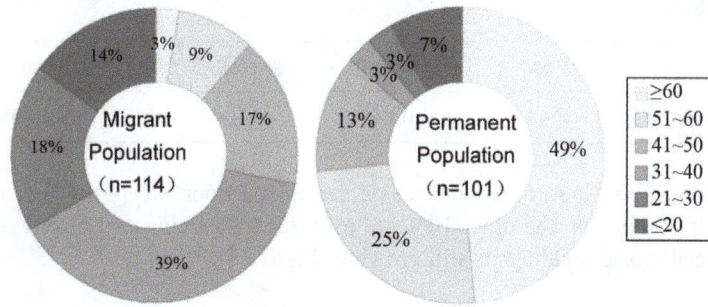

Figure 3. Age composition of migrant population and permanent population (n = 215).

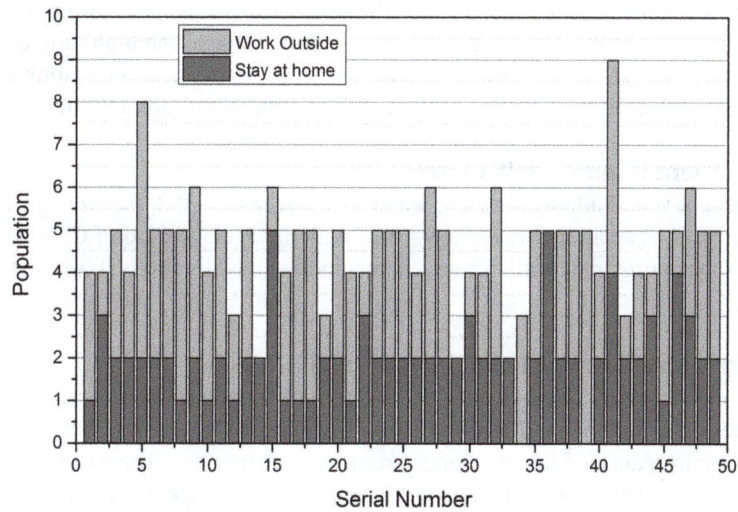

Figure 4. Family size and composition (n = 49 household).

Figure 5. Typical buildings built in different periods. (a) Built in early 1980s; (b) Built in 1980s and 1990s; (c) Built after 2000s; (d) Built after 2000s.

Table 2. Household income.

Household annual income	≤$5000	$5,000 - $10,000	$10,000 - $15,000	$15,000 - $20,000	≥$20,000
Percentage (n = 47)	4%	13%	49%	30%	4%

room, kitchen are on the first floor, bedrooms are on the second floor except the bedroom for elders. 90% rural buildings are for self-occupation, the other 10% are used for both living and business. The proportion of residential and commercial mixed buildings is still growing (**Figure 5**).

4.2.2. Floor Area

We calculated building area of each rural house according to Ligeng village plan and field measurement. The affiliated area, which was affiliated to the main building just for storing fire-woods or odds and ends that people always don't use it, was not calculated (**Figure 6**). The average floor area was 100.8 m^2, the average total building area was 199.5 m^2, and the building area per capita is 48 m^2. Due to the high ratio of migrant population, the building area per capita was up to 104.5 m^2. Because most of the permanent population are elders, the frequently used space was concentrated on the first floor, which was accounted for 60% (**Figure 6**).

4.2.3. Shape Coefficient Window to Wall Area Ratio

The shape coefficient of rural buildings is high, which is between 0.5 - 0.8, the average shape coefficient is 0.70. The window to wall ratio is small, between 0.06 - 0.17 with the average ratio of 0.10. The ratio of east and west orientation is lower than north and south. The average window to wall ratio is 0.07 and 0.06, the north and south is 0.13 (**Figure 6**).

4.2.4. Building Structure Building Envelope Thermal Bridge

There are brick-wood structure, brick-concrete structure and frame structure, respectively accounting for 2%, 94% and 4%. Because the material and construction of the building envelopes in different periods are quite different, the thermal performance of the envelopes are quite different too. The external walls were built up with rammed earth before early 1980s. After then, solid clay bricks took place and became the main wall material till now. Since 2000, the government announced to ban solid clay bricks in order to protecting cultivated land, and to spread new wall materials. Small-sized hollow concrete blocks, steam curing aerated concrete blocks, fired perforated bricks and so on, were used in rural buildings. The thermal bridge is quite different from different structures. The ratio of thermal bridge in rammed earth buildings and brick-concrete structure buildings built in 1980s to 1990s is about 10% to 15%, which is lower than the frame structure buildings built after 2000, with the ratio of 15% to 25% (**Figure 7**).

4.3. Household Appliances

Main appliances per household, power, energy efficiency label (EEL), etc. are shown as **Figure 8** and **Table 3**. Refrigeration equipments include electric fan and air conditioner. Average household electric fan is 2.73 units, and air conditioner is 1.18 units. Heating equipments include heater and electric blanket, the average household is 0.84 units and 0.92 units. The lower frequency used, the lower energy efficiency label, such as air conditioner and solar water heater. The higher frequency used, the higher energy efficiency label, such as refrigerator, washing machine and rice cooker. **Figure 9** shows the schedule of household appliances used.

4.4. Occupants' Activities

There are three typical period of time among a year in Ligeng village: normal times, farming season (in April) and Holidays. **Figure 10** shows the ratio of population at home to the total family population (H-F ratio) in 24 hours. Although, the activity of permanent population who lives in villages is not as fixed as people work in cities, however, their habits also have a relatively fixed schedule. April is the season of harvesting tea and bamboo. Because of far distance, farmers take lunch with them. As a result, the average H-F ratio from 6:30 to 4:00 is very low, only 5%. In holidays, the migration workers are back home. In our survey, we found that a small portion family working outside almost the whole year, only went back home for Spring Festival holiday.

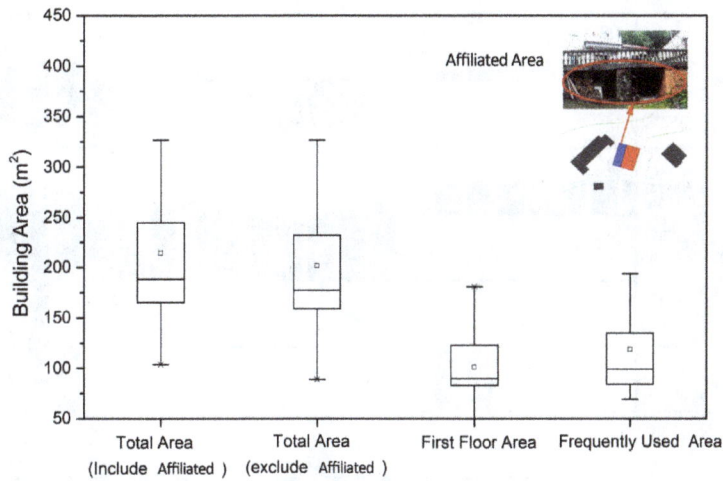

Figure 6. Total building area, floor area and frequently used area.

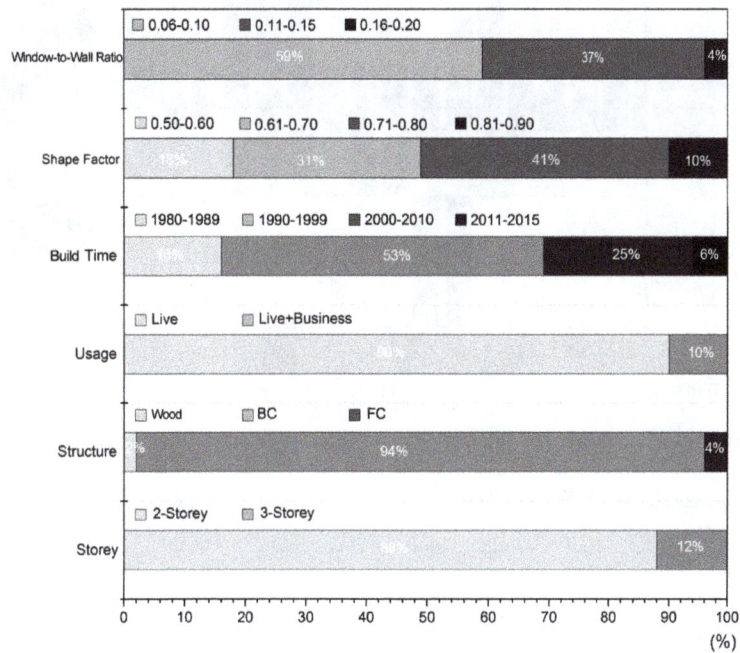

Figure 7. Fundamental state about rural buildings.

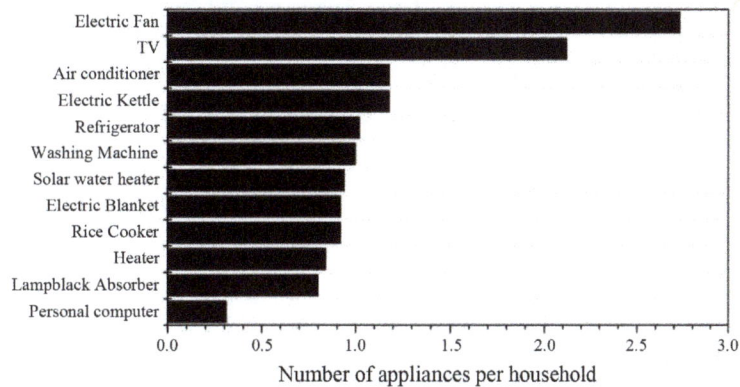

Figure 8. Average household appliances.

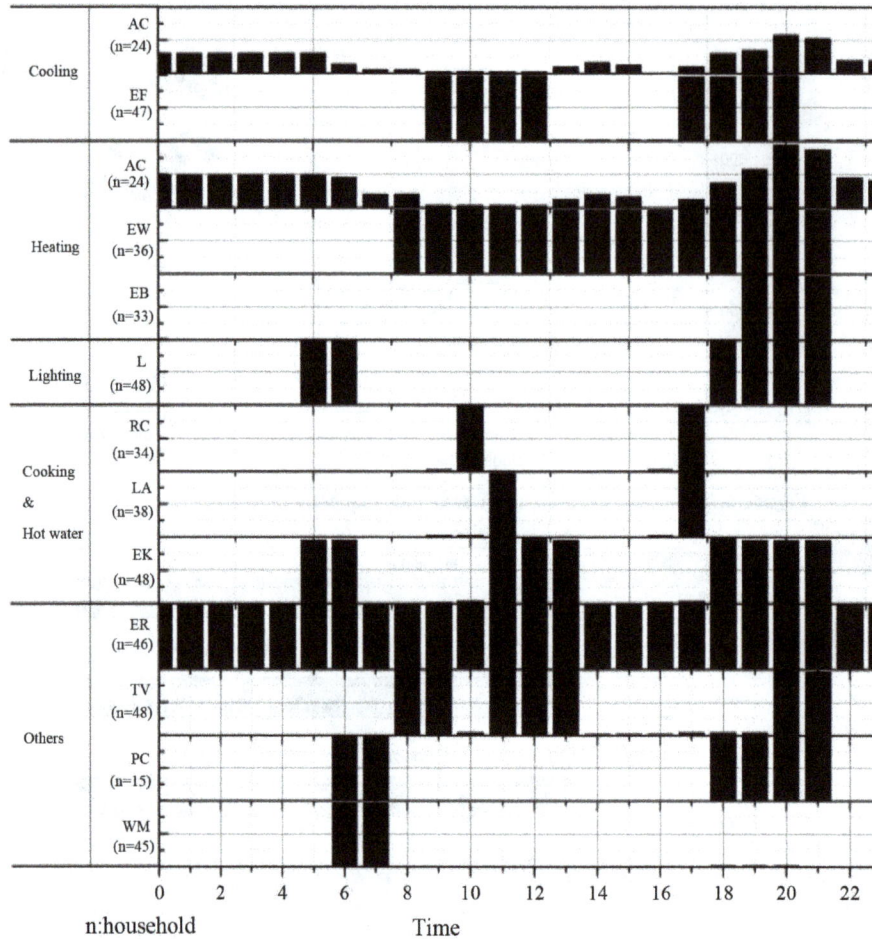

Figure 9. Schedule of household appliances.

Table 3. Power and energy efficiency label (EEL) of household appliances.

Type	Appliance	Power	EEL
Cooling	Air conditioner	1100 W	3/5
	Electric fan	50 W	——
Lighting	Energy saving bulb	50 W	——
Cooking & Hot water	Rice cooker	900 W	2/5
	Lampblack absorber	250 W	——
	Air conditioner	1500 W	
	Electric fan	1500 W	
Heating	Air conditioner	1100 W	3/5
	Heater	1800 W	——
	Electric blanket	100 W	——
Household appliance	Refrigerator	0.5 kwh/24h	1/5
	Washing machine	0.12 kwh/circle	2/5
	TV	200 W	——
	PC	200 W	——

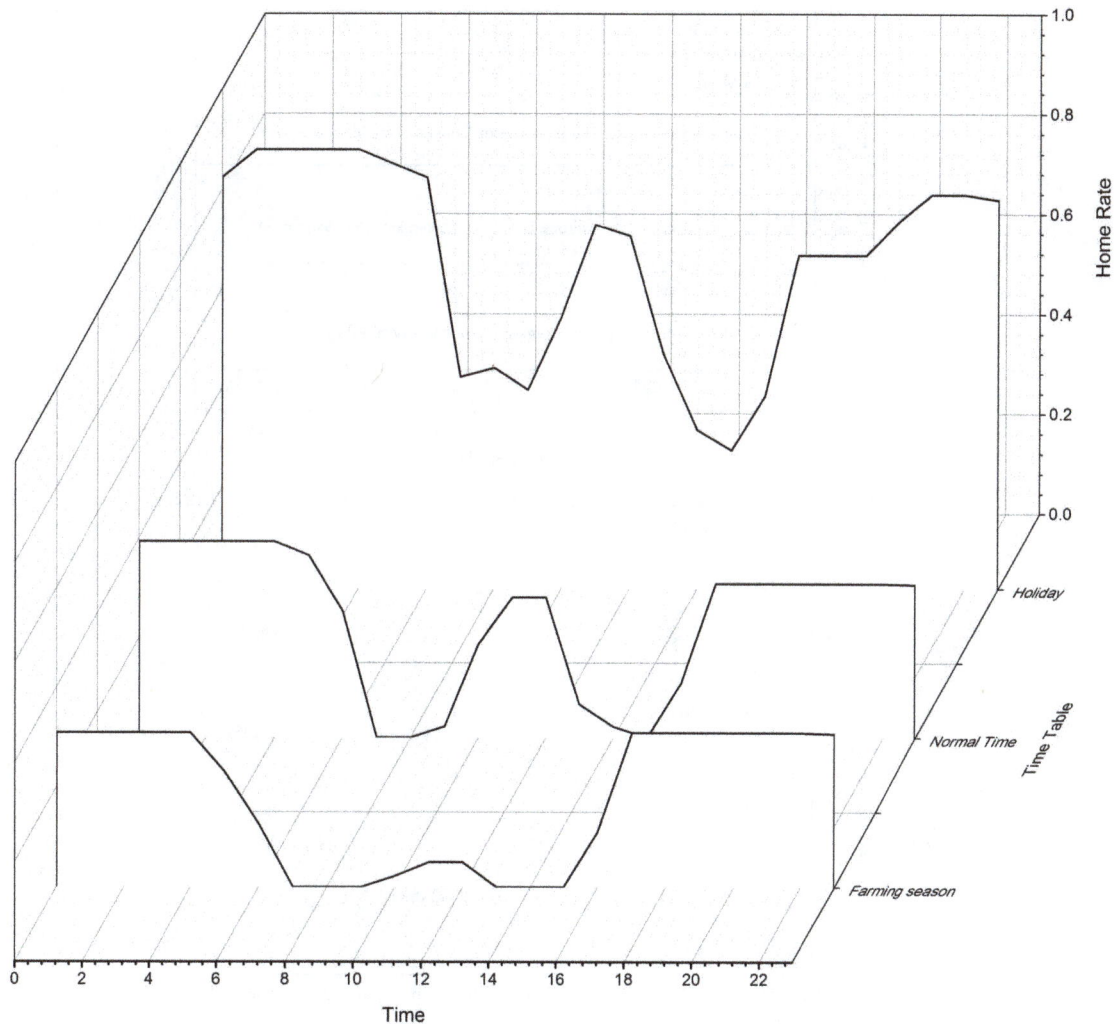

Figure 10. Schedule of human activity.

4.5. Household Energy Consumption

At present, according to the energy type used in rural buildings, there are four categories as follows: Electricity, LPG, firewood and solar energy. In addition, there are a very few farmers to use briquettes, which are not included in the statistics. Power supply bureau provided the total electricity consumption per year of the surveyed family from 2010 to 2014, and the electricity consumption per month from July 2014 to June 2015, according to number of electric meter. It shows that the total electricity consumption per year from 2010 to 2014 was on the rise, especially after 2011, the grow was more significant. It was calculated that the energy consumption in Ligeng village was only 6 kWh/(m²·a) (**Figure 11**). From July 2014 to June 2015, the average monthly household electricity consumption is 116 kWh. **Figure 12** shows that there were two peak periods of electricity consumption: in January, February and March which was heating period, the average monthly electricity consumption was 149.2 kWh; in August and September which was cooling period, the average monthly electricity consumption was 121.4 kWh. The maximum power consumption was in March, when is in the Chinese New Year holiday. Obviously, the electricity consumption in heating and cooling period was not balanced, the winter heating period consumption was significantly higher than the summer cooling consumption. In transition season, the consumption in spring (April, May, June, July) was lower than autumn (October, November and December).

LPG is carried in tanks, and the weight of gas is about 14 kg per carton. The average household using of LPG is about 1.5 barrels per year. Ligeng village is in the mountainous area, where is convenient to get firewood. Therefore, firewood remains one of the most important energy resource for cooking and heat water, with the

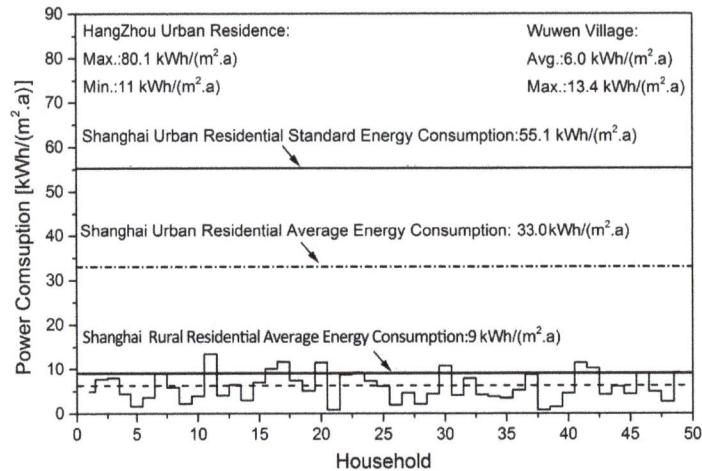

Figure 11. Comparison of building energy consumption between Ligeng village and other surveys.

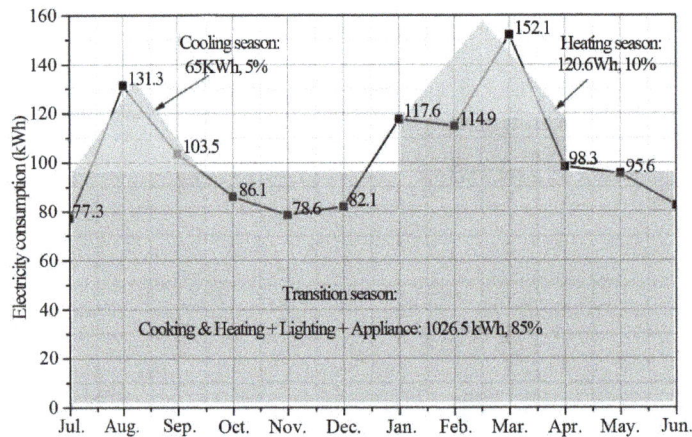

Figure 12. Energy consumption of heating, cooling and daily life (from July 2014 to June 2015).

usage of 1.78 tons per year. Moreover, solar water heater is widely used in local mainly to provide domestic water and shower. Most households answered that in summer, the solar water heater can meet the requirements of daily shower; but electric water heater is also needed in winter besides the solar hot water. The amount of LPG, firewood and solar energy is estimated according to the questionnaire statistics.

5. Building Energy Consumption

5.1. Household Energy Consumption

Different types of energy obtained from the survey, are converted according to the average low calorific value of the different fuel (**Table 4**). It was calculated that the energy consumption came from firewood was about 29.5 GJ, accounting for 83%; electricity was 4.2 GJ, accounting for 12%; LPG was 1.02 GJ, accounting for 3%. The usage of hot water for shower was difficult to survey. We supposed the flew rate of water flower was 0.04 L/s [26], and people took shower 3 times a week, 15 minutes each time. According to the questionnaire, electric water heaters were mainly used in winter. In the calculation, we supposed the total amount of heat supplied by the solar energy was 75%. So we can get the solar energy was 0.95 GJ, accounting for 2%.

5.2. Final Energy Consumption

Final energy consumption included heating, cooling, lighting, cooking and hot water, household appliances

(**Figure 13**). Each type of energy, power equipment was shown in **Figure 13**. Firewood, LPG and solar energy were for cooking and hot water. Electricity accounted for 12%, which covered five aspects of the final energy consumption of the country, was the most difficult part to be distributed.

Since the usage time of appliances was from respondents' estimation, and we didn't consider the impact of the depreciation of appliances, the exact proportion of hard statistics. It was difficult to get a precise proportion. Two methods were used to estimate the energy consumption.

Method one: Estimated from seasonal feature of the heating and cooling. As shown in **Figure 12**, the peak periods of summer and winter were mainly due to cooling and heating, which respectively took up 5% and 10%. Transition season accounted for 85%.

However, this method has two limitations. First, be merely able to divide heating, cooling and the daily life energy consumption. However, cannot be further subdivided the proportion of lighting, cooking and hot water, household appliances. Second, heating and cooling seasons happen to be the Summer holiday and the Spring Festival holiday in China, the P-F rate was higher than normal time. Therefore, the ratio of the actual energy consumption of cooling and heating would be smaller, the proportion of household energy use in daily life would be higher.

Method two: Based on the information obtained from field investigation——number of appliances per household, power, using time and using probability. It was found out that in the absence of heating and cooling

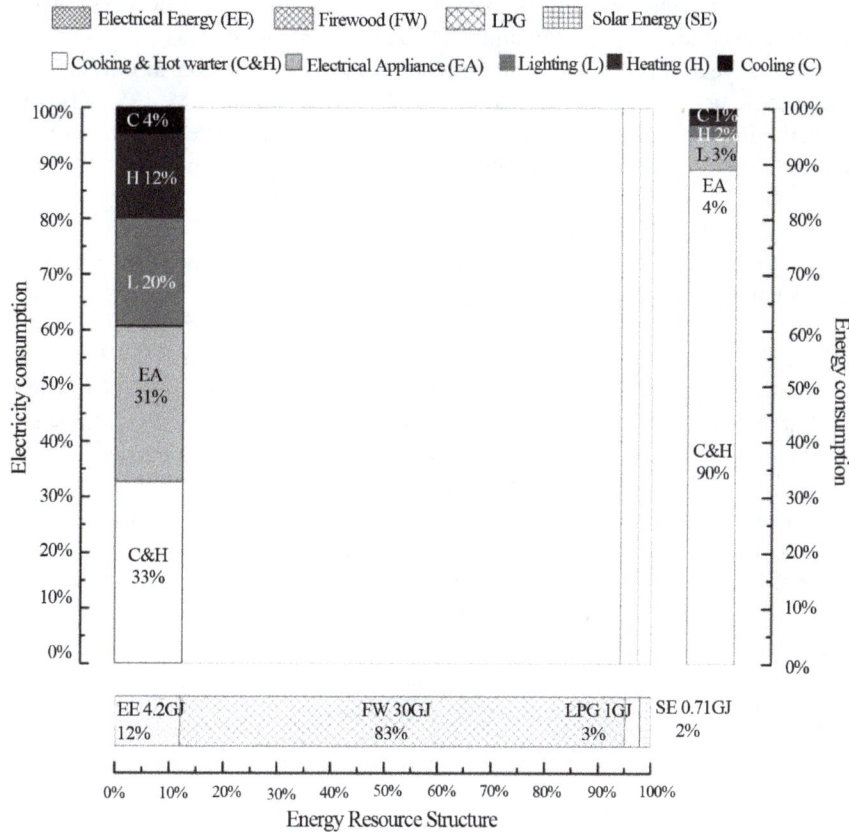

Figure 13. Energy resource structure and energy consumption.

Table 4. Conversion factors from physical unit to coal equivalent [1].

Energy	Average Low Calorific Value
Electricity	3600 KJ/kWh
LPG	50,179 KJ/kg
Firewood	16,726 KJ/kg

conditions, the temperature below 30°C in summer and higher than 8°C in winter in hot summer and cold winter area, most people think the indoor environment is bearable [8]. There is about 5°C temperature difference between indoor and outdoor. According to the typical meteorological year in Huzhou area, outdoor temperature which exceeds 35°C was totally about 15 days, and outdoor temperature below 5°C was totally about 15 days too. They are taken as the running days of air-conditioner, electric blanket and heater. Running time per day is counted by questionnaires. The using time of appliances such as TV, refrigerator, rice cooker and washing machine are relatively clear, and it would be determined in accordance with a higher probability of the emergence time (**Figure 9**). In the estimation process, we took no account of equipment depreciation. The total electricity consumption obtained by this method was larger than actual consumption. This is because of the survey error and the running time of appliances. By this method, in total electricity consumption, cooking and hot-water accounted for 33%, household appliances 31%, lighting 20%, heating 12% and cooling 4%. The heating, cooling and daily life electricity consumption calculated by the two methods are close (**Figure 13**).

From the energy resource types, energy consumption of cooling and hot water was 90%, household was 4%, lighting was 3%, heating was 2%, cooling was 1% (**Figure 13**).

5.3. Comparison with Other Surveys

According to statistical data in relevant statistical yearbooks [1]-[3], in commercial energy and non-commercial energy consumption in Zhejiang rural area in 2013, coal accounted for 1%, LPG accounted for 9% and electricity consumption accounted for 11%, non-commercial energy consumption was 79%, 47.5 GJ/per household/year and 15.5 GJ/per capita/year (**Figure 14**). The building energy consumption in our surveyed village, LPG accounted for 3%, electricity consumption accounted for 12%, firewood was 85%, 35.7 GJ/per household/year and 7.32 GJ/per capita/year. Due to the large migrant proportion, if we calculated according to the permanent population, it was 16.8 GJ per capita/year, which was slightly higher than the per capita consumption of Statistical Yearbook. By comparison, we can find that the proportion of electricity was close to the data of Statistical Yearbooks, but the difference of LPG and firewood were a slightly large. The main reason was that Ligeng village was far away from Cities, LPG were tanked which was expensive and not convenient to send. However, the rich vegetation resources in the mountain area provide ample firewood.

From July 2014 to June 2015, the electricity consumption was only 6 kWh/(m²·a). Sun *et al.* undertook a similar survey in Shanghai rural area in 2010. The electricity consumption was 9.0 kWh/(m²·a) [22]. In Shanghai rural area, the commercial energy usage has been reached to 90% in 2007 [28]. However, the electricity consumption was slightly higher than Ligeng village. This shows that the level of rural power consumption in the developed regions is relatively low. Zhu *et al.* surveyed 18 households urban residential electricity consumption in Hangzhou, the min. was 11 kWh/(m²·a) and the max. was 80.1 kWh/(m²·a) [29]. Jiang *et al.* investigated 1000 urban residential households electricity consumption in Shanghai, the results showed that the average electricity consumption was 33 kWh/(m²·a), and cooling and heating energy consumption accounted for 10 - 20 kWh/(m²·a) [30]. In the domestic and foreign studies, the research on the final energy consumption structure in rural area is lesser than urban. Ning *et al.* from the macroscopic view, and based on the statistical data of energy consumption of urban residents in China, they analyzed the characteristics of energy consumption of residential buildings from 1995 to 2010, and calculated the urban residential energy consumption was 5.59 GJ/per capita, with heating accounting for 25%, cooking and hot water 56%, household appliance 10%, cooling 6% and lighting

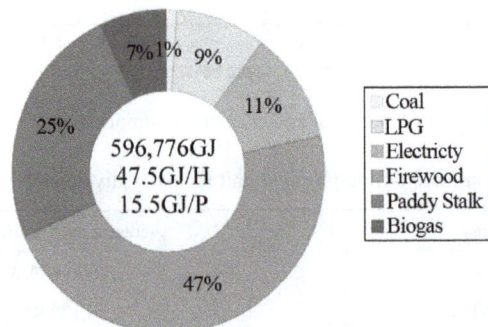

Figure 14. Energy resource structure in Zhejiang (2013).

3% in 2010 [27]. Li *et al.* selected two residential households and took a measurement of electricity, gas for a year to inspect the energy consumption. Finally, they put forward the results of energy consumption structure of Shanghai: 25% - 33% of gas, 11% - 13% of scouring bath, 19% - 21% of air conditioning, 5% - 6% of lighting, 8% - 10% of information and entertainment, 10% - 13% of refrigerators, and 15% - 20% of other appliances. Then they compared with the apartment houses energy consumption measured in 2002-2003 in Okinawa of Kitakyushu, Japan, and they found that energy consumed by each person of Okinawa was 1.424 times as that of Shanghai. From this, it can be seen that the rural building energy consumption in Zhejiang province is lower than urban residents, and far less than developed countries.

6. Analysis of the Influential Factors

There are many factors determining energy consumption in rural areas. Factors associated with buildings include shape coefficient, window to wall ratio and per capita floor area etc. Family-related factors include the permanent population, the population age structure (proportion of the population aged over 50 in the total household members). Appliance factors include the number of air conditioner, etc. These factors will be analyzed as follows:

6.1. Correlation with Income

We took the total electricity consumption in 2014 as the independent variable, and the average household income and per capita income as the dependent variable, then analyzed. **Figure 15** shows the correlation between total electricity consumption and the average household income, and the total electricity consumption and per capita income, and the determinant factor was 0.0115 and 0.015, which indicated no apparent correlation. Yutaka *et al.* surveyed the rural households in the fringes of Xian city, and they reached the same conclusions [16].

6.2. Correlation with Building Area

From the actual survey, it was found that there was a large leave unused area in rural buildings, the frequently used area is only 60% of the total building area. As first step, the correlation between total building area and total electricity consumption was examined. **Figure 16** shows the correlation was not correlative, because the determinant factor was only 0.1447. Secondly correlation between frequently used building area was also examined. The determinant factor was 0.378 and the results showed there was obvious correlation between them (**Figure 16**). From these results we concluded that total electricity consumption in rural areas appeared no correlation with total area, but had some dramatic correlation with frequently used area.

6.3. Correlation with Building Shape Coefficient and Window to Wall Ratio

Building shape coefficient and window to wall ratio are related to heating and cooling energy consumption.

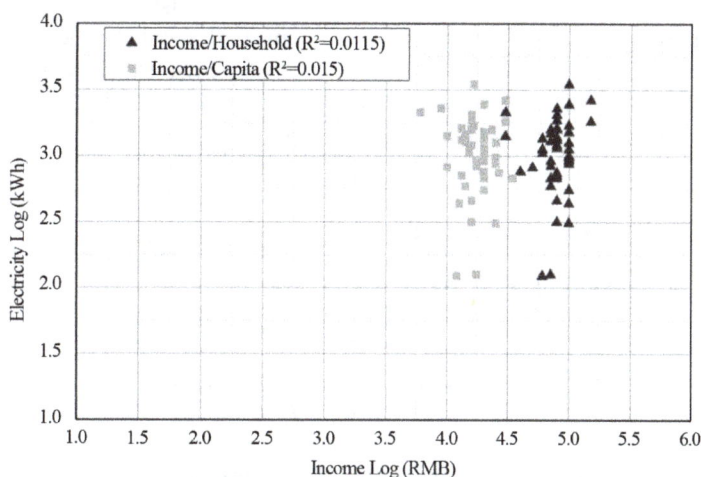

Figure 15. Correlation between income and electricity consumption.

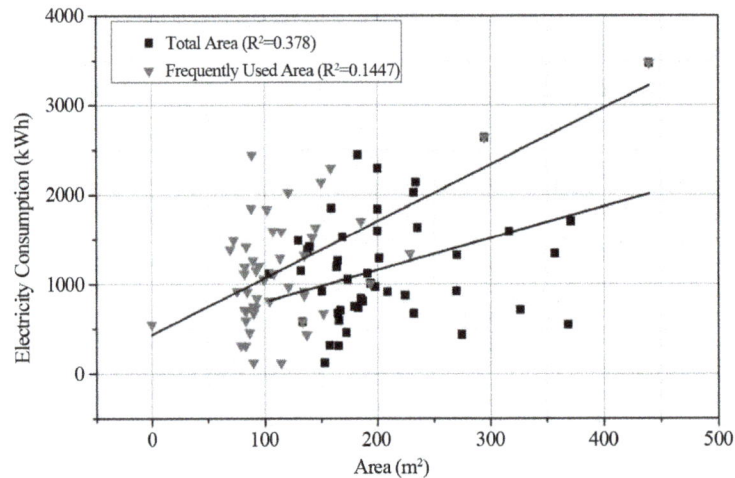

Figure 16. Correlation between total building area, frequent used area and electricity consumption per year.

Figure 17 shows that the correlation was not obvious, with determinant factor of 0.0228 and 0.0932. The reason is that only parts of the rural buildings adopted heating and cooling appliances in South rural area in China, coupled with a small proportion of the frequently used area, which led to no correlation with them. It indicated that in the energy-saving design of rural buildings in the southern region, it may be appropriate to relax the building shape coefficient and window to wall ratio requirements.

6.4. Correlation with Number of Air Conditioner

The number per household of appliances, such as refrigerator, TV and so on, were very close except air conditioner. The correlation between the number of air conditioner per household was examined, and the determinant factor was 0.4966 (**Figure 18**). The result showed an obvious correlation between them. However, from the survey, we found that the energy efficiency label of air conditioner in Ligeng village was low, most of them were three level which was the lowest. So the villagers should be encouraged to buy energy-saving air conditioners.

6.5. Correlation with Building Function

At present, there were several families deal in "agritainment", shops and agro-product processing. We chose those families, which has similar total building area, family size and permanent population, then divided into two groups according to different building functions: one group was just for living, the other group was for living and business (**Table 5**). Then the two groups were compared about the electric energy consumption per month. **Figure 19** shows that the electric energy consumption of the two groups were very close from November 2014 to April 2015. From July 2014 to October 2014, May and June in 2015, "L + B" group was about 48% higher than "L" Group, which was caused by village tourism.

7. Conclusions

The study investigated the rural energy consumption and the influential factors by taking Ligeng village as a model. From this study, the following conclusions are reached:

1) Ligeng village belongs to relatively wealthy village in Zhejiang Province; the electric energy consumption was 6 $kWh/(m^2 \cdot a)$, which was far less than urban residential household. Compared with the rural buildings around Shanghai, the ratio of commercial energy consumption was very high (above 90%), and the electricity consumption was close (9 $kWh/(m^2 \cdot a)$) to Shanghai rural buildings.

2) From energy consumption structure, the ratio of non-commercial energy consumption is higher than commercial energy consumption. Firewood accounted for 83%, electricity for 12%, LPG for 3% and solar energy for 2%.

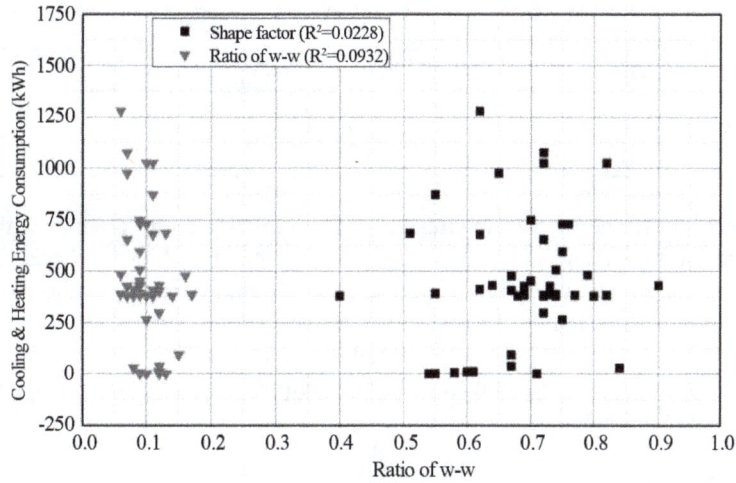

Figure 17. Correlation between cooling and heating energy consumption and building shape factor and ratio of w-w.

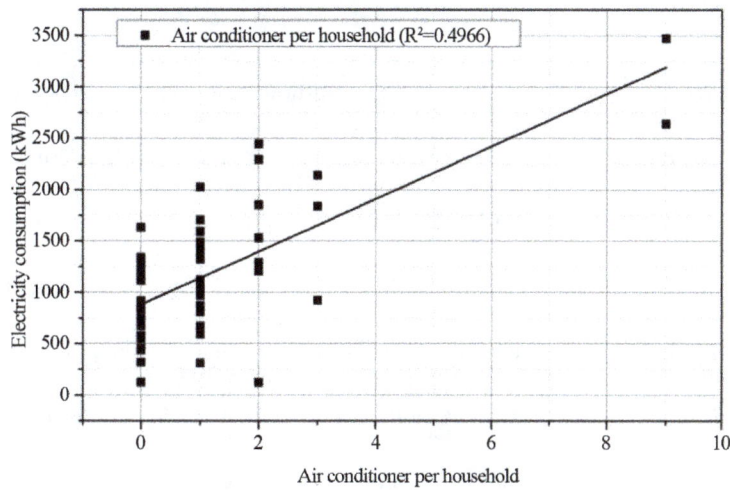

Figure 18. Correlation between electricity consumption and air conditioner per captia.

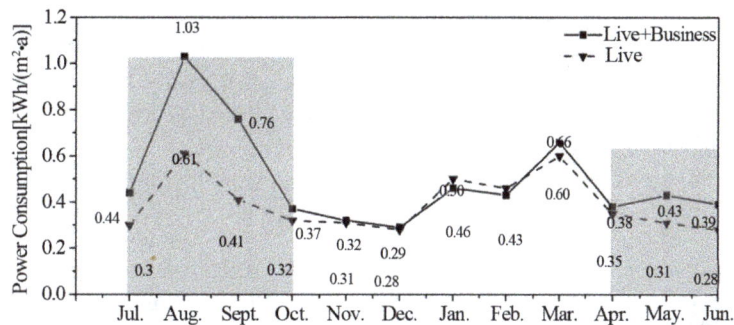

Figure 19. Power consumption compare between "L + B" group and "L" group.

3) At present, rural building energy consumption is still mainly to meet the basic daily needs. Energy consumption of cooking and hot water and household appliances are higher than others, respectively accounting for 33% and 31%, lighting for 20%, heating for 12%, and cooling for 4%.

4) Correlation analysis shows that frequently used area, and number of air-conditioner per household have

Table 5. "L + B" group and "L" group.

Group	Family Number	Total Building Area (m^2)	Family Size	Permanent Population
L+B	7, 10, 14, 24, 32	250 - 400	4 - 6	2
L	9, 23, 27, 35, 37, 40	250 - 400	4 - 6	2

obvious correlation with the energy consumption, and the correlation is not obvious with income, total building area, building shape coefficient and window to wall ratio. The building function has a greater impact on energy consumption. "L + B" group was about 48% higher than "L" Group, which was caused by village tourism.

To sum up the above arguments, the rural building energy consumption is low at present. However, the low energy consumption comes at the cost of uncomfortable and low economic activities. With the adjustment of rural industrial structure and the improvement farmers' living standard, rural building energy consumption will face tremendous pressure in the future.

Acknowledgements

The research was cooperated by the University of Kitakyushu (Japan) and Zhejiang University (China) and financially sponsored by the National Natural Science Foundation of China (Subject number: 51238011).

References

[1]   State Statistical Bureau, China (2013) China Energy Statistical Yearbook (2013-2000). China Statistical Publishing House, Beijing.

[2]   Zhejiang Statistical Bureau, China (2013) Zhejiang Statistical Yearbook (2013-2000). China Statistical Publishing House, Beijing.

[3]   Zhejiang Statistical Bureau, China (2013) Zhejiang Natural Resource and Statistical Yearbook on Environment (2013-2000). China Statistical Publishing House, Beijing.

[4]   Wang, X.H. (1994) Situation and Development of Rural Household Energy Consumption in China. *Journal of Nanjing Agricultural University*, **17**, 134-141.

[5]   He, B.J., Yang, L., Ye, M., Mou, B. and Zhou, Y.N. (2014) Over View of Rural Building Energy Efficiency in China. *Energy Policy*, **69**, 385-396. http://dx.doi.org/10.1016/j.enpol.2014.03.018

[6]   Luo, G.L. and Zhang, Y.M. (2008) Analysis on Rural Energy Consumption of China. *Chinese Agricultural Science Bulletin*, **24**, 535-540.

[7]   Wang, X.H. and Feng, Z.M. (2001) Study on Rural Household Energy Consumption in China: Consumption Level and Affecting Factors. *Transactions of the CSAE*, **17**, 88-91.

[8]   Guo, B.L. and Wang, Y.J. (2002) Present Situation and Prospects of China Residential Energy Consumption. *Rural Energy*, **105**, 4-7.

[9]   Energy Conservation Research Center of Tsinghua University (2013) 2012 Annual Report on China Building Energy Efficiency. China Building Industry Press, Beijing.

[10]   Wang, X.H., Hao, X.R. and Jin, L. (2014) Study on Rural Household Energy Consumption in China Based on Household Investigation from Typical Counties. *Transactions of the Chinese Society of Agricultural Engineering* (*Transactions of the CSAE*), **30**, 206-212.

[11]   Ning, X.Y., Zhang, X. and Gao, J. (2013) Analysis of Status Quo and Influencing Factors of Domestic Energy Consumption by Typical Villages and Towns in Different Climatic Regions. *Building Science*, **29**, 98-102.

[12]   Wang, P., Liu, Q. and Qi, Y. (2014) Factors Influencing Sustainable Consumption Behaviors: A Survey of the Rural Residents in China. *Journal of Cleaner Production*, **63**, 152-165. http://dx.doi.org/10.1016/j.jclepro.2013.05.007

[13]   Wang, X.H. and Hu, X.Y. (2010) Factors Influencing Rural Household Energy Consumption. *Transactions of the CSAE*, **26**, 294-297.

[14]   Zhu, L. (2014) Analysis of Residential Energy Utilization and Behavior Patterns in Hangzhou. Zhejiang University, Hangzhou.

[15]   Li, X., Lin, C., Wang, Y., Zhao, L.Y., Duan, N. and Wu, X.D. (2015) Analysis of Rural Household Energy Consumption and Renewable Energy Systems in Zhangziying Town of Beijing. *Ecological Modelling*, **318**, 184-193. http://dx.doi.org/10.1016/j.ecolmodel.2015.05.011

[16]   Tonooka, Y., Liu, J.P., Kondou, Y., Ning, Y.D. and Fukasawa, O. (2006) A Survey on Energy Consumption in Rural

Households in the Fringes of Xian City. *Energy and Buildings*, **38**, 1335-1342. http://dx.doi.org/10.1016/j.enbuild.2006.04.011

[17] Wu, Y.J., Xu, D. and Wang, X.Y. (2010) Investigation on Energy Efficiency of Rural Residential Housing in Jinzhou Area. *Journal of Liaoning Technical University* (*Natural Science*), **29**, 228-231.

[18] Yao, C.S., Chen, C.Y. and Li, M. (2012) Analysis of Rural Residential Energy Consumption and Corresponding Carbon Emissions in China. *Energy Policy*, **41**, 445-450. http://dx.doi.org/10.1016/j.enpol.2011.11.005

[19] Liu, W.L., Spaargaren, G., Heerink, N., Mol, A.P.J. and Wang, C. (2013) Energy Consumption Practices of Rural Households in North China: Basic Characteristics and Potential for Low Carbon Development. *Energy Policy*, **55**, 128-138. http://dx.doi.org/10.1016/j.enpol.2012.11.031

[20] Niu, H.W., He, Y.Q., Desideri, U., Zhang, P.D., Qin, H.Y. and Wang, S.J. (2014) Rural Household Energy Consumption and Its Implications for Eco-Environments in NW China: A Case Study. *Renewable Energy*, **65**, 137-145. http://dx.doi.org/10.1016/j.renene.2013.07.045

[21] Shan, M., Wang, P.S., Li, J.R., Yue, G.X. and Yang, X.D. (2015) Energy and Environment in Chinese Rural Buildings: Situations, Challenges, and Intervention Strategies. *Building and Environment*, **91**, 271-282. http://dx.doi.org/10.1016/j.buildenv.2015.03.016

[22] Sun, Y.-L., Lin, Z.-P. and Wang, X.-M. (2011) An Investigation of Envelope Situation and Simulation of Heating/Cooling Energy Consumption for Rural Residential Buildings in Shanghai. *Building Science*, **27**, 38-70.

[23] Li, Z.H., Sun, J. and Hiroshi, Y. (2009) Field Measurement and Analysis of the Residential Energy Consumption Structure of Shanghai. *Journal of Tongji University* (*Natural Science*), **37**, 384-389.

[24] Zhou, X.H., Zhou, X.Q. and Ma, J.L. (2011) Investigation on Energy Consumption and Analysis on Energy-Saving Potential of Rural Residential Buildings in Guangdong. *Building Science*, **27**, 44-47.

[25] Anji Statistical Bureau P.R. China (2013) Anji Statistical Yearbook. China Statistical Publishing House. http://www.ajtj.gov.cn/tjnj/2013/

[26] China's Urban and Rural Areas and Housing Construction Department (2012) Minimum Allowable Values of Water Efficiency and Water Efficiency Grades for Shower (GB28378-2012). China Building Industry Press, Beijing.

[27] Ning, Y.D., Cai, J.Y. and Ding, T. (2013) Urban Household Energy Consumption Structure in China. *Journal of Beijing Institute of Technology* (*Social Science Edition*), **15**, 26-33.

[28] Energy Conservation Research Center of Tsinghua University (2013) 2012 Annual Report on China Building Energy Efficiency. China Building Industry Press, Beijing.

[29] Li, Z. (2014) Analysis of Residential Energy Utilization and Behavior Patterns in Hangzhou. Zhejiang University, Hangzhou.

[30] Li, Z.H., Sun, J. and Hiroshi, Y. (2009) Field Measurement and Analysis of the Residential Energy Consumption Structure of Shanghai. *Journal of Tongji University* (*Natural Science*), **37**, 384-389.

9

High Proportion Renewable Energy Supply and Demand Structure Model and Grid Impaction

Xiaoxia Wei[1], Jie Liu[2], Tiezhong Wei[2], Lirong Wang[2]

[1]State Grid Energy Research Institute, Beijing, China
[2]Heilongjiang Grid Company, Harbin, China
Email: 20583853@qq.com

Abstract

In considering of high proportion of renewable energy supply in 2050, the accelerating of energy consumption gross, source and environment can affect the energy system restrict affection are stronger. Add wind and solar to electricity energy with large amount of energy source exploitation. The energy source amount per person is lower. Considering the renewable energy amount and supply, primary energy storage and structure problem is standing out. Before the wide spread of renewable energy, Using the high-carbon energy in China can pollute seriously. Chinese energy supply and demand problem is research key point. This paper researches Chinese energy supply and demand pattern system and evaluation methodology, gives out the inner and outer influencing elements. And evaluate Chinese energy supply and demand pattern from energy gross, structure, distribution and transportation. Use energy supply synthesize radar comparison chart in certain time period. From energy security, economy, clean and efficiency, analyze the benefit comparisons of Chinese energy supply and demand pattern. This energy supply and demand pattern model will give one certain theoretical analysis and practice reference to the further high proportion of renewable energy.

Keywords

High Proportion of Renewable Energy Supply, Inner and Outer Element, Power Grid Affection, Supply and Demand Pattern

1. Introduction

Energy is the fundamental element of modern civilization. From all countries development practice, the speed and degree of country and regional economic development is depending on modern energy supply guarantee system's ability and construction. As the maximal developing country, China attracted worldwide attention and most support from the energy system development. Energy is one of the fundaments to guarantee Chinese national economy sustain, stabilization, sound development, which is also one of the implement comprehensively, coordinately, sustainably. So strength our country energy supply and demand scientific analysis, its advanced management theory method research, have quite strong reality backgrounds (Sarak H. & Satman A. 2003. Ba-laehandra R. & Chan Du. 1999. Kumar A. & Bhattacharya S. C. & Pham H. L. 2003. Mackay R. M. & Probert S. D. 2001) [1]-[4].

Energy supply and demand is the key topic, confronting with new situation. High proportion of renewable energy supply has being considered in the ten to thirty years, affect the Chinese electricity energy supply and structure.

(a) In China, until 2030, coal usage will be the fundamental electricity energy in long period of time, coal transportation and coal power layout optimization are much more impendency. Parts of coal producing area are approaching to the science capacity. The receiving port power source construction is extruding power transportation space. Make the coal and power have little configuration space.

(b) In China, natural gas and import scale expect have rapid growth; shale gas is changing the world using gas condition. Distribution energy sources are changing the world power configuration, affect the power grid structure.

(c) Renewable energy's economic development and efficient utilization should be solved. In further, the renewable energy will be accelerated to the power generation energy supply. To wind, the abandon wind is one extrude problem. Wind basement exploit is suffocated. Disperse exploit cannot support goal. While to the solar energy, the large occupation land and low utilization time will affect the renewable energy efficiency and economy.

(d) International energy cooperation and global energy configuration promising are further infection. Energy consumption is approaching to the energy supply border. So it needs to layout Chinese power supply economically, stabilization, reliably under the global vision. And China is giving the new theory to establish global energy connection.

This paper analyzed the energy supply's new circumstances and key problems. Make the energy supply and demand structure as key line of current and future situations. Make the research in different views (Messner S. & Schrattenholzer L. 2000. Ediger V. S. & Tatlidil H. 2002. Cuaresma J. erespo & Hlouskova J. & Kossmeier S. 2004. Sharma D. P. & Chandramohanan Nair P. S. & Balasubramanian R. 2002) [5]-[7].

(a) Assessment indicator system of energy supply and demand. Describe energy supply pattern and evolution. Evaluate security, economy, clean and efficiency.

(b) Evolution characteristics of our energy supply and demand pattern. Research on Chinese energy development in analysis of the layout and project. And give out the long term energy changing pattern.

(c) Research on the power grid function and affection to the energy supply and demand.

2. Indicator System of Energy Supply and Demand Pattern

2.1. Key Element of Energy Supply and Demand Pattern

Energy supply and demand is focusing on whole country and different regions energy resource, producing, transportation, consumption, import and different energy in separating region balancing, trans-regional transportation condition. Evolution of energy supply and demand pattern is the track of different period of energy quality and quantity development. The key element of energy supply and demand pattern is gross, structure, distribution and transportation. Energy transportation includes variety, scale, direction and distance.

2.2. Indicator System of Energy Supply and Demand Pattern

Allocating the energy situation is one of important foundation to scientific judge energy management. This research area is lacking quantification description and evaluation method. Normally, it uses the perceptual knowledge and historical experience to judge energy distribution condition. The analysis method is urgent technical problem. This paper gives the energy distribution affection evaluation indicator system, carries out the demonstration evaluation. Strive to quantity and demonstrate energy distribution affection and breaking through.

Energy distribution affection and evaluation system include security, economy, clean and efficiency four aspects. Every aspect includes several support indexes. In order to embody power function to energy distribution, the support indexes include power sector and its index of correlation.

The support index and its choosing are following the authority theory. In the security evaluation aspect, the support indexes include energy reserves, foreign-trade dependence, coal railway transportation channel and its transportation use ration, coal-power transportation system reliability. In economic evaluation aspect, the support indexes include coal and coking industry and its ex-factory price indices of industrial product, oil and gas producer's price for manufactured products, fuel power purchase price indices, and power producer's price for manufactured products. In clean evaluation aspect, the support indexes include clean energy ration to primary energy consumption, sulfur dioxide discharge strength, sulfur dioxide discharge value, environmental pollution control investment ration to GDP, clean energy generate electricity ration and other indices. In the high effective aspect, the support indexes include energy loss per GDP, energy processing and transforming efficiency, power energy ration to terminal energy consumption (Persaud A. J. & Kumar U. 2001, Chow L. C. H. 2001) [8] [9] **(Figure 1)**.

2.3. Evaluation Methodology of Energy Supply and Demand Pattern

Energy distribution and affection evaluation can adopt the main element analysis method (PCA), through covariance matrix feature vector can reveal the common evolvement role, which can reach the index goal without considering man-made thinking.

The main element analysis (PCA) include three steps, the first step is construct the covariance matrix according to the random vector. The second step is solving the covariance value of matrix, and judge the contribution degree, confirming the main element. The third step is construct and solve the synthesize vectors using main element proper vector.

This research get one certain year as the base, made the base year's every prospect synthesize indexes evaluation score as ten points. Considering all aspects of synthesize indexes historical development and further trend, making the equal synthesize indexes to scale in one degree, get the historical and further degree evaluation description. This system adopts radar plot to descript the whole system synthesize evaluation and it can be viewed the relative development and changing trend from the line and scale **(Figure 2)**.

3. Relative Important Problem of Chinese Energy Distribution and Supply Demand

Energy supply and demand pattern is one big and comprehensive system. This system has different aspects. The inner elements affect the outer elements. This research combines the pattern time and space changing analysis, through the energy different elements (Bentzen J. & Engsted T. 2001. Finon D. 1976. Van der Voot E. 1982. Eltony M. N. & Al-Mutairi N. H. 1995. Ramanathan R. 1999) [10]-[14].

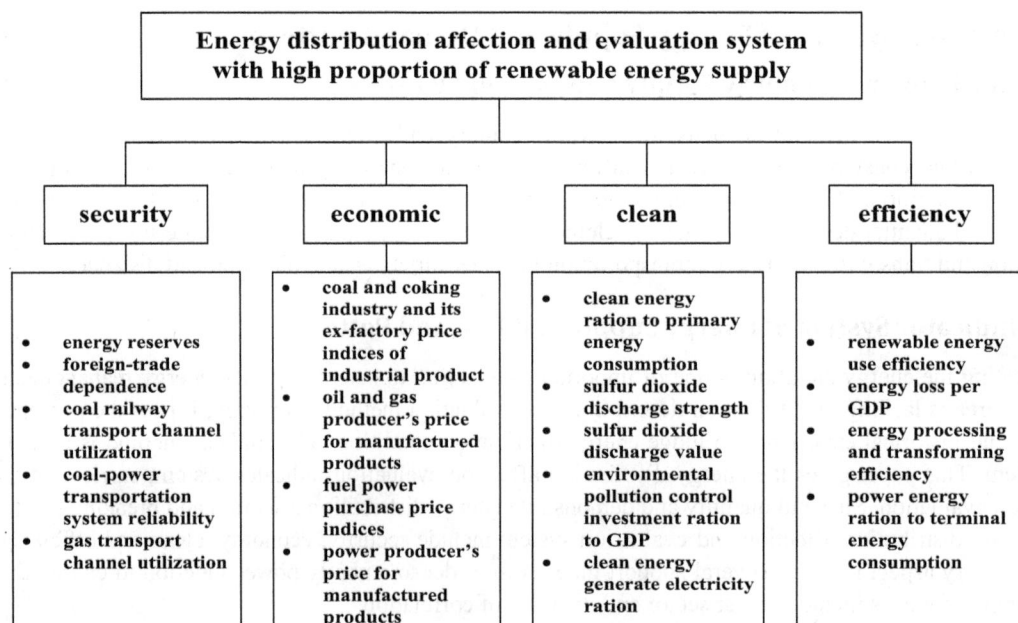

Figure 1. Energy distribution affection and evaluation system.

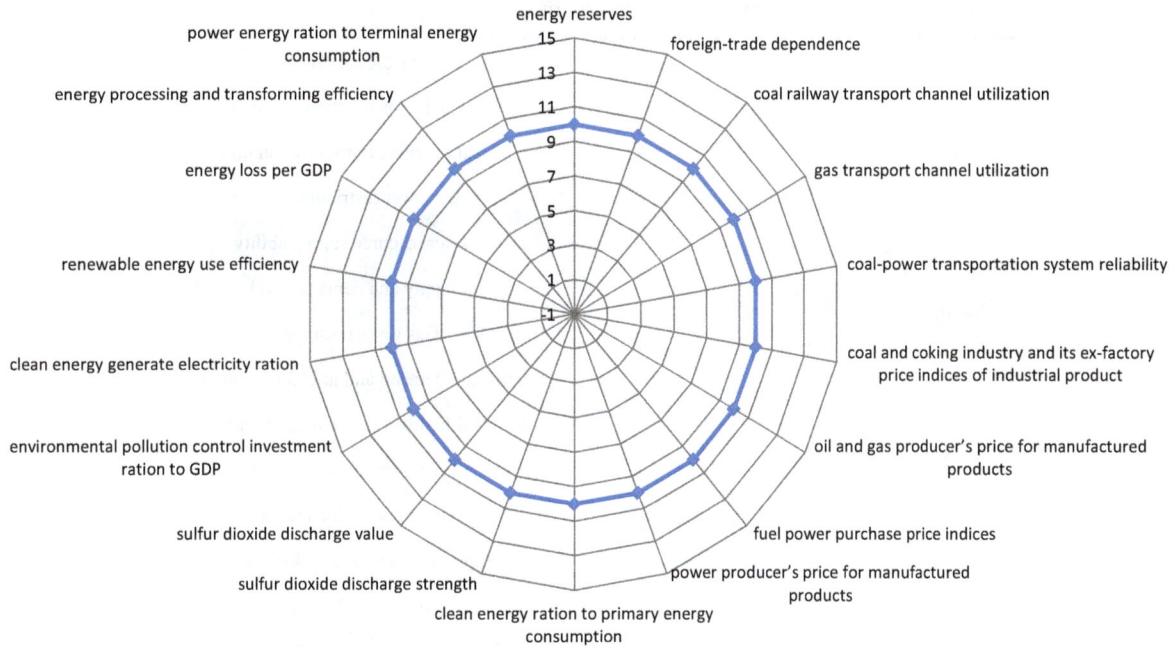

Figure 2. The comprehensive evaluation plot.

3.1. Energy Supply and Demand Inner Element

The inner elements of power supply and demand pattern include four parts. In this research, we consider coal, natural gas, water, nuclear, wind, solar and international power cooperation. These parts are considering the power resources, supply ability, power consumption and transportation (**Table 1**).

In China, the most popular energy resources are coal, natural gas and wind, to solar energy, which is using in the distributed area now. The solar is the main energy in the further usage.

Coal is the most important energy to China. To the world coal, Chinese coal reserve is low and difficult to exploit. If we consider the resource amount, mining condition, environment, water resource, transportation and security, China coal maximum exploitation amount is 41 billion tons/year. From 2009, Chinese coal import has increased 160%, and becomes the No.1 of the importing coal countries. In the further, railway and water transport will enlarge the scale. China will spread its coal move from the west to the east, which can satisfy the need of the economy. Also consider the Ultra-high voltage power grid, it can also increase the coal usage ratio.

Natural gas importing and producing are accelerating rapidly. It is changing the energy consumption condition. In china, natural gas can be imported through gas loop and LNG, from 2010 to 2014, the natural gas transportation is reaching 1370 billion steres, gas pipeline is 4.4 ten thousand kilometers, reserve gas ability is 210 billion steres/year. Big gas power generation and distribution gas system is the main type of gas usage. And from the using type, the distribution is most effective. From the cost of generation, the combined cycle is lowest. From the savage of generation, single loop has the fast adjustment.

Water is influenced by weather and season condition, according to the survey and draw data, chinses water installed capacity will be saturation in 2035. Then part of water installed capacity will be the basis capacity of the power supply. Pumped-hydro energy storage will be peak load regulation.

Nuclear is the restricted energy in certain period of exploitation time, China is focusing on its serious affection to human and environment, and its will be basis energy capacity.

Wind generation is facing two key problems, one is abandon wind and the other is lack of fund. Actually, in china, we have several different wind exploitation and usage methods, such as basement plus local consumption, distribution exploitation, basement plus combine output and basement plus wind as the main output. Wind generation is most popular energy usage in china, we adopt demand side management, power adjustment, optimize power operation and strengthen the output passage way to increase the wind energy consumption.

Solar generation will be the main renewable energy in the next thirty year, it has wide power supply. Now it is transforming into the thermoelectricity.

Table 1. The inner element of power supply and demand pattern.

Coal	Sustainable supply ability
	Coal transportation ability
	Coal to power transportation ratio
	Coal power distribution space
	Water resource supply ability
Natural gas	Supply and demand trend
	Gas generation type
Water	Supply and demand and in 2035 to saturation
	Pumped-hydro energy storage regulation
Nuclear	Stable development with conservatism
Wind	Development model and choose
	Programmer target and distribution
Solar	Development model and choose
	Programmer target and distribution
International power cooperation	Resource around of surrounding countries
	Developing potential of international energy basement
	Prospect cooperation of international power

International power cooperation of China is near to the Russia, Mongolia, Kazakhstan and other countries, which have wide spread area and plenty of source can be used by China. Russia has plenty of coal, oil and natural gas reserves. Mongolia has coal, wind, solar reserves. Kazakhstan has amount of fossil energy. In the further, China will make high voltage transmission to transport energy, which will reduce the discharging and energy import.

3.2. Energy Supply and Demand Outer Element

Economy affects energy usage, and imposing the energy development. Energy usage guarantees Chinese long-term of economy accreting. At the same time, Chinese fast economy increasing is the main impetus. To the investment, energy consumption and investment have opposite direction, the investment and its policy will affect the energy usage. The industrial structure and household consumption can determine the energy demand structure, with the industrial structure adjustment, the synthesize indexes will be changed, which can determine the regional economy development. In the further, the east part of China is still the most develop part and its energy consumption will be ahead.

Society part includes population gross, urbanization and whole people energy saving consciousness. Energy consumption has positive relation with population gross, changes the usage type. The urban is the main energy consumption places.

Policy part includes guidance quality and low carbon environmental protection property. The energy strategy in China is now focusing on energy management, reserve, market competition and clean energy development, which are very important to country energy strategy. The constraint, inducing and stimulating policies are inducing the further energy development and its trend. The low carbon and environmental policy can also improve and enhance energy consumption structure (Samimi R. 1995. An Bentzen & Tom Engsted. 2001) [15] [16] (**Table 2**).

International part includes energy market fluctuations price and finance crisis. From the world energy market view, the low energy price can affect the produce profit and less the replacement chasing trend. The high price will accelerate the industry producing cost and salesmen price, and it will affect the energy consumption. The

Table 2. The outer element of power supply and demand pattern.

	GDP
	Investment
Economy	Resident consumption
	Industrial structure
	Regional economic development
	Total population
Society	Urbanization process
	National energy saving awareness
Policy	Oriented policy
	Low carbon, environmental protection policy
International	International energy market price fluctuations
	Finance crisis

finance crisis will also push the real economy, lead to its recession, reduce the energy demand (Wankeun Oh& Kihoon Lee. 2004. George Hondroyiannis. 2004. Khalifa H. Ghali& EI-Mutairi M. I. T. 2004) [17]-[19].

4. Prediction of Energy Supply and Demand Pattern Evolution Trend in 2020

4.1. Energy Gross Evolution

In the further, Chinese energy power consumption demand will accelerate fast. The 12th Five-Year Plan, the annual average increases 5.0%, The 13th Five-Year Plan, the average will be 3.4% in prediction.

The power consumption demand is synchronized with energy consumption demand. At the same time, with the fast development of renewable energy, power is standing out in energy system consumption demand, the power consumption speed increase is faster than energy consumption speed.

4.2. Energy Structural Evolution

(1) With the fast acceleration of energy supply structure and its clean adjustment, natural gas and non-fossil energy and power ration has fast acceleration.

From the primary energy consumption structure, coal consumption to energy consumption gross proportion will be 57% in 2020, and its decreasing amplitude is 12%. Natural gas consumption ratio will accelerate 9% until 2020. The non-fossil energy ration will reach to 15% in 2020. The renewable energy of electricity power generation to primary energy will be nearly to 80%.

From the power installation structure, coal installation ration will reduce to 54.1% in 2020. The gas power generation installation ratio is 5.7% in 2020. The clean installation ratio will reach to 40%.

(2) Energy supply structure will have obvious diversification trend, in the further, energy supply will combine with civil fossil energy, civil clean energy and foreign import to China.

In statistics from 2010 to 2020, China's new primary energy consumption is 16.6 billion tons standard coal. Chinese fossil energy newly increased supply is 16.1 billion tons standard coal. And the newly increased clean energy supply is 4.4 billion tons standard coal. The foreign import newly increase energy supply is 6.1 billion tons standard coal.

(3) Energy distribution electrification degree will increase, power is the prominent in energy distribution.

From the energy conversion, non-fossil energy is mainly be transformed into electricity power. Power generation energy ratio to primary energy is rapid increasing fast. In 2020, power generation energy ratio to primary energy is 50%.

4.3. Energy Distribution Evolution

In the further, the energy distribution intensification in China will be increased fast, big-scale energy basement to energy production gross will also accelerate.

In 2020, Chinese coal production is most spread in the north parts, such as Shanxi, Shanxi, Neimeng and Xinjiang. The production mount will be 27 billion tons. The oil spreads in north-east, north-west oil production areas. Natural gas spreads in south-west and north-west energy basement. The newly accelerating hydro-power is in south-west basement. And wind installment will be in tri-north parts of China. The nuclear power is spreading in east part.

In 2020, Chinese big synthesize energy basement production ratio to whole primary energy production is 73%.

4.4. Energy Transportation Evolution

In China, energy transportation harmonization degree will improve rapid, which will form the reasonable division of labor and complement each other's advantage, and form one energy synthesize system.

Railway capacity and coal transportation scale is matching. In 2020, Chinese tri-west parts railway outward transport coal is about 12 billion tons. The use ration is 60% and ample is big enough, which can fit the coal transport seasonal fluctuation and outward changing.

The highway coal transport will be controlled in reasonable degree. In the further ten years, china can take place of the coal highway transportation about 2 billion tons, railway transportation about 1 billion tons.

China will form coal-power energy transportation combination system. In 2020, from the Chinese energy basements outward transport, power transportation will reach 23%, railway coal transportation will be 72%, highway coal transportation will 5%. That will form the reasonable division of labor and complement each other's advantage energy transportation system. And the power grid will have great impact to wind and hydro-power output.

5. Evaluation of Chinese Energy Supply Pattern Evolution to 2050

From security, economy, clean and efficiency, evaluate energy supply pattern evaluation trend. The result showed that security, clean and efficiency will be promoted rapidly, and clean is outstanding. Consider the clean energy development and its environmental benefit, transportation structure will be adjusted. Then the distribution economic degree will be promoted.

5.1. From 2020 to 2050 Installed Capacity

According to the energy development prediction, Chinese renewable energy will be 80% to the primary energy in 2050. The fossil energy will be the fundamental load. Wind and solar will be the electricity power supply. From 2020 to 2050, the renewable energy will take place of most part of the fossil energy. The coal capacity will be 78,669 kilo watt, wind capacity will reach to 120,000 kilo watt, solar capacity will be 150,000 kilo watt (**Figure 3**).

From **Figure 4**, from 2015 to 2050, the proportion of renewable energy to primary energy is accelerating from 10% to 77.1%. The proportion of renewable energy electric quantity to gross generation is from 21.7% to 83.2%. The proportion of generation energy to primary energy is from 46.4% to 92.7%. The proportion of electric energy to terminal is from 25.6% to 67.4%.

5.2. Analysis of Energy Supply Pattern Evolution Trend

Make the 2014 as the basic year, analyze the 2020 to 2050 per ten-year as period which can get four series dates. In the energy distribution, economy, clean, efficiency elements are increasing in the further, and the security is getting weak because of renewable energy volatility (**Figure 5**).

The decline of security is mainly because of sufficient railway, import risk coefficient decline, energy diversification degree increment. The improvement of economy is mainly because of environment benefit accelerating, considering the green GDP conception and its economical efficiency. The improvement of clean is mainly because of non-fossil energy ratio increase, and pollution emissions strengthen losing. The improvement of efficiency is mainly because of energy loss per GDP, energy consumption elastic coefficient decline and electricity to terminal power consumption ratio increasing.

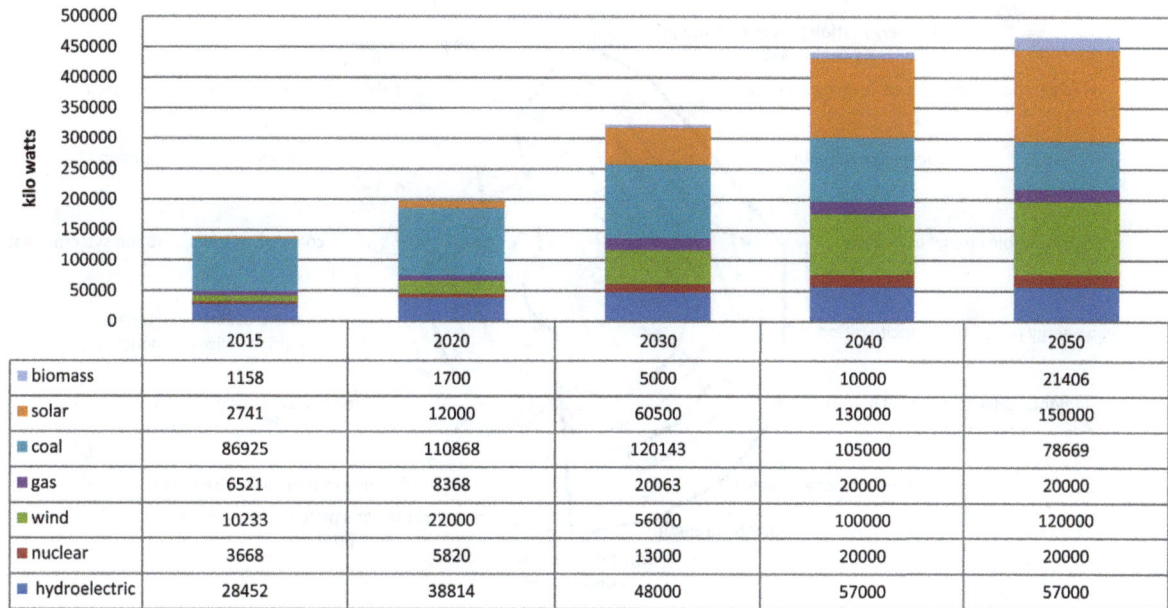

	2015	2020	2030	2040	2050
▪ biomass	1158	1700	5000	10000	21406
▪ solar	2741	12000	60500	130000	150000
▪ coal	86925	110868	120143	105000	78669
▪ gas	6521	8368	20063	20000	20000
▪ wind	10233	22000	56000	100000	120000
▪ nuclear	3668	5820	13000	20000	20000
▪ hydroelectric	28452	38814	48000	57000	57000

Figure 3. Energy capacity prediction from 2015 to 2050.

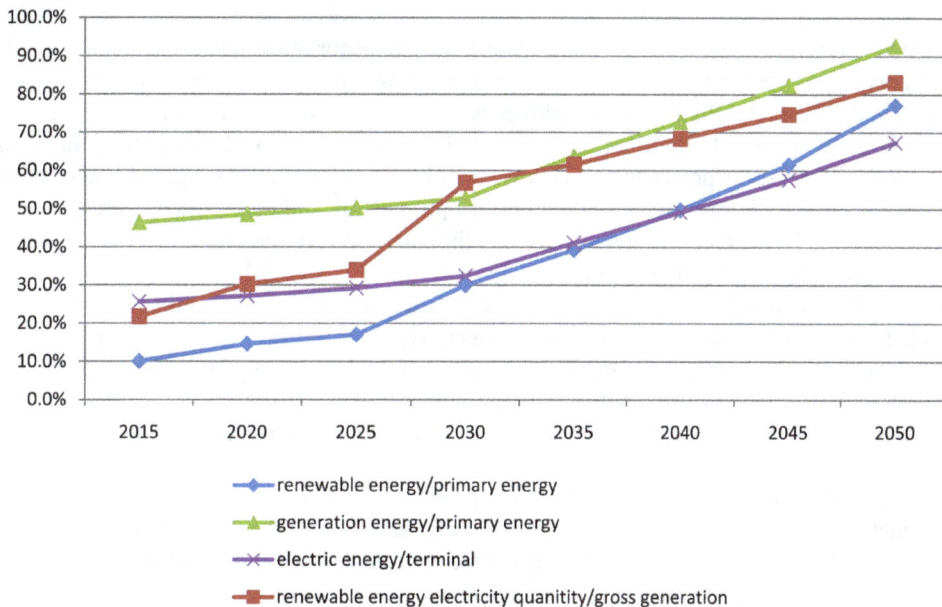

Figure 4. Energy proportion comparison from 2015 to 2050.

6. Power Grid Impact to Energy Development and Supply Pattern

In China, the power grid function has four aspects in new development. (1) Satisfy energy and power large scale long distance transportation. (2) Satisfy receiving areas multiple energy high efficient integration and coordinated operation. (3) Satisfy new energy development and new type power-using service needing. (4) Satisfy the power market platform material base needing. In the further, power grid will be pivotal role in energy exploit, transform, transportation, consumption, security, service and market trading system.

(1) Satisfy energy and power large scale long distance transportation. China is still in industrialization, informatization, urbanization and agricultural modernization rapid development stage. The power demand will continue to increase. The east and midland will restrain as the power load center in long period of time, while the new coal power, hydro-power, wind and solar power are mainly in the west and north areas. Power demanding

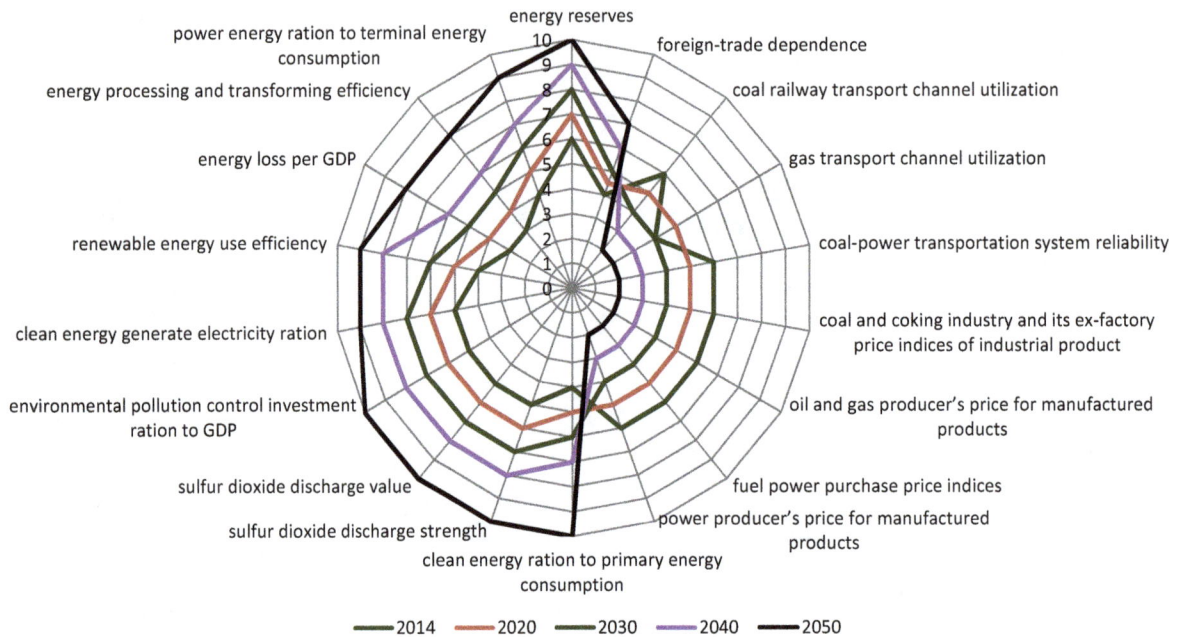

Figure 5. Energy supply pattern evolution trend from 2020 to 2050.

and generation is reverse. Long-distance, large-scale energy transportation is the most important characteristic in Chinese energy supply pattern evolution. Chinese west power transport to east and north power transport to south will be still. The energy basement of hydro-power in south-west, coal power in west and north, wind in tri-north and solar in west, will be transported to east and mil-part of china. Russia, Mongolia, Central Asia and Southeast Asia will transport power to China. Grid will be the most important part Chinese energy synthesize transportation system. As improvement of coal power optimization and distribution, the clean non-fossil energy will be transported in large scale and long distance, which will be very important to Chinese energy power large scale transportation in energy synthesize system.

(2) Satisfy receiving areas multiple energy high efficient integration and coordinated operation. With power trans-regional scale rapid improvement, the development region will have large ratio import power. Power system comprehensive degree will accelerate. The importing of coal power, hydro-power, reproducible energy and locale coal power, gas power, hydro-power, storage power, nuclear power, reproducible energy combine into coordination system, and it requires high demanding, such as big area, strong grid, dynamic balance and security high level.

(3) Satisfy new energy development and new type power-using service needing. Wind power, solar power and other new energies have randomness and intermittent characteristics. Their controllability and predictability is lower than original fossil power generation. Large scale exploitation will bring big challenge of power grid control and coordination. So the power grid is needing advanced automatics, coordination control and storage technique to implement the new energy accuracy control and high efficient utilization. At the same time, with the wide spread of electric car and electric household appliance, the energy quantity will face the higher requirement. Intelligentize will be the further needing of grid.

(4) Satisfy the power market platform material base needing. The whole country power market platform operation needs whole power grid connection, which can carry out the fundamental effect of larger degree, wider scale to give play to energy source distribution. Promote the non-fossil energy large scale exploitation usage and make larger economic benefit, power save benefit, source environmental benefit, security benefit and social benefit. In order to adapt the big change of power grid function, the further power grid must be strong and intelligent. With the breaking through of high voltage transportation, intelligent power grid construction can be carry forward synchronization. And it will form high voltage as the center, all or different levels harmonious development forming intelligent grid.

In China, energy development carries out production and consumption concentration ratio reaching to high level, energy producing and consumption showing reverse distribution, trans-regional demand reaching long distance. And these characteristics will change energy distribution configuration, coal-transport cooperation, propel intelligent grid construction, drive energy structure adjustment and energy development changing, realize energy security, economy and clean.

6.1. Energy Supply Security

Promote clean energy consumption and fuel oil replacement, which can decline foreign-trade dependence. In 2020, electric car will replace fuel oil 700 million tons. In 2050, electric car will totally replace the traditional vehicles.

Promote energy transportation system diversified development, reduce transportation risk, and accelerate energy security level. In 2020, the basements will get coal to transportation ratio is 4:1. In 2050, the reliability of coal transportation will be 1:1 (**Table 3**).

6.2. Energy Clean Development

Promote clean energy consumption. In 2020, trans-region power contribution degree to clean energy consumption is 27%, and in 2050, the ration will reach to 77%.

Reduce pollution emission of development area. In 2020, the sulfur dioxide is 75million tongs/year. And to 2050, the dioxide emission will be zero (**Table 4**).

6.3. Energy Distribution Efficiency

Promote the power transportation and clean energy usage efficiency. In 2020, trans-regional power can lack abandoning abandon water and electricity is 320 billion kilowatt-hours, ratio of trans-regional wind power to total power is 40%. In 2050, generation energy to primary energy is 92%.

Promote ratio power to terminal energy consumption. In 2020, generating power ratio to primary energy consumption will accelerate 8.2%, and in 2050, the ratio will be 67.4% (**Table 5**).

6.4. Energy System Economy

In 2020, trans-regional power can save cost 670 billion yuan/year. From the green GDP, it will reduce whole country environmental loss 30 billion yuan/year. In 2050, with high proportion of renewable energy connecting to the power system, we do not needing to accelerate the environment investment (**Table 6**).

7. Conclusion

Energy is the most important to country development, and the speed and degree of country and regional economic development is depending on modern energy supply guarantee system's ability and construction. This

Table 3. Grid effect of power security supply.

classify		2015	2020	2050
Security	Foreign-trade dependence	Limit of electric car scale	Electric car ownership is500 millio, replace fuel oil 700 million tons.	Electric car totally replacement
		Limit of trans-region of power energy volume exchange	Rapid improve grid construction, replace coal importing 5000 million tons.	None coal importing
	Reliability of coal-transportation	3:1	4:1	1:1

Table 4. Grid effect of power clean supply.

classify		2015	2020	2050
clean	Ratio of clean energy to primary energy consumption	Ratio of clean energy to primary energy is 9%	Ratio of clean energy to primary energy is 15%	Ratio of clean energy to primary energy is 77%
	Sulfure dioxide emission	Changjiang area is 45 tons	Changjiang area is 75 tons	none

Table 5. Grid effect of power efficiency supply.

classify		2015	2020	2050
efficiency	Energy process transition	Line loss rate 6.5%	Line loss rate 6.2%, Power appliance usage ratio will be accelerating	Generation energy to primary energy is 92%
		Clean energy has little trans-regional exchanging power, the north abandon wind ratio is exceeding 20%.	Less the hydro-power in southwest is 320 billion kilowatt-hour, reduce the abandon wind power is billion kilowatt-hour.	Hydro-power is stable, and wind power can be stored
	Power ratio to terminal energy	20.9%	27.5%	67.4%

Table 6. Grid effect of power economy supply.

Classify		2015	2020	2050
Economy	Power industrial products ex-factory price	Coal transportation to receiving port generating power has low economy	Save the cost of generating power cost is 450 billion yuan	Renewable energy replacement
	Green GDP	No space in development	Lack environment loss is 30 billion yuan/year	No needing to add investment

paper researches Chinese energy supply and demand pattern system and evaluation methodology with high proportion of renewable energy supply, gives out the inner and outer influencing elements. And evaluate Chinese energy supply and demand pattern from energy gross, structure, distribution and transportation. Give the energy supply synthesize radar comparison chart in certain time period. From the energy security, economy, clean and efficiency, analyze the benefit comparisons of Chinese energy supply and demand pattern. This energy supply and demand pattern model will give one certain theoretical analysis and practice reference.

References

[1] Sarak, H. and Satman, A. (2003) The Degree-Day Method to Estimate the Residential Heatingnatural Gas Consumption in Turkey: A Case Study. *Energy*, **28**, 929A-939A. http://dx.doi.org/10.1016/S0360-5442(03)00035-5

[2] Balaehandra, R. and Chandru, V. (1999) Modelling Electricity Demand with Representative Load Curves. *Energy*, **24**, 219-230. http://dx.doi.org/10.1016/S0360-5442(98)00096-6

[3] Kumar, A., Bhattacharya, S.C. and Pham, H.L. (2003) Greenhouse Gas Mitigation Potential of Biomass Energy Technologies in Vietnam Using the Long Range Energy Alternative. *Energy*, **28**, 627-654. http://dx.doi.org/10.1016/S0360-5442(02)00157-3

[4] Mackay, R.M. and Probert, S.D. (2001) Forecasting the United Kingdom's Supplies and Demands for Fluid Fossil-fliels. *Applied Energy*, **69**, 161-189. http://dx.doi.org/10.1016/S0306-2619(01)00003-4

[5] Messner, S. and Schrattenholzer, L. (2000) MESSAGE-MACROO: Linking an Energy Supply Model with a Macroeconomic Module and Solving It Iteratively. *Energy*, **25**, 267-282. http://dx.doi.org/10.1016/S0360-5442(99)00063-8

[6] Cuaresma, J.C., Hlouskova, J. and Kossmeier, S. (2004) Forecasting Electricity Spot-Prices Using Linear Univariate Time-Series Models. *Applied Energy*, **77**, 87-106. http://dx.doi.org/10.1016/S0306-2619(03)00096-5

[7] Sharma, D.P., Chandramohanan Nair, P.S. and Balasubramanian, R. (2002) Demand for Commercial Energy in the State of Kerala, India: An Econometric Analysis with Medium-Range Projections. *Energy Policy*, **30**, 781-791. http://dx.doi.org/10.1016/S0301-4215(01)00138-0

[8] Persaud, A.J. and Kumar, U. (2001) An Eclectic Approach in Energy Forecasting: A Case of Natural Resources Canada,s(NRCans) Oil and Gas Outlook. *Energy Policy*, **29**, 303-313. http://dx.doi.org/10.1016/S0301-4215(00)00119-1

[9] Chow, L.C.H. (2001) A Study of Sectoral Energy Consumption in Hong Kong (1984-97) with Special Emphasis on the Household Sector. *Energy Policy*, **29**, 1099-1110. http://dx.doi.org/10.1016/S0301-4215(01)00046-5

[10] Ediger, V.S. and Tatlidil, H. (2002) Forecasting the Primary Energy Demand in Turkey and Analysis of Cyclic Patteras. *Energy Conversion and Management*, **43**, 473-487. http://dx.doi.org/10.1016/S0196-8904(01)00033-4

[11] Finon, D. (1976) Un model energetique pour la France. Centre National de la Recherche Scienti-fique, Paris.

[12] Van der Voot, E. (1982) The EFOM 12C Energy Supply Model with in the EC Modeling System. *Omega*, **10**, 507-523.

http://dx.doi.org/10.1016/0305-0483(82)90007-X

[13] Eltony, M.N. and Al-Mutairi, N.H. (1995) Demand for Gaspoline in Kuwait: An Empirical Analysis Using Cointegration Techniques. *Energy Economies*, **17**, 249-253. http://dx.doi.org/10.1016/0140-9883(95)00006-G

[14] Ramanathan, R. (1999) Short- and Long-Run Elasticities of Gasoline Demand in India: An Empirical Analysis Using Cointegration Techniques. *Energy Economies*, **21**, 321-330. http://dx.doi.org/10.1016/S0140-9883(99)00011-0

[15] Samimi, R. (1995) Road Transport Energy Demand in Australia: A Integration Approach. *Energy Economics*, **17**, 329-339. http://dx.doi.org/10.1016/0140-9883(95)00035-S

[16] Bentzen, J. and Engsted, T. (2001) A Revival of the Autoregressive Distributed Lag Model in Estimating Energy Demand Relationships. *Energy*, **26**, 45-55. http://dx.doi.org/10.1016/S0360-5442(00)00052-9

[17] Oh, W. and Lee, K. (2004) Energy Consumption and Economic Growth in Korea, Testing the Causality Relation. *Journal of Energy Modeling*, **26**, 973-985.

[18] Hondroyiannis, G. (2004) Estimating Residential Demand for Electricity in Greece. *Energy Economies*, **26**, 319-334. http://dx.doi.org/10.1016/j.eneco.2004.04.001

[19] Ghali, K.H. and EI-Mutairi, M.I.T. (2004) Energy Use and Output Growth in Canada: A Multivariate Integration Analysis. *Energy Economies*, **26**, 225-238. http://dx.doi.org/10.1016/S0140-9883(03)00056-2

10

The Prospects for Renewable Energy through Hydrogen Energy Systems

Marc A. Rosen

Faculty of Engineering and Applied Science, University of Ontario Institute of Technology, Oshawa, Ontario, Canada
Email: Marc.Rosen@uoit.ca

Abstract

The prospects for renewable energy are enhanced through the use of hydrogen energy systems in which hydrogen is an energy carrier. As easily accessible fossil fuel supplies become scarcer and environmental concerns increase, hydrogen is likely to become an increasingly important chemical energy carrier. As the world's energy sources become less fossil fuel-based, hydrogen and electricity are expected to be the two dominant energy carriers for the provision of end-use services, in a hydrogen economy. Thus, hydrogen energy systems allow greater use of renewable energy resources. In this paper, the role of hydrogen as an energy carrier and hydrogen energy systems, and their economics, are described and reviewed.

Keywords

Renewable Energy, Hydrogen, Hydrogen Economy

1. Introduction

Many energy sources exist, e.g., fossil fuels, uranium, renewable energy resources. Fossil fuels are the world's main sources and are also energy carriers. Most renewable energy sources (solar energy, wind, falling water, tides, waves, etc.) and uranium must first be converted to an energy carrier (commonly electricity) before use. As fossil fuels become harder to obtain and environmental impacts (climate change, stratospheric ozone depletion, acid precipitation, smog, etc.) increase, renewable energy sources will likely be increasingly sought. But renewable energy sources cannot act as energy carriers and presently are mainly used to produce the energy carrier electricity. This limits the prospects for renewable energy since societies cannot operate effectively with only electricity. They also need chemical fuels for processes such as transportation [1]-[3].

Thus, eventually it will be necessary to produce chemical fuel, either directly from non-hydrocarbon energy sources or from the electricity they can generate [4]. Many researchers feel that hydrogen is the most logical choice as a future chemical fuel. A "hydrogen economy," with hydrogen and electricity acting as complementary energy carriers, has been envisioned for decades [1] [4] [5]. Although some feel that renewable energy resources can supply all energy requirements, others believe the potential is limited due to their challenges, e.g., intermittency, but hydrogen energy systems appear capable of improving the prospects for renewable energy.

The objective of this paper is to describe the prospects for renewable energy provided by hydrogen energy systems, by reviewing hydrogen energy and hydrogen energy systems and their economics. The use of hydrogen as an energy carrier and hydrogen energy technologies, including those for hydrogen production, utilization, storage and distribution, are described.

2. Hydrogen as an Energy Carrier

Hydrogen is a useful energy carrier for numerous reasons:
- Hydrogen can be produced from fossil fuels, uranium and renewable energy sources. The hydrogen is obtained by splitting water in the last two cases.
- Hydrogen can be used as a fuel (and feedstock) in industrial, transportation, residential and commercial activities, and for electricity generation in devices such as fuel cells.
- Hydrogen can be stored in large quantities in a variety of forms, unlike electricity.
- Hydrogen can be transported in many ways to (road, rail, ship, pipeline, etc.).
- Hydrogen use is environmentally benign, the main output of its oxidation being water [6].

3. Hydrogen Energy Systems and Technologies

Hydrogen energy systems are mainly comprised of technologies for the production, utilization, storage and distribution of hydrogen [4] [7].

3.1. Production

The main energy sources for hydrogen production are shown in **Figure 1**. The main processes for hydrogen production include fossil fuel-based processes (e.g., steam reforming of natural gas, catalytic decomposition of

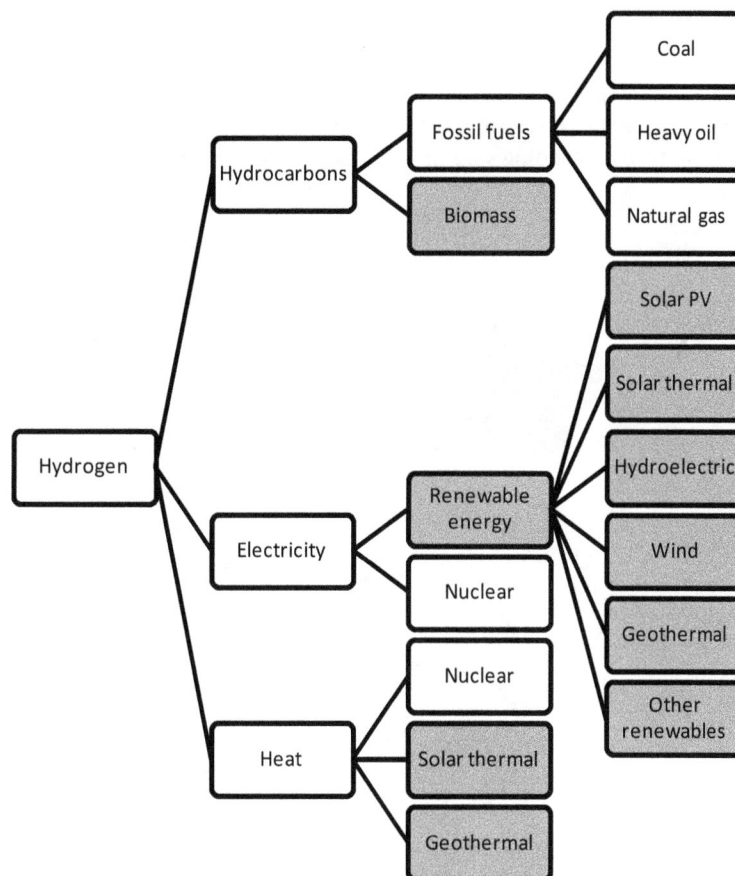

Figure 1. Main energy sources for hydrogen production, with renewables highlighted.

natural gas, partial oxidation of heavy oil, coal gasification), and water splitting technologies (e.g., water elec-trolysis, thermochemical water decomposition, photo-chemical, electrochemical and biological processes). The latter group can be driven by renewable energy. Most hydrogen is produced today by steam reforming of natural gas. For hydrogen production processes using hydrocarbons such as fossil fuels and biomass, the hydrogen is derived from the hydrogen in the hydrocarbon itself and water. The hydrogen is derived from the hydrogen in water for the hydrogen production processes driven by electrical and thermal energy, via the chemical reaction $H_2O \rightarrow H_2 + O_2$.

Photochemical, photoelectrochemical and photobiological processes are at the early research stage, but water electrolysis and thermochemical water decomposition have greater potential:

• Water electrolysis is a mature technology in which hydrogen and oxygen are produced by using electricity to split water. Water electrolysis is used at present on a small scale, often in specialized processes that require high purity hydrogen. Water electrolysis achieves a hydrogen concentration over 99.5% compared to 97% - 98% for fossil fuel-derived hydrogen. High-temperature electrolysis, which utilizes steam, is under devel-opment [4] [8]. The near-term option for hydrogen production from renewable energy is likely water elec-trolysis.

• Thermochemical water decomposition (or splitting) consists of a sequence of chemical reactions for which the net reaction is water decomposition [9]-[12]. Many such cycles have been proposed, potentially driven by heat from nuclear power plants [13] or high temperature solar thermal facilities. Two examples are the copper-chlorine cycle, which requires heat at up to 550°C, and the sulphur-iodine cycle, which requires heat at about 900°C. Hybrid thermochemical cycles that use electricity and heat to decompose water are also pos-sible. Thermochemical water decomposition is only at the research and development stage, but it is a sig-nificant future candidate for large-scale hydrogen production.

The main processes for hydrogen production are illustrated in **Figure 2**, where three categories presented: processes based on fossil fuels, processes not based on fossil fuels, and processes that integrate multiple hydro-gen production processes.

3.2. Storage

One reason hydrogen complements electricity as an energy carrier is that it can be stored in bulk over long times, in many forms and using many technologies. For example, hydrogen can be stored as a cryogenic liquid (at 20.3

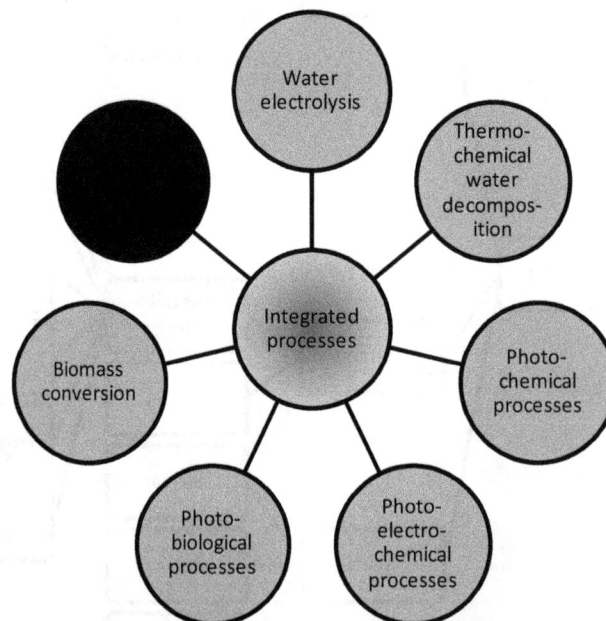

Figure 2. Main hydrogen production processes, including those based on fossil fuels (dark shading), non-fossil fuels (light shading) and in-tegrated processes (center gradient shading).

K) in insulated dewers (tanks), using metal hydride technologies, which allow large quantities of gaseous hydrogen to be thermally adsorbed and desorbed from the surface of certain metals, and as a compressed gas in cylinders or in underground reservoirs and caverns.

3.3. Distribution

Several technologies for the distribution of hydrogen exist. Hydrogen can be transported in bulk by pipeline as a gas, or possibly, for short distances, as a liquid; truck or rail as a highly compressed gas in cylinders; and truck or rail or ship as a liquid.

3.4. Utilization

Hydrogen can be used as an energy carrier in various ways:
- a fuel for producing electricity in fuel cells [14], which combine hydrogen and oxygen electrochemically to produce electricity and water;
- a supplement to natural gas, which can be added directly in natural gas distribution networks;
- a feedstock in the manufacturing of synthetic fuels, especially methane and ammonia;
- a fuel for urban motor vehicles, locomotives, marine vessels and aircraft, using fuel cells and/or combustion engine technology [1] [15]; and
- a fuel for space heating, in appropriate circumstances.

At present, hydrogen is used extensively as a chemical feedstock in industry, mainly in refinery processes, and petrochemical and chemical production. But in a hydrogen economy, hydrogen will be used for energy services as the prime chemical energy carrier, with the sector most likely to rely on hydrogen being transportation [1] [3].

4. Hydrogen Energy Systems Economics

Economic assessments of hydrogen energy systems typically depend on many factors and assumptions. Such assessments must account for costs of producing, storing, distributing and utilizing hydrogen, all of which are significant in a hydrogen economy. Details on these follow:
- *Production*: Costs for producing hydrogen via conventional and non-conventional methods depend on factors such as feedstock, labour and capital costs, and are often a significant part of the costs of hydrogen energy systems.
- *Transport and storage*: Hydrogen transport and storage costs vary by method. Usually, hydrogen transport by pipeline as a compressed gas is often least expensive for large volumes and distances, while small quantities of gaseous hydrogen are transported most economically over short distances in tube tanks by truck. The transport cost is much lower for liquid hydrogen than compressed gaseous hydrogen, but liquefaction increases the hydrogen cost by about 60%. Costs are normally significantly less for hydrogen transport and storage than for production.
- *Utilization*: Several hydrogen uses are expected to be economic in the future, with a main focus likely to be in transportation. The earliest large-scale use of hydrogen from water electrolysis is expected to be in upgrading of refinery products, where hydrogen produced by steam-methane reforming can be substituted. Hydrogen may also be used in biomass-to-methanol conversion.

5. Conclusions

As accessible fossil fuel supplies become scarcer and environmental concerns increase, hydrogen is likely to become an important chemical energy carrier. Hydrogen energy systems can thereby enhance the prospects for renewable energy. There are many commercial processes for producing hydrogen from fossil fuel and non-fossil fuel sources (including renewables). Technologies for the storage and distribution of hydrogen exist. Technologies are developing for utilizing hydrogen as an energy carrier, especially in transportation. The technologies needed for hydrogen energy systems are undergoing much research and development.

Acknowledgements

The Natural Sciences and Engineering Research Council of Canada provided support.

References

[1] Scott, D.S. (2007) Smelling Land: The Hydrogen Defense against Climate Catastrophe. Canadian Hydrogen Association, Ottawa, Canada.

[2] Balat, M. (2008) Potential Importance of Hydrogen as a Future Solution to Environmental and Transportation Problems. *International Journal of Hydrogen Energy*, **33**, 4013-4029. http://dx.doi.org/10.1016/j.ijhydene.2008.05.047

[3] Turgut, E.T. and Rosen, M.A. (2010) Partial Substitution of Hydrogen for Conventional Fuel in an Aircraft by Utilizing Unused Cargo Compartment Space. *International Journal of Hydrogen Energy*, **35**, 1463-1473. http://dx.doi.org/10.1016/j.ijhydene.2009.11.047

[4] Muradov, N.Z. and Veziroglu, T.N. (2008) "Green" Path from Fossil-Based to Hydrogen Economy: An Overview of Carbon-Neutral Technologies. *International Journal of Hydrogen Energy*, **33**, 6804-6839. http://dx.doi.org/10.1016/j.ijhydene.2008.08.054

[5] Gnanapragasam, N.V., Reddy, B.V. and Rosen, M.A. (2010) Feasibility of an Energy Conversion System in Canada Involving Large-Scale Integrated Hydrogen Production Using Solid Fuels. *International Journal of Hydrogen Energy*, **35**, 4788-4807. http://dx.doi.org/10.1016/j.ijhydene.2009.10.047

[6] Lubis, L.I., Dincer, I., Naterer, G.F. and Rosen, M.A. (2009) Utilizing Hydrogen Energy to Reduce Greenhouse Gas Emissions in Canada's Residential Sector. *International Journal of Hydrogen Energy*, **34**, 1631-1637. http://dx.doi.org/10.1016/j.ijhydene.2008.12.043

[7] Holladay, J.D., Hu, J., King, D.L. and Wang, Y. (2009) An Overview of Hydrogen Production Technologies. *Catalysis Today*, **139**, 244-260. http://dx.doi.org/10.1016/j.cattod.2008.08.039

[8] Turner, J., Sverdrup, G., Mann, M.K., Maness, P.-C., Kroposki, B., Ghirardi, M., Evans, R.J. and Blake, D. (2008) Renewable Hydrogen Production. *International Journal of Energy Research*, **32**, 379-407. http://dx.doi.org/10.1002/er.1372

[9] Rosen, M.A. (2010) Advances in Hydrogen Production by Thermochemical Water Decomposition: A Review. *Energy—The International Journal*, **35**, 1068-1076. http://dx.doi.org/10.1016/j.energy.2009.06.018

[10] Lewis, M.A., Masin, J.G. and O'Hare, P.A. (2009) Evaluation of Alternative Thermochemical Cycles—Part I: The Methodology. *International Journal of Hydrogen Energy*, **34**, 4115-4124. http://dx.doi.org/10.1016/j.ijhydene.2008.06.045

[11] Lewis, M.A. and Masin, J.G. (2009) Evaluation of Alternative Thermochemical Cycles—Part II: The Down-Selection Process. *International Journal of Hydrogen Energy*, **34**, 4125-4135. http://dx.doi.org/10.1016/j.ijhydene.2008.07.085

[12] Andress, R.J., Huang, X., Bequette, B.W. and Martin, L.L. (2009) A Systematic Methodology for the Evaluation of Alternative Thermochemical Cycles for Hydrogen Production. *International Journal of Hydrogen Energy*, **34**, 4146-4154. http://dx.doi.org/10.1016/j.ijhydene.2008.11.118

[13] Naterer, G.F., Suppiah, S., Stolberg, L., Lewis, M., Ahmed, S., Wang, Z., Rosen, M.A., Dincer, I., Gabriel, K., Secnik, E., Easton, E.B., Lvov, S.N., Papangelakis, V. and Odukoya, A. (2014) Progress of International Program on Hydrogen Production with the Copper-Chlorine Cycle. *International Journal of Hydrogen Energy*, **39**, 2431-2445. http://dx.doi.org/10.1016/j.ijhydene.2013.11.073

[14] Thomas, C.E. (2009) Fuel Cell and Battery Electric Vehicles Compared. *International Journal of Hydrogen Energy*, **34**, 6005-6020. http://dx.doi.org/10.1016/j.ijhydene.2009.06.003

[15] Lapeña-Rey, N., Mosquera, J., Bataller, E. and Ortí, F. (2010) First Fuel-Cell Manned Aircraft. *Journal of Aircraft*, **47**, 1825-1835. http://dx.doi.org/10.2514/1.42234

11

Review on People's Lifestyle and Energy Consumption of Asian Communities: Case Study of Indonesia, Thailand, and China

Didit Novianto*, Weijun Gao, Soichiro Kuroki

The University of Kitakyushu, Kitakyushu, Japan
Email: *u3dbb002@eng.kitakyu-u.ac.jp, weijun@kitakyu-u.ac.jp, kuroki@kitakyu-u.ac.jp

Abstract

This paper focuses on the residential housing performance and people's lifestyle in terms of energy use. The research has to be continuously study in order to change the Asian people's lifestyle moving toward the low energy consumption. The research began with collecting data by questionnaires which already distributed to some cities of some countries in Asia. In this paper, we conducted review of energy behavior in housing of three different countries, Indonesia, China, and Thailand. Through questionnaire surveys distributed to more than 600 households, we revealed the housing unit characteristics, household characteristics, ownership of domestic electrical appliances, use of indoor thermal equipment, and monthly energy use. Based on questionnaire results from all households in the seven districts, we conducted statistical analyses to find the major factors influencing the energy use in households. Finally, the research results will be used to indicate the energy use and develop an idea for energy conservation in Asian countries. For further studies, the researchers also try to discover the new way of changing the people's lifestyle in terms of energy consumption.

Keywords

Lifestyle, Housing, Energy Use, Survey, Asia

1. Introduction: Energy Consumption Situation in Each Country

Indonesia: Indonesia is rich in natural resources including the petroleum, natural gas, and coal. About 35% of the petroleum mining products exported to other countries. On the other hand, Indonesia's economic growth af-

---

*Corresponding author.

fects the increasing of domestic energy demand. However, these energy resources are limited and need hundreds of years of production. From 2000 until 2009, Indonesia's oil export decreased 48% as an impact of oil production decrease [1]. According to the National Energy Outlook of Indonesia, in the period 2000-2009 energy consumption in Indonesia increased from 709.1 million SBM (Barrels of Oil Equivalent/BOE) to 865.4 million SBM, or increased by 2.2% per year. Until the end of 2011, the largest sector of final energy consumption is still dominated by industrial and then followed by household sector and transportation sector, each are 37%, 36%, and 21% [2]. Even so, with the national population booming, there is possibility for the household sector will increase significantly.

China: Energy consumption of China has been increasing rapidly due to the recent economic growth and development. This leads to serious environment problems such as air pollution and acid rains. Meanwhile the building sector accounts for large parts of energy consumption. It is almost 30% of total energy consumption in 2007 [3]. Especially in recent years, people's demand for life quality triggered drastic annual increase of energy consumption on urban areas in China. In order to estimate the future trend of residential indoor environment and energy consumption in China it's necessary to understand the actual conditions of the usage of facilities the indoors thermal conditions and quality in different zones. In China, the space heating energy consumption is about 40% in the urban area on average though there is great difference between the south region where people use individual heating equipment and the north region where district heating is generally used. However, the energy consumption is comparatively large in the villages of China because firewood and crops are used as the energy source.

Thailand: The National statistical office of Thailand has conducted a survey of household energy consumption by 2009 and found out the comparison of the energy cost of the households across the country in 2008 and 2009 has increased from 1568 baht (51 USD) to 1818 baht (60 USD) or increase the percentage of 15.9 per year. Particularly oil, biodiesel and renewable energy, although it is rarely used but has been increased 50%, gas was increased 33.7% and diesel increased by 25.7%. The electric consumption was found increases of 11.6 over the year 2008, which has just in 2.1%. From 2008 to 2009, the monthly energy cost of residential sector has increased 10.6% (from 83 USD to 92 USD) [4].

Responding this issue, some development countries innovate with environmentally hi-technologies, such as wind power, solar panel, and nuclear power plant. The technologies were proven in reducing consumption of oils and coals, but still low in affordability and not a bit of technologies would lead to another disaster like the damage of nuclear power plant in *Fukushima*, Japan, 2011. Rather than just reducing consumption, the technologies were also increase the energy supply, which means people can use it without limits as their needs increase. From these issues, we are thinking about how to change the community towards energy efficient by changing their lifestyle. By using the energy as necessary will significantly able to save the environment and human life, with the combination of the passive technologies.

2. Study of Literature

In response to the issue of increasing energy demand in the housing sector, several studies on residential energy use have been conducted by many researchers in Asian and Western countries. Some researches about trends of energy use and its relationship with households attributes in Japan were reviewed. *Nakagami et al.* [5] reported that the energy demand is expected to increase continuously as well as the increase of housing appliances ownership in Japan. They also surveyed residential energy consumption and its indicators in 18 countries, both developed and developing countries, and revealed that in the Western Countries, household energy consumption shows a trend toward saturation, but in the Asian Countries it is likely that household energy consumption will continue to rise. *Kagajo et al.* [6] analyzed the household energy consumption trends based on the standard models of family pattern, aging society, and its life schedule. *Nomura et al.* [7] conducted a study on the inter-relationship between household electricity consumption and family member ages. *Tsurusaki et al.* [8] found that space-heating, space-cooling, lighting and usage of electronic entertainment equipment are the major influences on household energy consumption in the cold climate area, Hokkaido, Japan. *Fong et al.* [9] pointed out the lifestyles in terms of family pattern, employment status; employment sector, gender, and age do have a significant impact on household energy consumption, later the lifestyles were categorized as indirect lifestyles. *Yu et al.* [10] made a questionnaire and simulation in northeast, middle and east of China, and generated an equation to calculate the energy used for district heating of households in Shenyang, Dalian, Beijing, and Luoyang. *Chen et al.*

[11] using Quantification Theory 1, investigated the influential factors on energy consumption in seven big cities of China in the summer time and presented a comparative study of the results. *Lam et al.* [12] also made the research on residential energy consumption with five different building types in Hong Kong, and evaluated the end use values. *Long et al.* [13] conducted an investigation on the electricity use of air conditioners in Shanghai, and got the mean monthly energy consumption values in the air conditioning seasons and transition seasons. *Tso et al.* [14] studied domestic energy structure in Hong Kong and found the breakdown of end use and the daily properties of energy loads by calculations, and finally revealed the influential factors of electricity use in summer and winter. *Yoshino et al.* [15] carried out the investigation to 240 houses during the winter from 1998 to 2000 by means of questionnaire in terms of the space heating and indoor thermal environment of residential buildings in three big cities of China. *Ogawa et al.* [16] carried out research to investigate the characteristics of building energy standard of the residential building in China and classified the climate zones based on thermal environment design, specified new standard of sunlight, lighting, and ventilation. *Wei et al.* [17] analyzed the current condition of urban and rural residential energy consumption of Jilin, China by establishing a model of residential energy demand, which can predict energy demand based on demand types and environmental load until 2020. *Quyang et al.* [18] identify the changes in the occupants' behavior of residential housing in Hangzhou, China, and predicted that residential electricity use will increase continually in the near future due to improved standards of living and a greater dependency on electric appliances, also more than 10% of household electricity use can be conserved by informing occupants of energy-saving measures to improve their behavior. *Hubacek et al.* [19] examined the contribution to CO_2 emissions (I) of population growth (P), affluence (A) (representing different lifestyles and consumption patterns) and CO_2 intensity (T) representing technology. *Kang et al.* [20] developed energy load prediction equations which can be easily used to estimate the energy consumption of multi-residential buildings in Korea by using Orthogonal Array to carry out simulation and investigated the relative importance of each energy factor with ANOVA. *Yoo et al.* [21] identified the changes in occupants' lifestyles using national statistics survey data, and then estimations were made by connecting each occupant activities to the corresponding residential appliances.

In the Western societies, numerous researches about residential energy use and family lifestyle were also reviewed in this paper. *Wood et al.* [22] emphasized the importance of time-series electricity use analysis of each residential appliance in a particular house, in order to reflect the occupant's characteristics and how those characteristics impact electricity use in the UK. *Diamond et al.* [23] characterized the physical environment and identified the forces that would be influential in making implications for the future energy use in US buildings. *Brounen et al.* [24] analyzed the extent to which the use of gas and electricity is determined by technical specification of dwelling as compared to demographic characteristics of the resident in Dutch homes, revealed that the gas consumption is determined principally by structural characteristics, while electricity consumption varies more directly with household composition, income, and family structure. *Haas et al.* [25] investigated the impact of the thermal quality of buildings, consumer behavior, heating degree days, building type (single- or multi-family dwellings) on residential energy demand for spaces heating have been investigated in Austria. *Linde'n et al.* [26] based on a survey of 600 Swedish households, revealed those behavioral patterns that are efficient and those that need to be improved for energy conservation, after that several policy instruments for change were identified in the study and they include combinations of information, economic measures, and administrative measures and more user friendly technology as well as equipment with sufficient esthetic quality. *Olexsak et al.* [27] compiled 274 measurements of observed changes in electricity demand caused by Earth Hour events in 10 countries, the events reduced electricity consumption by 4% on average, with a range of +2% (New Zealand) to −28% (Canada). *Leighty et al.* [28] investigated the changes in behavior and technology induced by a transient crisis which can permanently lower electricity use in Juneau, Alaska by 25% and concluded persistent electricity savings appear to be the result of reduced thermostat settings, continued CFL bulb use, keeping fewer lights on, unplugging appliances when not in use, use of power-saving settings on appliances, and showering behavior (both shorter duration and fewer).

Several studies have obtained the major influential lifestyle factors on household's energy use and proposed some strategies to change the behavior of occupants, but the sample mostly were randomly taken, especially in terms of household's types. In this study, we conducted surveys in Asian big cities: Shanghai (China), Jakarta (Indonesia), and Bangkok (Thailand) to more than a thousand respondents in order to grasp the house energy situation in developing and developed countries. In this paper, we selected some local cities of three countries and targeted the younger nuclear families aging between 20's to 40's (married couples with children, or single parent with children) as representative for future energy demand. The large number of feedback and unique database are new contribution to this research field, since it is very difficult for researcher to conduct such kinds of questionnaire surveys in different countries and especially in Japan due to privacy issues and strict rule of local government. Furthermore, we also contribute to the discussion on energy use pattern by types of houses and energy sources.

3. Survey on Lifestyle and Housing Energy Consumption

The research framework can be seen in **Figure 1**. The great differences of the climate and architecture characteristics among three countries lead to residential indoor environment and energy usage structures in different countries having their own characteristics. For that reason a survey was carried out for these three countries from August 2011 to December 2012 among more than 4500 households in total of all countries. Questionnaire distribution and collection were done through local cooperative school with their teacher's assistance. Questionnaire has seven parts: building characteristics, family income and outcome, building appliances, family characteristics, life style, interior air quality condition, and effort on energy consumption reduction. The investigated contents are shown in **Table 1**. The types of buildings involved in the survey range from low-rise to high-rise building, the monthly energy consumption data of gas & electricity and field measurement data of humidity & temperature in a year were obtained also.

In Indonesia the survey was conducted from September 2011 to October 2011 among 500 households in the transition time between rainy season and dry season respectively, which were selected in urban area. Almost of the data were collected from three big cities in Java Island, are Jakarta, Bandung, and Semarang City. In China the survey was conducted from September 2011 to December 2011, between summer and autumn, which were selected in Shanghai area, among more than 800 households. In Thailand the survey data was collected from Bangkok City in the same period with Indonesia survey.

3.1. Basic Information

Table 2 shows the building characteristics among all investigated households and its comparison of three different countries. As for the architecture floor area, 32.7% households from Indonesia and 34% households from Japan are between 121 m^2 and 200 m^2, which is the largest one in the country. The 91 m^2 - 120 m^2 is dominating in China, probably is an impact of economic growth in Shanghai, the land also getting higher. While in Indonesia there is 2% households are have less than 15 m^2 of floor area, less than the minimum standard of Indonesian housing, which is 36 m^2. For the building type, in Indonesia, 94% buildings investigated are detached house

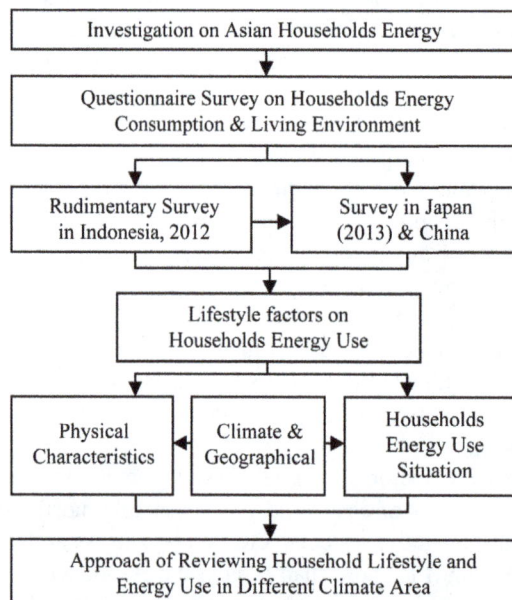

Figure 1. Research framework.

Table 1. Questionnaire contents.

Item	Content
Building characteristics	Housing area, housing type, housing status, housing structure, housing construction year
Family income and outcome	Annual income, electricity bill consumption, water bill consumption, gas bill consumption
Building appliances	Heating and cooling system type, kitchen equipment, entertainment equipment, cleaning & showering equipment, lighting type, transportation type, etc.
Family characteristics	Number of occupants, work type
Life style	Heating & cooling period, heating & cooling time, number of staying persons
Interior air quality condition	Sense of thermal comfort, satisfaction of environment, environmental conditioning method
Effort on energy consumption reduction	Environmental problem, waste management, passive technology usage, etc.

Table 2. Building characteristic.

Building characteristic		Indonesia	Thailand	China
Architecture area (m²)	<15	2		
	15 - 45	3.3	7	
	46 - 60	5.7	9	3.1
	61 - 90	19.3	11	16.6
	91 - 120	15.7	17	38.8
	121 - 200	32.7	34	25.8
	201 - 250	10	12	8.9
House type	Detached house	94	75	25.8
	Hi-rise apartment	0.7	25	72.2
	Low-cost flat	3.3		
	Shop house	2		2
House floor level	1 F	47.3	27	7.2
	2 F	47	73	16.5
	3 F or more	5.7		76.3
Construct. year (year)	< 5	9		36.6
	5 - 10	30.3	49	35.1
	11 - 20	39	33	19.5
	21 - 30	10.7	16	5.2
	31 - 40	5.7	2	3.4
	41 - 50	2.3		0
	51 - 60	2.7		0.3
	60 >	0.3		
Structure type	Concrete	43.3	90	29.1
	GRC	4.2		66.1
	Brick	26.7	10	3.3
	Steel	0.8		1.2
	Wood	21.7		0
	Bamboo	3.3		0.3
House status	Buy	91	84	29
	Rent	9	16	71

which is the largest one. 3.3% buildings are low cost apartment which received government's subvention. 2% buildings are shop houses; houses with commercial space under it. 0.7% buildings are apartment or mansion without subvention. Thailand situation has similarity on the detached house domination, it is about 75%, but the apartment number is much bigger than Indonesia, about 25%. China situation is different, the house building development growth vertical, 72.2% is hi-rise apartment. As for floor level, there are no significant differences between 1 floor and 2 floors in Indonesia, but in Thailand and China almost of the building are higher than two-storey. In Indonesia most of buildings were built between 1990 and 2000 accounting for 39 %, Thailand is 49% were built around year 2005. The newest building dominate in China, about 36% were built after year 2005. In Indonesia and Thailand, most of buildings use concrete as main structure each is s about 43.3% and 90%. GRC is really popular in China, reach 66.1%, followed by concrete. While about 21.7% and 3.3% are each using wood and bamboo as the main structure which is the traditional building style (*Javanese* and *Sundanese* traditional house) in Indonesia. As for house status, 91% and 84% households in Indonesia and Thailand choose to buy their house. Probably they feel more secure if the families buy the houses also the land price still affordable for most of citizen. Only 9% and 16% households rent the houses. Different with China, more than 70% householder rent their house, as the impact of vertical development.

3.2. Energy Use

Figure 2 shows the electricity and water consumption value comparison between Indonesia and Thailand. All the energy use converted to calorific value in Mega Joules. The gas consumption is difficult to track and make the comparison because householder in the country like Indonesia and Thailand consume gas from the LPG tube which they bought every three weeks or 1 month depend on their need. But in the China, some area already consumes gas from the pipe. From the **Figure 2**, the comparison between Thailand and Indonesia has slightly different in electricity price, but it has similar shape in the respondent number as shown in black parabolic line. Although, there is growth of consumption in Indonesia from point 2000 MJ/year, means there are some household using the electricity bigger than the average (orange parabolic line). In the other hand, China has different consumption graphic pattern, it has wider variation of user's lifestyle makes the average energy consumption is bigger than the other two countries.

3.3. Household Appliances

In order to find an idea of future energy conservation method, the influence factors of residential energy consumption has to be analyzed, also the reasons which result in the differences of energy consumption quantities between high and low energy use family group need a further analysis. Housing appliances user comparison between Indonesia, Thailand, and China can be shown from **Figures 3-9**. The research investigate almost all of the electric equipment in residential housing, but in this paper the only the housing appliances that consume high energy to be described, such as refrigerator, microwave oven, television, computer, and type of stove.

Figure 3 shows that more than 94% China families and about 60% Thailand families have more than 1 unit refrigerator in their house. As **Figure 5**, the microwave oven not really popular in Indonesia and Thailand, but

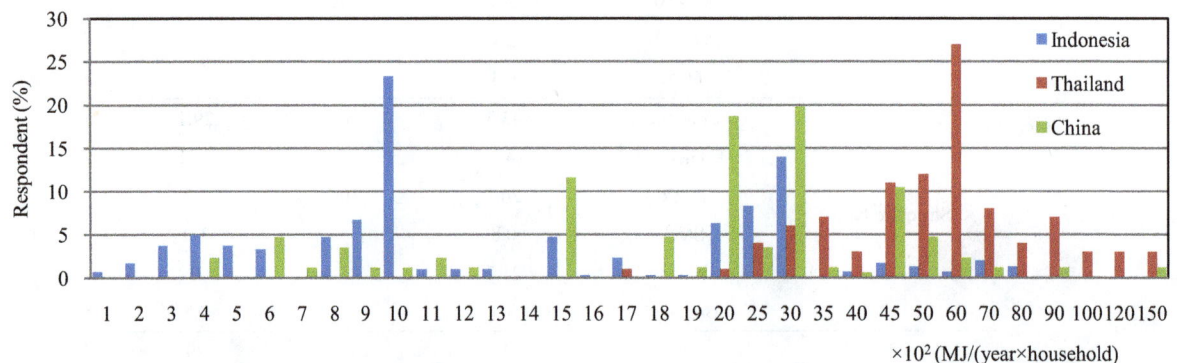

Figure 2. Electricity consumption by countries.

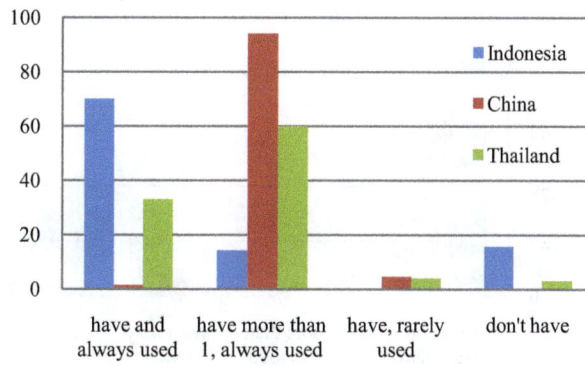

Figure 3. Refrigerator.

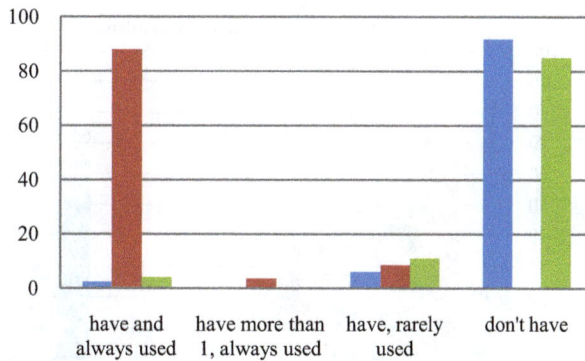

Figure 4. Microwave.

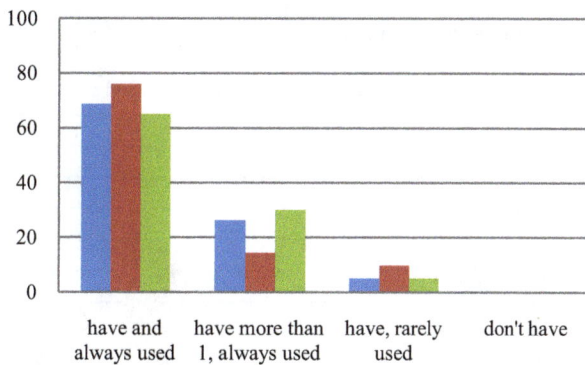

Figure 5. Television.

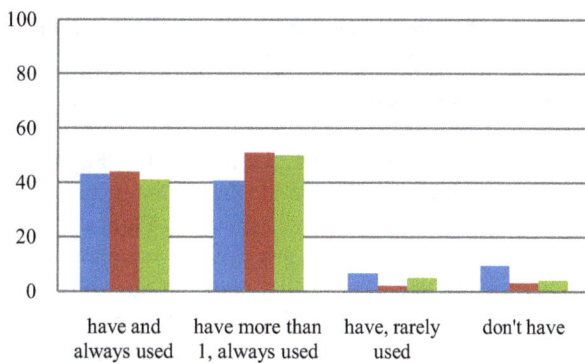

Figure 6. Computer.

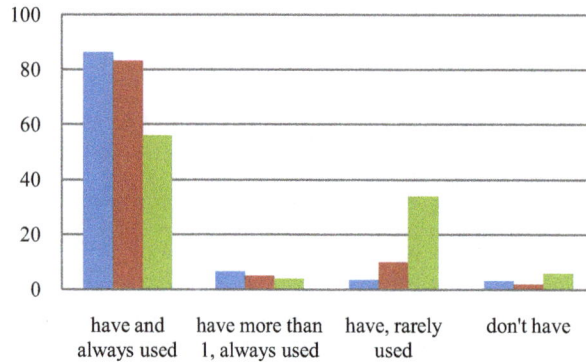

Figure 7. Gas stove.

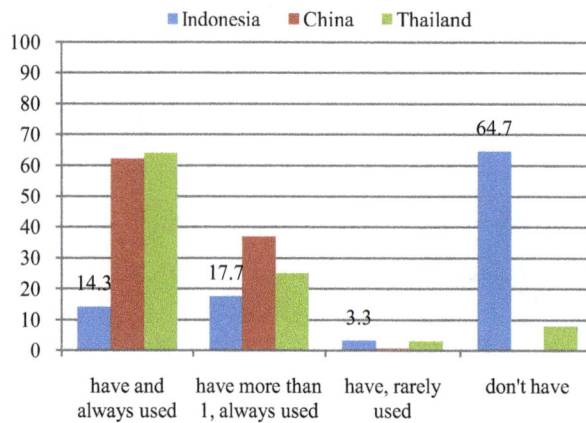

Figure 8. Air conditioner.

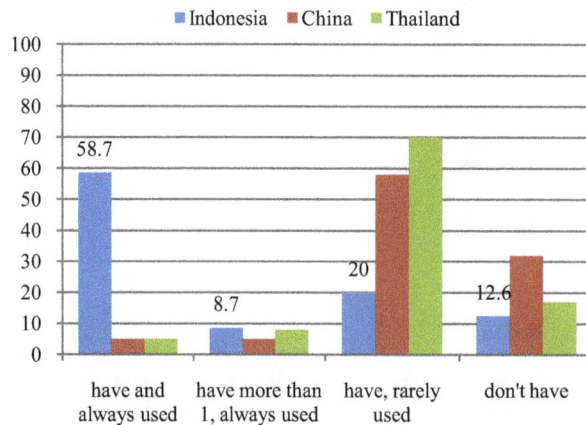

Figure 9. Electric fan.

more than 87% families in China have and always use the microwave. **Figure 5** shows the similarity of the television composition user, until nowadays, televisions are being the primary need of families. Like television, computer and laptop have been modern people lifestyle, shown in **Figure 6**, almost all countries have more than 1 unit computer in the families. **Figure 7** shows the gas stove user composition, Thailand is higher at non-user families in gas stove, probably because the Thai people prefer to buy food rather than cook the food by themselves, different with Indonesian and Chinese who still like to cook for the family.

Indonesia and south part of Thailand are still in the tropical climate zone, mean it only have two main seasons, dry season and rainy season. Even so, the average temperature in Thailand is higher than Indonesia, the result shows there are more than 60% families who have and always used the air conditioner in their houses. Different with those two countries, China has been divided five architecture thermal zones, the very cold zone, the cold zone, the moderate zone, the hot summer & cold winter zone and the hot summer & warm winter zone, its make the insulation and air conditioning system in China are the important thing to reach the thermal comfort. In the other hand, there are more than 64% of families who did not install the air conditioning system and still comfortable with the use of electric fan. Although, there is growth of air conditioner user in the Jakarta area, it is not only because the Jakarta temperature but also because of the people's lifestyle growth.

4. Influential Factors on Housing Energy Use

Multiple Regression Analysis (MRA) with *Quantification Theory I* was used to weigh the contribution of independent variables (lifestyle types) to the dependent variable (total energy use per households). In this method, the qualitative and quantitative variables of yearly energy use can be introduced into models and be analyzed together. Firstly, the correlation analyses conducted to more than 40 variables and only the factors that have positive correlation were taken for the next stage of analyses. The significant test was taken to judge that to what extent the Partial Correlation Coefficient (PCC) was large enough, the factors will have effect to the energy use. The bigger Significance Probability (SP) has the less PCC. It means the factor has a bigger influence on the housing energy use. In order to get valid results, the regression analysis was conducted on all predicted factors. As a result, 24 factors were revealed with a high correlation score. In this case, there were 24 variables used as independent variables. They were: 1) *House size*, 2) *Family size*, 3) *Household age*, 4) *House type*, 5) *Residential year*, 6) *Electricity current*, 7) *Daily stay hours in home*, 8) *Wakeup/sleep schedule*, 9) *Sleeping hours per day*, 10) *Toilet and kitchen lighting*, 11) *Electric light pattern*, 12) *Bath method*, 13) *Bath length per person*, 14) *Cooking schedule*, 15) *House energy type*, 16) *TV ownership*, 17) *Solar PV ownership*, 18) *Computer ownership* 19) *LED lamp ownership*, 20) *Electric fan ownership*, 21) *AC ownership*, 22) *Oil heater ownership*, 23) *Elec. heater ownership*, 24) *Gas heater ownership*.

Then the regression was conducted once again to include 24 factors above with the filtering method correlation score up to 95%. It means if SP less than 0.05, the factor has big influence on energy use. The score of each variable was used to analyze the influence extent of all the categories of qualitative variables (e.g. housing characteristic, family characteristic, and lifestyle characteristic), and quantitative variables (e.g. appliances ownership) on the dependent variable (yearly energy use).

In the other hand, the data filtering was conducted once again on 600 households. Only data with 100% validity was selected. According to the assumption that significance probability was smaller than 0.05 (or in other words has correlation score bigger than 95%), the factor had a big influence on energy use. As a result, the influence extent of the factors on energy use can be established in the following order in **Table 3**.

Based on the survey results in Indonesia and through the application of Quantification Theory I, households energy use in Indonesia are strongly influenced by the ownership of AC, number of people stay (family member), the ownership of refrigerator, the ownership of electric fan, and the ownership of electric stove, and the ownership of gas stove.

In China, the family member, the yearly income, the land size, the housing floor number, the sleep length of the householder, the ownership of washing machine, and the housing floor size are the major factors influencing the household's energy use.

Table 3. Influential factors on energy use by countries.

Countries	Empirical equation	R^2
Indonesia	$y = -9.29 + 8.06[\text{AC}] + 3.03[\text{people stay}] + 3.24[\text{refrigerator}] - 2.88[\text{fan}] + 5.07[\text{electric stove}] - 2.63[\text{gas stove}]$	0.612
China	$y = 739[\text{Family member}] + 359[\text{income}] + 985[\text{landarea}] - 880[\text{floornumber}] - 1266[\text{sleeplength}] - 513[\text{washingmachine}] - 513[\text{floorarea}] - 820$	0.734
Thailand	$y = 3.82[\text{floorarea}] + 3.47[\text{AC}] + 2.13[\text{residenceperiod}] + 1.98[\text{TV}] + 1.44[\text{electriccurrent}] + 1.28[\text{cookingperiod}] - 16.515$	0.791

While in Thailand, the floor area, residence period of stay, TV ownership, the electricity current contract choices, and daily cooking period have biggest influence on energy use in households.

Figure 10 shows the correlation between surveyed results and predicted energy use from regression models of three countries. The coefficient determination (R^2) was 0.612, 0.734, and 0.791, indicating that the regression model is reasonably well fitted with investigated values. Therefore, the empirical regression equation was found to have a considerable predictive power.

5. Conclusions

The following conclusions can be made through this survey and comparison to the survey conducted from September 2011 to December 2011 among three countries:

1) Indonesian and Thai residential building has a lot of similarity in the physical architecture style and in occupant's behavior, probably because the similarity in the climate condition and economic situation. The slight difference between these two countries is the air conditioner use, 89% of houses installed and always use air conditioner in their houses, while Indonesia only 32% households always use the air conditioner in their house. Compare with Thailand and China, energy consumption cost in Indonesia is lower, it could be the electricity and

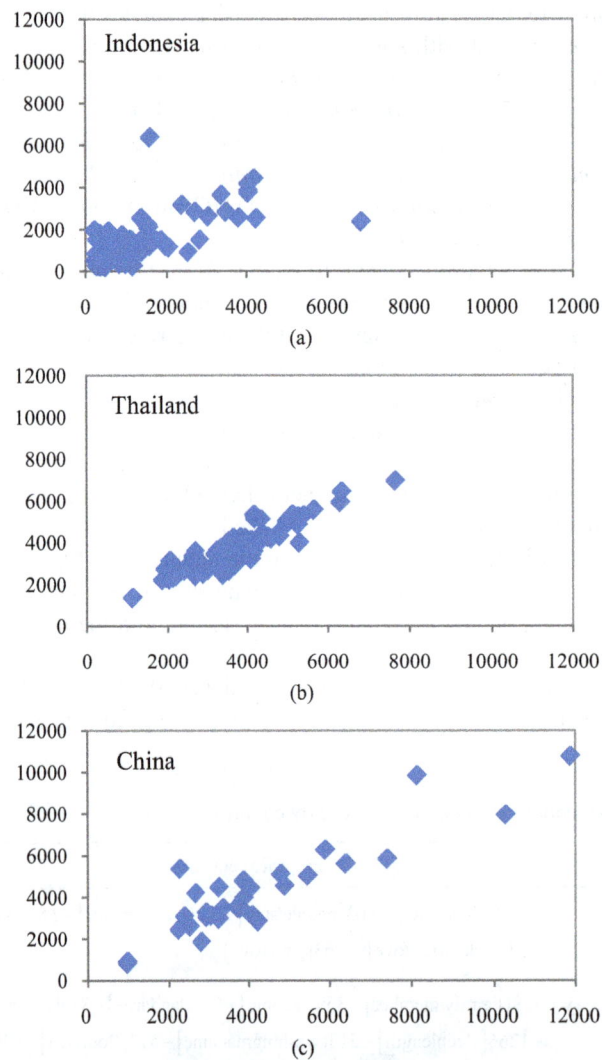

Figure 10. Correlation between x-axis-surveyed value energy use (MJ/household) and y-axis-predicted value (MJ/household).

water tariff also lower in exchange rate. Although there are some families group consume bigger than the average. Even so, the energy consumption per capita is much lower than others, probably there is just little share consumption in space cooling and almost the people don't need the heating system.

2) Beside the climate situation, the housing appliances are very important influence factors on energy consumption in these 3 countries. There is a big possibility of family who doesn't have such kind of appliances now will follow the richer family's life standard. Not only families in one country, but also there is also possibility of country's life standard change follow the developed country. If Indonesian follows the Chinese and Thai lifestyle, it would have bigger energy consumption average because of the human population.

3) The architecture design of the housing building is also an important factor in energy conservation. From the result data, in Thailand and Indonesia, energy consumption increases with the popularity of air conditioners. To prevent the increase of energy consumption, it is necessary to design open spaces between rooms for effective ventilation, to use window openings for night time ventilation, to construct roof insulation and air tightening in rooms with air conditioner. Also, the appropriate structure material choice, building orientation, and space configuration will impact to the occupant's behavior lifestyle to consume less energy. The housing design which can maximize the natural lighting and ventilation or insulation system will reach the indoor thermal comfort without consuming much energy.

Acknowledgements

This research was partly supported and funded by Ministry of Education, Culture, Sports, Science and Technology (MEXT) Japan, under the contract No. 24560724. Authors would also like to thank Mr. Alan Meier, (Energy Efficiency Center, UC Davis) for his advices to improve the quality of this paper. Also thanks to Mr. Ian Jarvis (The University of Kitakyushu) for checking the English.

References

[1]   (2009) Ministry of Energy and Mineral Resources of Republic Indonesia. http://www.esdm.go.id

[2]   Ning, Y.D., Tonooka, Y. and Zhao, X., *et al.* (2008) Analysis on Trends of Urban Housing Energy Consumption in Shanghai. *Proceedings of the Fifth International Conference on Building Energy and Environment*, Dalian, 412-419.

[3]   (2008) The World Fact Book. CIA.

[4]   National Statistical Office, The Survey of Household Energy Consumption by 2008-2009, Thailand.

[5]   Nakagami, H. (1997) Appliance Standards in Japan. *Energy and Buildings*, **26**, 69-79. http://dx.doi.org/10.1016/S0378-7788(96)01014-6

[6]   Kagajo, T. and Nakamura, S. (1997) Analysis of Household Energy Consumption Trends Based on the Standard Models of Family Pattern. *Energy Economics*, 11-29.

[7]   Nomura, N. and Ohya, H. (2000) Interrelationship of Household Electricity Consumption and Family Member Ages. *Journal of the Japan Institute of Energy*, **80**, 727-735. http://dx.doi.org/10.3775/jie.80.727

[8]   Tsurusaki, T., Murakoshi, C. and Yokoo, M. (2000) Measurement and Analysis of Residential Energy Consumption in Hokkaido. *Proceeding of the Conference on Energy, Economy and Environment*, 417-422.

[9]   Fong, W.K., Matsumoto, H., Lun, Y.F. and Kimura, R. (2007) Influences of Indirect Lifestyle Aspects and Climate on Household Energy Consumption. *Journal of Asian Architecture and Building Engineering*, **6**, 395-402.

[10]   Yu, L., Watanabe, T., Yoshino, H. and Gao, W. (2008) Research on Energy Consumption of Urban Apartment Buildings in China. *Journal of Environmental Engineering*, **73**, 183-190. http://dx.doi.org/10.3130/aije.73.183

[11]   Chen, S., Yoshino, H. and Li, N. (2010) Statistical Analyses on Summer Energy Consumption Characteristics of Residential Buildings in Some Cities of China. *Energy and Buildings*, **42**, 136-146. http://dx.doi.org/10.1016/j.enbuild.2009.07.003

[12]   Lam, J.C. (1996) An Analysis of Residential Sector Energy Use in Hongkong. *Energy*, **21**, 1-8. http://dx.doi.org/10.1016/0360-5442(95)00089-5

[13]   Long, W.D., Zhong, T. and Zhang, B.H. (2003) Situation and Trends of Residential Building Environment Services in Shanghai. *Proceedings of the 4th International Symposium on Heating, Ventilating and Air Conditioning*, Beijing, 9-11 October 2003, 493-498.

[14]   Tso, G.K.F. and Yau, K.K.W. (2003) A Study of Domestic Energy Use Pattern in Hongkong. *Energy*, **28**, 1671-1682. http://dx.doi.org/10.1016/S0360-5442(03)00153-1

[15]   Yoshino, H. and Lou, H. (2002) Indoor Thermal Environment of Residential Buildings in Three Cities of China. *Jour-*

nal of Asian Architecture and Building Engineering, **1**, 129-136. http://dx.doi.org/10.3130/jaabe.1.129

[16] Ogawa, Y., Gao, W., Zhou, N., Watanabe, T., Yoshino, H. and Ojima, T. (2005) Investigation on the Standard for Energy and Environmental Design of Residential House in China. *Journal of Asian Architecture and Building Engineering*, **4**, 253-258. http://dx.doi.org/10.3130/jaabe.4.253

[17] Wei, X., Xuan, J., Yin, J., Gao, W., Batty, B. and Matsumoto, T. (2006) Prediction of Residential Building Energy Consumption in Jilin Province, China. *Journal of Asian Architecture and Building Engineering*, **5**, 407-412. http://dx.doi.org/10.3130/jaabe.5.407

[18] Quyang, J., Gao, L., Yan, Y., Hokao, K. and Ge, J. (2009) Effects of Improved Consumer Behavior on Energy Conservation in the Urban Residential Sector of Hangzhou, China. *Journal of Asian Architecture and Building Engineering*, **8**, 243-249. http://dx.doi.org/10.3130/jaabe.8.243

[19] Hubacek, K., Feng, K. and Chen, B. (2012) Changing Lifestyles towards a Low Carbon Economy: An IPAT Analysis for China. *Energies*, **5**, 22-31. http://dx.doi.org/10.3390/en5010022

[20] Kang, H.J. and Rhee, E.K. (2012) A Development of Energy Load Prediction Equation for Multi-Residential Buildings in Korea. *Journal of Asian Architecture and Building Engineering*, **11**, 383-389. http://dx.doi.org/10.3130/jaabe.11.383

[21] Yoo, J.H. and Kim, K.H. (2014) Development of Methodology for Estimating Electricity Use in Residential Sectors Using National Statistics Survey Data from South Korea. *Energy and Buildings*, **75**, 402-409. http://dx.doi.org/10.1016/j.enbuild.2014.02.033

[22] Wood, G. and Newborough, M. (2003) Dynamic Energy-Consumption Indicators for Domestic Appliances: Environment, Behavior and Design. *Energy and Buildings*, **35**, 821-841. http://dx.doi.org/10.1016/S0378-7788(02)00241-4

[23] Diamond, R. (2003) A Lifestyle-Based Scenario for US Buildings: Implications for Energy Use. *Energy Policy*, **31**, 1205-1211. http://dx.doi.org/10.1016/S0301-4215(02)00172-6

[24] Brounen, D., Kok, N. and Quigley, J.M. (2012) Residential Energy Use and Conservation: Economics and Demographics. *European Economic Review*, **56**, 931-945. http://dx.doi.org/10.1016/j.euroecorev.2012.02.007

[25] Haas, R., Auer, H. and Biermayr, P. (1998) The Impact of Consumer Behavior on Residential Energy Demand for Space Heating. *Energy and Buildings*, **27**, 195-205. http://dx.doi.org/10.1016/S0378-7788(97)00034-0

[26] Linden, A.L., Kanyama, A.C. and Eriksson, B. (2006) Efficient and Inefficient Aspects of Residential Energy Behavior: What Are the Policy Instruments for Change? *Energy Policy*, **34**, 1918-1927. http://dx.doi.org/10.1016/j.enpol.2005.01.015

[27] Olexsak, S.J. and Meier, A. (2014) The Electricity Impacts of Earth Hour: An International Comparative. *Energy Research & Social Science*, **2**, 159-182. http://dx.doi.org/10.1016/j.erss.2014.04.014

[28] Leighty, W. and Meier, A. (2011) Accelerated Electricity Conservation in Juneau, Alaska: A Study of Household Activities That Reduced Demand 25%. *Energy Policy*, **39**, 2299-2309. http://dx.doi.org/10.1016/j.enpol.2011.01.041

12

New Robust Energy Management Model for Interconnected Power Networks Using Petri Nets Approach

Mouhammad Alanfaf Mohamed Mladjao[1,2], El Abbassi Ikram[2], Darcherif Abdel-Moumen[3], El Ganaoui Mohammed[1]

[1]University of Lorraine, IUT Henri Poincaré de Longwy, LERMAB Longwy, Nancy, France
[2]ECAM-EPMI, LR2E-Lab, Cergy-Pontoise, France
[3]ECAM-EPMI, Quartz-Lab, Cergy-Pontoise, France
Email: mohamed-mladjao.mouhammad-al-anfaf@univ-lorraine.fr, mohammed.el-ganaoui@univ-lorraine.fr, i.elabbassi@ecam-epmi.fr, m.darcherif@ecam-epmi.fr

Abstract

The problematic of energy management, particularly in terms of resources control and efficiency, has become in the space of a few years an eminently strategic subject. Its implementation is both complex and exciting as the prospects are promising, especially in relation with smart grids technologies. The deregulation of the electricity market, the high cost of storage, and the new laws on energy transition incite some significant users (collectivities, cities, regions, etc.) to form themselves into local producers in order to gain autonomy and reduce their energy bills. Thus, they may have their own sources (classic and/or renewable energy sources) to satisfy their needs and sell their excess production instead of storing it. In this idea, the territorial interconnection principle offers several advantages (energy efficiency, environmental protection, better economic balance). The main challenge of such systems is to ensure good energy management. Therefore, power distribution strategy must be implemented by matching the supply and the demand. Such systems have to be financially viable and environmentally sustainable. This allows among others to reduce the electricity bill and limit the systematic use of the national power network, typically using non-renewable sources, and thereby support sustainable development. This paper presents an original model for aid-decision in terms of grid configurations and control powers exchanged between interconnected territories. The model is based on Petri nets. Therefore, an iterative algorithm for power flow management is based on instantaneous gap between the production capability (photovoltaic, wind) and the demand of each user. So, in order to validate our model, we selected three French regions: the PACA region, the Champagne-Ardenne region and the Lorraine region. Due to their policy, their geographical and climatic features, we opted for two renewable sources: "wind" and "photovoltaic". The numerical simulations are performed using the instantaneous

productions of each region and their energy demand for a typical summer day. A detailed economic analysis is performed for two scenarios (with or without interconnections). The results show that the use of renewable energy in an interconnection context (*i.e.* pooling), offers serious economic and technical advantages.

Keywords

Smart Grid, Energy Management, Interconnection, Economic Analysis, Pv System, Wind Energy Conversion System (WECS)

---

1. Introduction

"The world is at a critical crossroad", the temperature of the land and ocean surfaces has increased globally by almost $1°C$ (from the early 20th century) and in parts of Africa, Asia, north and South America, the increase is up to $2.5°C$. We are moving towards a warming of $5°C$ - $6°C$ if nothing is done [1]. Carbon emissions from energy use dominate the total greenhouse gas emissions of most countries. One of the major challenges of the energy transition is the integration and management of a large volume of renewable sources of energy in the economic system and the electrical network. Indeed, renewable energy is growing at a rapid pace in the world and is becoming an important segment of the energy industry. Inserting an ever larger share of renewable energy in the energy mix poses new challenges compared to traditional energy sources [2].

With an abundant potential still underexploited, photovoltaic and wind energy are advantageous economically and environmentally. However, they have a common weakness: their dependence on climatic hazard. It makes necessary to rethink the structure of electrical networks and energy markets, as well as changes in network management methods. There are several studies on optimal energy management and control models, within the framework of environmental and economic issues. Most of these studies are directed to microgrids [3]. We can classify control techniques applied to the energy management and control of microgrids by their problem formulation (sometime their cost function objective) and optimization methods used to resolve it [4]. For example, Tazvinga *et al.* [5] present an optimal energy management model of a solar photovoltaic-diesel-battery hybrid power supply system. The model proposed minimizes both fuel costs and battery wear costs, and finds the optimal power flow taking into account photovoltaic power availability, battery bank state of charge and load power demand. Kriett and Salani propose [6] a generic mixed integer linear programming model to minimize the operating cost of a residential microgrid. Sortomme and El-Sharkawi [7] model the load and generation of two microgrids with wind farms to implement optimal power flow using particle swarm optimization on a system of two microgrids for a 48 hour period. It was shown that, the costs could be reduced by 14% and the system load peaks could be shaved by over 10 MW. Riffonneau *et al.* [8] propose an optimal power flow management with predictions. Lagorse *et al.* [9] proposed in their work a distributed energy management solution by multi-agent systems means as an application for hybrid energy sources.

In addition to the deregulation of the electricity market which has as general goals: to reduce the government involvement in the electricity supply sector, to introduce competition in electricity generation and selling, and to increase the demand side participation [10]; and the high cost of storage means, users (local authorities for example) may have their own sources of renewable energy and sell their excess production instead of storing it.

In the case of France, due to rapid industrialization and high population growth rate, energy consumption is rising exponentially, in recent years. Thereby French government is making a steady change towards energy efficiency and alternative sources of energy. Several declarations have been issued in recent years emphasizing concerns and commitment of regional powers to achieve sustainable development. Energy Strategy 2030 introduced by European Union aims to reduce the domestic 2030 greenhouse gas at least 40% compared to 1990. This 2030 policy framework aims to make the European Union's economy and energy system more competitive, secure and sustainable and also sets a target of at least 24% for renewable energy and energy savings by 2030 [11]. This is an attractive driver for European countries to adopt solutions that reduce overall energy consumption. In the same spirit, considering the rapid rise in power demand in the country, French government are now looking to diversify their energy mix from their primary energy source to a greater reliance on renewable energy. French energy efficiency ranking is expected to get a major boost due to the development of large renewable energy projects in their different regions. It is in this context that the law on the energy transition has recently been adopted by the French parliament [12]. Whence, the efficiency of energy production and management patterns requires improvement.

Many regions in France have set dynamic strategic direction to achieve immediate reduction in carbon emissions. The interconnection between these regions can contribute to the improvement of energy efficiency, environmental protection and a better economic balance. It's an asset for the global energy future. However, the development and operation of such systems involve many challenges. One of them is to ensure good energy management between regions which must be able to share instantly power flow, as well as between regions and the national network. Therefore, a management system is necessary. On the other hand, the complementarity that has the photovoltaic and wind in daily energy output leads to a solution that will not require storage. That is hybrid system connected to the national network without storage [13]. Yamegueu *et al.* [14] present an experimental study of a PV/diesel hybrid system without storage by taking four different daily constant loads to simulate its behavior. It has been verified through this study that the functioning of a PV/diesel hybrid system is efficient for higher loads and higher solar radiations. Such a system has a great advantage, because it significantly reduces the initial cost of a traditional hybrid system provides with a storage device. Modelling of such a system is achieved by modelling the components that constitute its [14]. Ai *et al* [15] presented complete set of match calculation methods for optimum sizing of PV/wind hybrid system by developing a set of match calculation programs with adopting the more practical mathematical models for characterizing components of the system. Diaf *et al.* [16] presents a methodology to perform the optimal sizing of an autonomous hybrid PV/wind system by finding the configuration, among a set of systems components, which meets the desired system reliability requirements. The reliability criterion used is the levelized cost of energy. Evans *et al.* [17] present simplified methods to design and estimate overall performance of photovoltaic Systems. Nema *et al.* [18] review the current state of the design, operation and control requirement of the stand-alone PV solar-wind hybrid energy systems with conventional backup source *i.e.* diesel or grid. They also highlight the future developments, and conclude that the hybrid energy system combining variable speed wind turbine and PV array generating system may be integrated to supply continuous power to the load with optimal design of hybrid controller. Various optimization techniques for modelling and design of PV/wind hybrid system, by modelling wind turbine output [19], photovoltaic panel power output [20] and optimizing it has been reported in the literature [21]. Zhou *et al* [22] present a state of the simulation, optimization and control technologies for the stand-alone hybrid solar-wind energy systems. They continued that research and development effort in this area is still needed.

The production of electricity from renewable energy is continually increasing. Some authors like Gurkaynak and Khaligh [23] and Ruiz-Romero *et al.* [24] have shown that its integration to the national network associated with a management system is more than necessary. The studies of Lu *et al.* [25] and Wang *et al.* [26] propose models based on Petri nets (PN) that is following operation modes for power management of a multi-source system. In this sense, a generic multi-source/multi-load model was developed for sustainable city, using Petri nets for power system optimization [27]. Indeed, PN is powerful tools for modeling and control for control of power systems [2].

In regards to interconnection between local networks, it has attracted considerable interest only recently, because of losses in long distribution lines and also development of the Renewable.

Energy Sources (RESs) are used as alternate source for power generation and nearer load. Operating grid along with distribution generation is become very complicated. So the electrical network should be converted into local networks interconnected through the main network. Thus, Zhu *et al.* [28] described environmental influences and potential benefits of the interconnection of regional power grid in China by increasing the production of energy from natural sources. Edmunds *et al.* [29] considered the technical benefits of additional energy storage and electricity interconnections in future British power systems. It appears that the increased level of interconnection and energy storage provides significant technical benefits in future British power systems. The work by Minciardi and Sacile [30] follows this research approach. They present a model to support optimal decisions in a network of cooperative grids formalized as an original discrete and centralized problem and defined as cooperative network of smart power grids problem. In this framework, each grid is referred to as a "smart power micro grid" (or simply as a "micro grid"), the connection to the local energy provider as a connection to the "main grid," and the overall set of micro grids connected among them and to the main grid is defined as the "network." the control variables are the instantaneous flows of power in the network of grids. They use graph theory to model this system. Two case studies are compared, the first one supposes that each micro grid is independent, while the second assumes that all micro grids are connected among them by a lattice network. In the latter case, it is supposed that a unique decision maker (DM) can decide the optimal strategy to control the storage level and the power flows in all micro grids, following a cooperative approach as described by the CNSPG problem. Hooshmanda *et al.* [31] have investigated the problem of power flow management for a network of cooperating microgrids within the context of a smart grid and present a power flow management method in a model predictive control framework. Hammad *et al.* [32] focused on the development of an autonomous distributed framework for cooperation amongst a set of microgrids without support from a traditional centralized grid. This work proposes a distributed algorithm that supports autonomous operation, enhances cost efficiency, and increases reliability of the overall system. This algorithm is based on coalition formation games. They also develop an autonomous distributed framework for cooperation amongst a set of grid-independent microgrids to improve the overall microgrid network reliability. Thus, they study the ability of proposed global storage to facilitate cooperation amongst the microgrids within the network. Their main goal was to improve individual microgrid demand-supply balance by cooperation, as well as to use better renewable Distributed Energy Ressource (DER) by exchanging surplus between microgrids [33]. Numerical results demonstrate the benefits of employing this cooperative model especially for high penetration levels of renewable DERs. Ouammi *et al.* [34] present a centralized control model based on a linear quadratic Gaussian problem, to support optimal decisions in the control of the power exchanged for a smart network of microgrids. It considered grid interconnections for additional power exchanges and incorporated storage devices, various distributed energy resources, and loads. The proposed model is evaluated through a case study in the Savona district, Italy, consisting of four microgrids that cooperate together and connected to a main grid. Even, some hypotheses that have been introduced to simplify the model, it demonstrated that the cooperation among grids has significant advantages and benefits to each single grid operation in terms of integrating a common strategy to face shortage or excess of power production due to the intermittent behavior of RESs. Recently, Yu Wang *et al.* [35] investigate a hierarchical power scheduling approach to optimally manage power trading, storage and distribution in a smart power grid with a macrogrid and cooperative microgrids. They developed two algorithms, one about online power distribution in the macrogrid, and the other about distributed cooperative power scheduling algorithm.

This paper presents an original model for optimization and control of interconnections between users using a special modelling tool, Petri nets and an iterative algorithm for flow management based on instantaneous gap between the power production capability and demand of each user.

The aims of the model is to support optimal decisions in the control of the power exchanged for a smart interconnected network composed by local networks via the main grid. The model is applied to three French lands: PACA, Champagne-Ardenne and Lorraine. The three regions are connected via the national network as shown on **Figure 1**. Each region has its own renewable sources of energy production. First, the system sources are presented before explaining the principle of interconnection used for energy management. Then, the simulation of the model and its energy management are detailed, and simulation results are presented and analyzed. Two scenarios are considered (with or without interconnections). Finally, a detailed economic analysis is performed.

1.1. Conventional Grid and the Need of Smart Grid

Over hundred years, when electricity was first used, local authorities designed and built independent systems to supply electricity large towns or a part of them. Then, connection of all of these independent systems was made to become national grid, in the goal to provide better supply security and to reduce costs. It allowed larger and more efficient power stations to be built far from where the power was consumed. As a result, the previously independent local systems gradually lost their generation and became the distribution systems we have today [36]. The main difference between the classic systems (mainly with fossil-fired generators) and Smart Grid is the way that production and demand are kept in balance.

The electricity system design and operating strategies used in classical system is operational for decades and used around the world. However, electricity supply sector should make changes to achieve government's carbon reduction targets. All components of the system have an important role in this achievement, but one of the most important is the replacement of fossil-fired generation with low or zero carbon generation technologies [37]. However, these generation technologies are uncontrollable sources of energy and operator will need to find new means of balancing production and consumption for system stability and ensure that there is no overload in the grid causing loss of supply as more intermittent generation and connected load can cause.

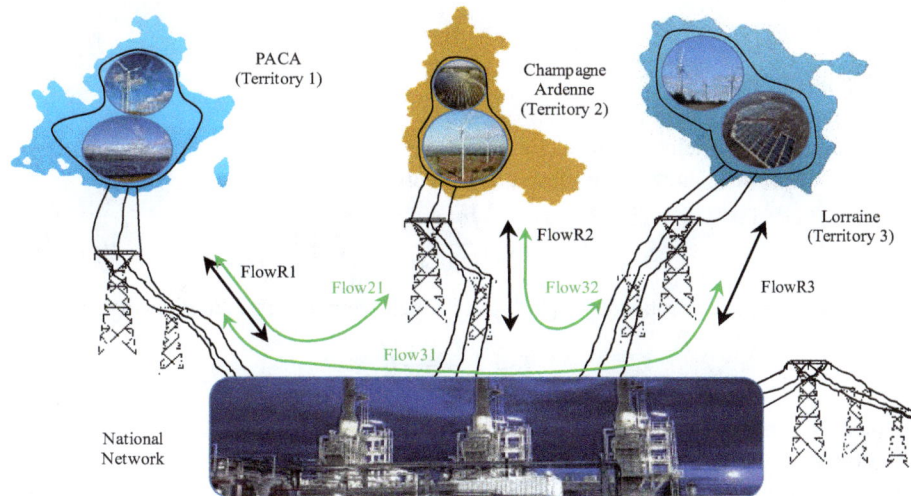

Figure 1. Schematic representation of the interconnection of the three considered regions.

1.2. The Need for Interconnection

Greater interconnection between independent local networks could be helpful. Flow of power could be in both directions. Intelligent monitoring and control will provide the lowest cost means asked to meet this permanent balance. The surplus will complement each other's needs. It will help the network companies to meet more quickly their consumer requirements and reduce cost efficiently.

2. Hybrid PV/Wind System

The wind and solar energy are omnipresent, freely available, and environmental friendly. In recent years, the combination of these renewable energy sources (hybrid PV/wind system) become attractive and viable alternative of oil-produced energy to meet electricity demands. Components of such renewable hybrid energy system consists of both energy sources, a power conditioning equipment, a controller and an optional energy storage system. With the complementary characteristics between solar and wind energy resources, a hybrid PV/wind system without storage presents a good alternative to satisfy the electricity needs for locations connected to national grid, especially during peak consumption.

Because of the stochastic aspect of both solar and wind energy, the main aspects to be taken into account in the design of such a PV/wind hybrid system is the reliable supply of consumption in various weather conditions and the cost of kWh of energy. The efforts of research and development in solar and wind energy technologies, and other should continue in the way to improve their performance, to establish accurate prediction techniques of production and their integration into the national network.

2.1. Photovoltaic System Model

In this paper, a mathematical model for estimating the power output of PV modules is used. Basically, the output power generated from a PV generator, can be calculated using the following formula [16]:

$$P_p(t) = A\eta\eta_{pt}I \tag{1}$$

A: the area of a single module used in a system (m^2)
I: the global irradiance incident (W/m^2).
η: the efficiency of the PV generator.

As the operation and the performance of a PV generator is interested in its maximum power, the models describing the PV module's maximum power output behaviors are more practical for PV system assessment. With a perfect maximum power point tracker, the efficiency of power tracking equipment $\eta_{pt} = 1$ [16].

All the energy losses in a PV generator, including connection losses, wiring losses and other losses, are assumed to be zero.

η is defined as follows [20]:

$$\eta = \eta_r \left[1 - \beta \left(T_c - T_r \right) \right] \tag{2}$$

T_c: the temperature of PV cell (°C),
T_r: the reference cell temperature (°C)
β: the cell temperature coefficient of efficiency (°C^{-1}) respectively.
η_r: is the PV generator efficiency at reference temperature T_r.

$$\eta_r = \frac{P_{p\max}}{1000\,A} \tag{3}$$

Based on the energy balance proposed by [17], the PV cell temperature T_c is given as follows:

$$T_c = T_a + \frac{\alpha\tau}{U_L} I \tag{4}$$

T_a: the ambient temperature (°C),
α : the absorptance coefficient of PV cell.
τ : the transmittance coefficient of PV cell respectively.

Based on the energy balance, it is shown that U_L and $\alpha\tau$ are related to the nominal operating cell temperature NOCT as follows [16]:

$$\frac{\alpha\tau}{U_L} = \frac{I_{T,NOCT}}{NOCT - T_{a,NOCT}} \tag{5}$$

$I_{T,NOCT}$: the global irradiance incident at NOCT conditions[1] (800 W/m^2),
$NOCT$: nominal operating cell temperature (°C),
$T_{a,NOCT}$: the ambient temperature at NOCT conditions (20°C)
U_L: the overall heat loss coefficient (W/(m^2. °C)).

If the number of photovoltaic modules installed and connected to the national network is N_{pvi}, then the total power produced for each territory i is: $P_{pv_i}(t) = N_{pv_i} \times P_p(t)$.

2.2. Wind Turbine Model

Different types of wind generators usually have different power output performance curves. However, for a specific wind generator, a model should be developed according to its power output performance curve, which is given by the manufacturer. Different models for predicting the performance of wind turbines have proposed in the literature [19].

In this paper, the wind generator power output is estimated through characteristic equations of a wind turbine developed by Ai et al. [15] by fitting its actual power curve using method of least squares. The instantaneous wind power output from the turbine can be predicted from the wind power equation discussed here as under, following power curve of wind turbine [38]. The power curve of the wind turbine is separated to sub-functions as illustrated.

$$\left. \begin{array}{ll} P_e(t) = 0 & \text{for } v < v_c \\ P_e(t) = a_1 v^2 + b_1 v + c_1 & \text{for } v_1 \leq v \leq v_2 \\ P_e(t) = a_2 v^2 + b_2 v + c_2 & \text{for } v_2 \leq v \leq v_3 \\ \quad \vdots & \\ P_e(t) = 0 & \text{for } v > v_f \end{array} \right\} \tag{6}$$

$$v_c < v_1 < v_2 < v_3 < \cdots < v_f$$

V, v_c and v_f are the wind speed (m/s), the cut-in speed (m/s) and cut-out (or furling) speed of the wind turbine (m/s) respectively. $a_{1,2,\cdots,k}$; $b_{1,2,\cdots,k}$; $c_{1,2,\cdots,k}$, are the coefficients of quadratic equations.

If the number of wind turbines installed and connected to the national network are N_{wti}, then the total wind power produced for each territory is: $P_{wt_i}(t) = N_{wt_i} \times P_e(t)$.

---

[1]$NOCT$ conditions: I_T, $NOCT = 800$ W/m^2, Ta, $NOCT = 20$°C wind speed = 1 m/s and $\eta = 0$.

2.3. PV/Wind Model

The energy interconnection can be justified on the basis of energy independent, economic interest of the territories and the reduction of greenhouse gas emissions.

It is assumed that for regions, the energy delivered by the power grid is still available, but limited by the information of the substation that connects the PV/Wind power system of each region. Keeping in mind the objective of providing energy independence of the regions, it was decided to focus on the exchange between regions over the region-national network exchange.

The total energy generated and needed ($W_{gen_i}(t)$, $W_{dem_i}(t)$), over a period of 24 hours for each user (region) can be defined in terms of wind and photovoltaic energy generated and power needed $P_{dem_i}(t)$ as follows [39]:

$$W_{gen_i}(t) = \sum_0^{24}\left[(\Delta S)\left(P_{gen_i}(t)\right)\right] = \sum_0^{24}\left[(\Delta S)\left(N_{wt_i}P_{wt_i}(t) + N_{pv_i}P_{pv_i}(t)\right)\right] \qquad (7)$$

$$W_{dem_i}(t) = \sum_0^{24}\left[(\Delta S)\left(P_{dem_i}(t)\right)\right] \qquad (8)$$

$P_{gen_i}(t)$ is total PV/wind power generated, t is the time (time of day) and ΔS is the time between samples (one hour).

To have a balance between power production capability and demands in a given period of time, the gap $\Delta P_i(t)$ must have an average of zero. Note that the positive values of $\Delta P_i(t)$ indicates the availability of production and negative values indicates insufficient production capability [40].

$$\Delta P_i(t) = P_{gen_i}(t) - P_{dem_i}(t) \qquad (9)$$

3. Interconnection Model

Main element of the model is the instantaneous gap between the power production capability and the demand. The principle of the interconnection system is shown in **Figure 2**.

Two cases arise. Whether an energy insufficiency or sufficiency. In our experiences, the production capability of each user is calculated using mathematical models of wind turbines and photovoltaic panels for a day, on the basis of hourly weather data available. Then using the instantaneous power demands data from each user, the gap between power production capability and demands over 24 hours can be calculated.

3.1. Flow Management Strategy

One of the main problems of the interconnection model is related to the control and supervision of the energy

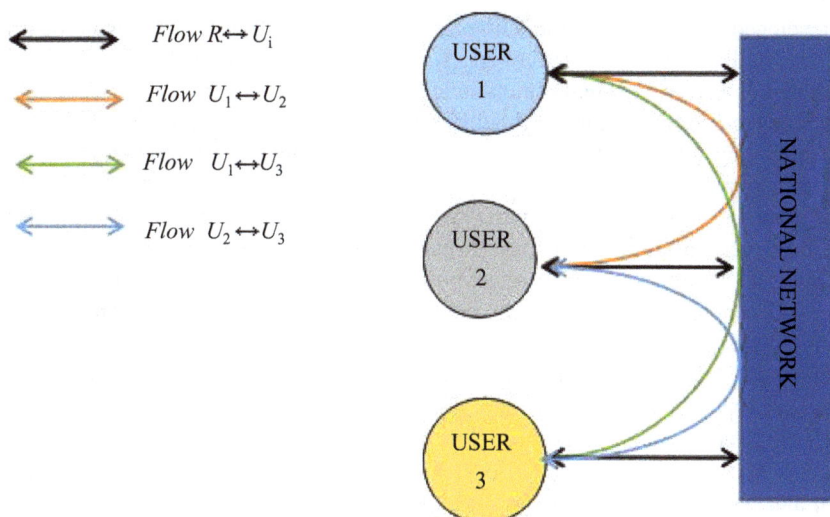

Figure 2. Topology of the system model.

distribution. The dynamic interaction both between users and between users and national network can lead to, critical problems of stability and power quality. Indeed, the absence of storage and the problem of intermittent power generation can lead to important losses. The decentralized injection permits a decrease of the network load if the energy produced is immediately consumed locally. Otherwise, it can quickly lead to overloading of the distribution network. Also, decentralized systems (PV, Wind) influence on the quality of the voltage and may even cause non-compliant energy returns to the upstream network. Whence, the vital role that management system can play on such system.

To assure continuous supply of the load demand, managing the flow of energy throughout the proposed model is to be done. The main objective of the energy flow and management is to supply the load with its full demand. The goal of this study is to design an effective power management system for interconnection system between users based on their gap between renewable hybrid power system production and needs.

The following operating strategy is employed:
- The use of electric power generated by the photovoltaic arrays and wind turbine generators has priority in satisfying electricity demand over that provided by national network.
- If the total electric power generated by the photovoltaic array and wind turbine generators is higher than the demand, the additional electric power sold to users in demand via the national network. If there is not users in demand, power sold to national network.
- If the total electric power generated by the photovoltaic arrays and wind turbine generators is less than the demand, electric power will be bought from others user's additional electric priority. Then from national netw-ork, if demand still cannot be satisfied.

To achieve this operation strategy, an iterative algorithm is developed in order to control and manage the power flows exchanged according to the gap between production capability and the demands for each user.

The algorithm uses the characteristic data of the demands $P_{demi}(t)$, and daily weather data from location of user i. It allows to calculate the wind power generation $P_{wti}(t)$, and photovoltaic power generation $P_{pvi}(t)$. Their sum equal to the wind/PV production capability $P_{geni}(t)$. Also, it enable to calculate the gap $\Delta P_i(t)$ between the production capability $P_{geni}(t)$, and power demands $P_{demi}(t)$ instantaneously.

Several configurations are possible depending on the number of users. In this work, three users are considered $i=1, 2, 3$.

For each user there are 8 configurations that can exist:
- $\Delta P_i(t)$ can be lower to zero ($\Delta P_i(t) < 0$), it is given the index 0;
- $\Delta P_i(t)$ can be greater than or equal to ($\Delta P_i(t) \geq 0$), it is given the index 1;

The configurations [000, 001, 010, 011, 100, 101, 110, and 111], are assigned to the coefficients [$\Theta 0$, $\Theta 1$, $\Theta 2$, $\Theta 3$, $\Theta 4$, $\Theta 5$, $\Theta 6$, $\Theta 7$], respectively.

3.2. Control of Interconnection System Based on Petri Nets

The model is developed using a special modeling tool, Petri nets. It composed by two blocks (**Figure 3**).
- The block witch generated the sign of $\Delta P_i(t)$ with an iterative algorithm,
- A block that describes Petri nets simulations. It consists in two subnets:
- Subnet modeling configurations of the different territories at a given time.
- Subnet modeling the regulation of flows exchanged through the interconnection user/user via the national network and user/national network.

Several configurations are possible depending on the number of users.

PN1 depicts how the configuration change (**Figure 4**) and PN2 represents the Petri Nets that regulates the power flow (**Figure 4**), it selects the appropriate power exchange to reach the demand. The transition from one place to another is dependent on the sign of gap between the production capability and power demands.

For PN1, eight places (P0, P1, P2, P3, P4, P5, P6 and P7) representing eight states. They are linked by twenty-one transitions. Each state corresponds to a configuration. Each configuration is characterized by the presence of a token in the corresponding place. For example, a token in the place P0, means the system is on configuration 000. This state occurs when the gap between production capability and needs of each territory becomes negative. Indeed, transitions fired if all conditions are met. PN1 can be explained as **Table A.1**.

For each user and according to the configuration, when needs become higher than the production capability, it buys surplus from other users via the national network. If the latters could not meet its request, it bought directly from the national network own production.

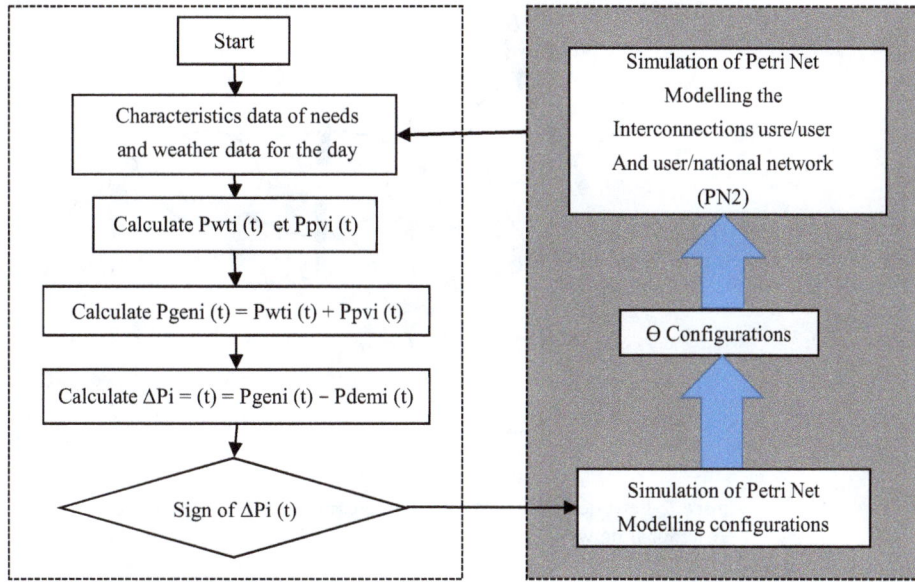

Figure 3. Blocks scheme of the model.

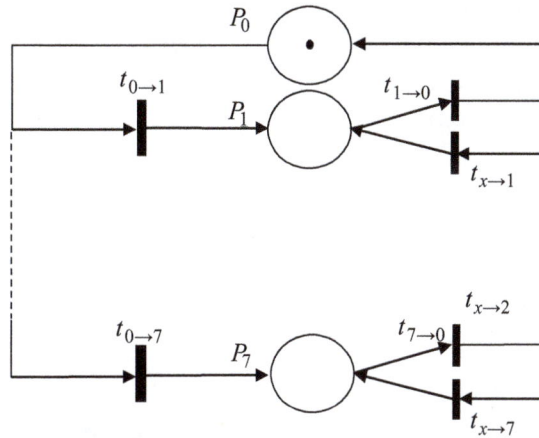

Figure 4. Petri Net (PN1) modelling configurations.

P_{U1}, P_{U2} and P_{U3} places correspond to the three user states (**Figure 5**). To describe the interconnections user/ user and user/national network, two states are defined for each user (PN2).

As described above, either there is sufficiency or insufficiency. Indeed, according to α configuration at time t, and the sign of the sum of the gaps. $\sum_{i=1}^{3}\Delta P_i(t)$, transitions can be fired. The presence of a token in a place corresponding to the user, characterizes the state of sufficiency of the latter. Otherwise, characterizes its insufficiency. Thus, different flows can be traded, according to different conditions as **Table A.2**. The Petri net control is composed of arcs inhibitors, their role is to prevent the presence of more than one token in places P_{U1}, P_{U2} and P_{U3} [2].

4. Cost Modelling

The objective of the economic analysis is to show the economic advantages of the model. For this, the levelised cost of energy for hybrid wind/PV power system is estimated as follows [41]:

$$LCE = \frac{TAC}{E_l} \tag{10}$$

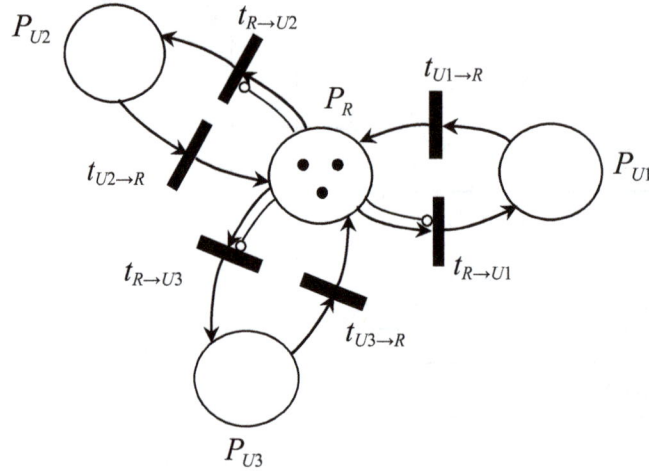

Figure 5. Petri net (PN2) modelling interconnections user/user and user/national network.

where TAC and E_l represent, respectively, the total annualized cost, and the annual total energy. The total annualized cost is calculated by taking into consideration the present value of costs (PVC) and the capital recovery factor (CRF) [42]:

$$TAC = PVC \cdot CRF \tag{11}$$

The capital recovery factor can be expressed as follows [43]:

$$CRF = \frac{\gamma (1+\gamma)^N}{(1+\gamma)^N - 1} \tag{12}$$

where γ is the annual discount rate, and N the system life in years (generally 25 years for PV/wind systems) [41].

The present value of costs for hybrid system consists of the sum of C_{pv}, and C_{wind}. What are the sum of present value of capital and maintenance costs of the PV system and the sum of present value of capital and maintenance costs of the wind turbines in system life, respectively [44]:

$$PVC = C_{pv} + C_{wind} \tag{13}$$

According to the studied system, the present value of costs is composed of the initial cost (IC), the present value of maintenance cost (MC) and the present value of replacement cost (RC) [42]. The initial cost of system components comprises the AC (acquisition costs) and the installation costs. Installation cost of system components is taken as a fraction of AC (ε) [45]. It is usually taken as 50% of the PV system and 25% of the wind turbines [39].

The present value of maintenance costs of the hybrid system is expressed as:

$$CM = Cm0 \cdot IC \tag{14}$$

where $Cm0$ represents the maintenance cost in the first year. It can be expressed as a fraction m of the initial cost:

$$PVC = C_{pv} + C_{wind} \tag{15}$$

In this work the maintenance cost is considered the same as first year. m is equal to 1% of the acquisition cost for the PV system, and 5% of the acquisition cost for the wind turbines [42].

In the context of cost of electricity minimization in the French electricity sector. The following paragraphs present a comparative basis of two *scenarii*:

▪   A baseline scenario (BS) that does not include the interconnect option and considers the current state of operation centered on national network of regional electrical systems.

▪   An interconnection scenario (IS) favoring the interconnection of territories.

5. Interconnection Model

5.1. Conventional Grid and the Need of Smart Grid

In this paper "the PACA region" is taken as user 1, "the Champagne-Ardenne region" as user 2 and the "Lorraine region" as user 3.

The production capability is calculated from weather data.

Ten percent of the hourly average daily demand for each region for a summer day (June 30, 2014) are considered. The gap between production capability and demand is calculated over a 24 hours period as shown in **Figure 6**.

According to **Figure 6**, five configurations are occurred:

Configuration 000 (parts a, c and g) shows that all regions are in energy insufficiency ($\Delta P_{i=1,2,3}(t) < 0$). This occurs in the interval from 00 to 8 am which the PV/wind production capability is low than needs, causes less favorable weather conditions. And from 19 h to 22 h, interval for which the needs for electricity is highest in France, and the weather conditions become less favorable to the production, especially photovoltaic. Therefore regions have a status of buyers because their production is not sufficient to meet 10% of their needs as shown in **Figure 7**, **Figure 8**. They buy from the national network, and *flowNationalNetwork/PACA*, *flowNational- Network/Champagne*, *flowNationalNetwork/Lorraine* are negative and are obtained using (Equation. B.1).

- Configuration 010 (part h) shows that PACA and Lorraine regions are energy insufficiency ($\Delta P_{i=1,3}(t) < 0$), unlike the Champagne Ardenne region what is oversupplied ($\Delta P_{i=2}(t) < 0$) as shown in **Figure 5** and **Figure 6**.

In this case the Champagne Ardenne sells its surplus to the other regions, and *flowChampagne/PACA*, *flowLorraine/Champagne* are obtained using (Equation. B.6). Note that the sign of the sum of the energy gaps is negative, which explains the dissatisfaction of the needs of regions in energy insufficiency. To meet them, they also buy from the national network, and *flowNationalNetwork/Champagne*, *flowNationalNetwork/Lorraine* are negative as shown in **Figure 7** and **Figure 8**.

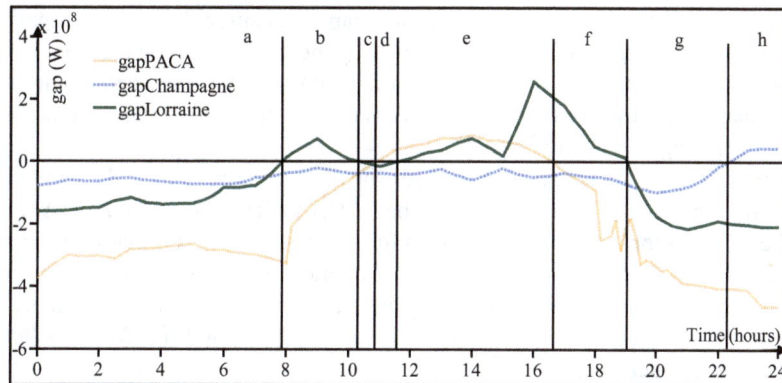

Figure 6. Gaps between PV/Wind production capability and needs for each region.

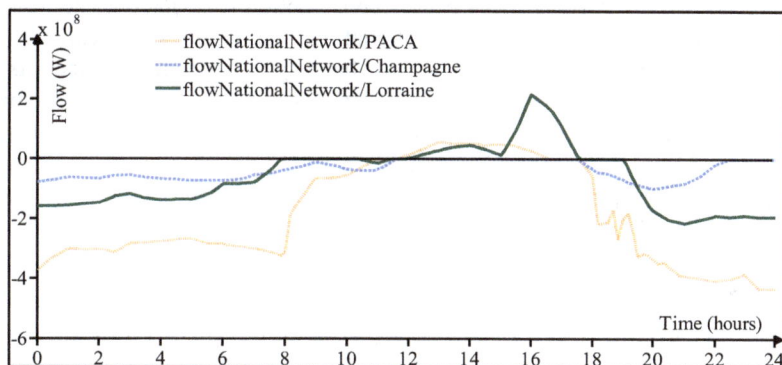

Figure 7. Flow exchanged between the national network and the three regions.

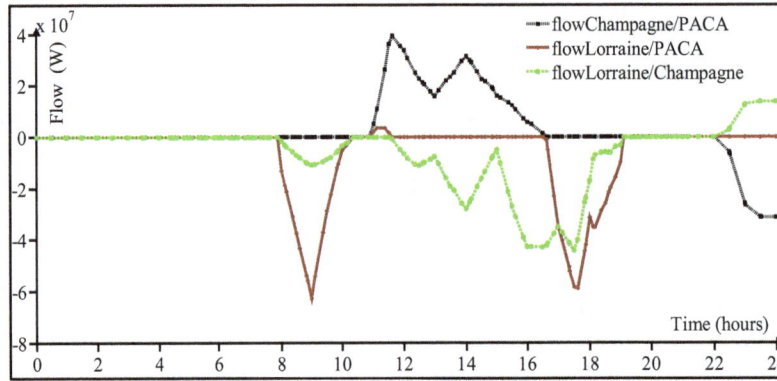

Figure 8. Flow exchanged between the three regions.

- Configuration 001 (parts b and f) shows that the PACA region and Champagne Ardenne are in energy insufficiency ($\Delta P_{i=1,2}(t) < 0$), unlike the Lorraine region that is in situation of oversupply ($\Delta P_{i=1,2}(t) > 0$).

In this case the Lorraine Region sells its surplus to the other regions in need, and *flowLorraine/Champagne*, *flowLorraine/PACA* are obtained using (Equation. B.7) for the parts where the sum of energy gaps is negative, and using (Equation. B.2) for the parts where the sum of gaps is positive. For the positive part, to meet their needs, both regions in energy insufficiency buy from national network, *flowNationalNetwork/PACA*, and *flow-National-Network/Champagne* are negative as shown in **Figure 7** and **Figure 8**. Otherwise, it's the Lorraine region that sells to national network, the remaining surplus after meeting the needs of other regions.

- Configuration 101, which is shown in part (e) shows that the PACA and Lorraine regions are oversupply ($\Delta P_{i=1,3}(t) > 0$), unlike the Champagne Ardenne which is in energy insufficiency ($\Delta P_{i=2}(t) < 0$).

In this case both regions sell their surplus to Champagne Ardenne region, and *flowChampagne/PACA*, *flow-Lorraine/Champagne* are obtained using (Equation. B.3), when the sign of the sum of their gaps is negative, and using (Equation. B.6), when the sign of the sum of their gaps is positive. When it is negative, the production capability of both regions in energy sufficiency can't meet le needs of the region in energy insufficiency. Therefore, this latter buys from the national network, *flowNationalNetwork/Champagne* is negative as shown in **Figure 5**, **Figure 6**. Otherwise, these are both regions in oversupply that sell to the national network, the rest of the surplus after meeting the region in insufficiency.

- Configuration 100, which is shown in part (d) shows that the regions and Champagne Ardenne Lorraine are in energy insufficiency ($\Delta P_{i=2,3}(t) < 0$), unlike the PACA which is oversupplied ($\Delta P_{i=1}(t) > 0$).

In this case the both regions buy to PACA its surplus as shown in **Figure 6**, **Figure 7**, and *flow Champagne/PACA*, *flowLorraine/PACA* are obtained using (Equation. B.4). Note that the sign of the sum of the gaps is negative, which explains the dissatisfaction of regions in energy insufficiency. To meet their needs, they also buy from the national network as shown in **Figure 6** and **Figure 7**, and *flowNationalNetwork/PACA*, *flowNationalNetwork/Lorraine* are negative.

With storage means cost and environmental issue, regionsmay sell their excess production by the interconnection with others regions. It is clear from **Figure 7** and **Figure 8** that, the interconnection can significantly reduce regional dependence on non-renewable energy consumption from national network and the need of storage. It should become more attractive as costs of national network supplies increase. Indeed, from 10 pm to 12 pm there is not flow bought from national network to Champagne Ardenne, its excess is sold to both regions, PACA and Lorraine. Around 10:30 am to 7 pm, even its gap is negative as shown in **Figure 7**. Its needs are satisfied by excess from PACA and Lorraine as shown in **Figure 7**. Same case is occurred between 8 am and 10:20 am and around 4:30 pm to 6 pm for PACA as shown in **Figure 7**. Its needs are satisfied by excess from Lorraine and Champagne. This proves that the interconnection of regions can replace the centralized system on national network and reduce the need of storage.

5.2. Cost Analysis

For each PV system the acquisition costs of photovoltaic panel are about 1.42 €/W. For the inverter it is approximately 0.11 €/W, are considered. For each wind system, acquisition costs are about 0.41 €/W. The levelised cost of energy (€/kWh) for each region is defined as **Table 1**, taking into account the production percentages by renewable energy as **Table 2** and **Table 3**. Price per kWh for EDF in optional base is 0.1403 €.

a.   Source: Observ' ER

b.   LCE for PV power sale PV is 0.11688 € and for wind power it is 0.082 €.

To investigate the feasibility and economic advantages of the interconnection of regions, two *scenarii* are developed:

-    A baseline *scenarii* (BS) and

-    An interconnection *scenarii* (IS) favoring the interconnection of regions in the context of a possible minimization of the cost of electricity and sustainable development of the French electricity sector.

The following paragraphs present a comparative basis of both *scenarii*. **Figure 9** shows the gap results between buying and selling costs of both *scenarii* for each region. Notice that only 10% of demands are considered.

Results on **Figure 9** show pertinently that the interconnection scenario (IS) is very economically advantageous over the classic model.

For PACA, it is about 42% of gain over the baseline scenario (BS). Then for Champagne Ardenne, it is about 156% of benefices over the baseline scenario (BS). Finally 53.53 % of gain for Lorraine.

The use of such a model can be interesting, if in the future the leaders of the regions and the national network find mutual agreement for the integration of such a system in the country's energy environment.

6. Conclusions

This work presents an original interconnection model for power control and management between territories (cities, departments, regions). Through their own energetic resources (hybrid PV/Wind systems with or without storage) and their connection to the national network, the users (territories) can be both "producers and consumers" of energy. This means that the power flow could be in both directions. A control scheme using Petri Nets has been proposed for controlling and managing the production capabilities and demands of each user. An original algorithm is developed for operation strategy. It is based on the sign of output gap between energy production capability and demands. This leads to the statements and configurations of users. Each user can have two statements: energy insufficiency or sufficiency. A detailed economic analysis has been performed based on two scenarii: A baseline scenario (BS) and an interconnection scenario (IS) favouring the interconnection.

Table 1. Levelised cost of energy (LCE) for the three regions[a].

Regions	LCE established by the government in France[b] (€)	LCE calculated (€) (using model)
Champagne-Ardenne	0.0830	0.0288
Lorraine	0.0869	0.057
PACA	0.0865	0.1563

a) Source: observ' ER, b) LCE for PV power sale PV is 0.11688 € and for wind power it is 0.082 €.

Table 2. Installed and produced wind turbine capacity, and its percentages depending on total PV/Wind production for each region[a].

Regions	Installed Wind turbine capacity in MW	Total Wind production	%
Champagne-Ardenne	1113	2451 GWh	97
Lorraine	716	1225 GWh	86
PACA	45	116 GWh	13

a) Source: observ' ER.

Table 3. Installed and produced PV capacity, and its percentages depending on total PV/Wind production for each region[a].

Regions	Installed PV capacity in MWp	Total PV production	%
Champagne-Ardenne	80	82 GWh	03
Lorraine	198	202 GWh	14
PACA	616	811 GWh	87

a) Source: observ' ER.

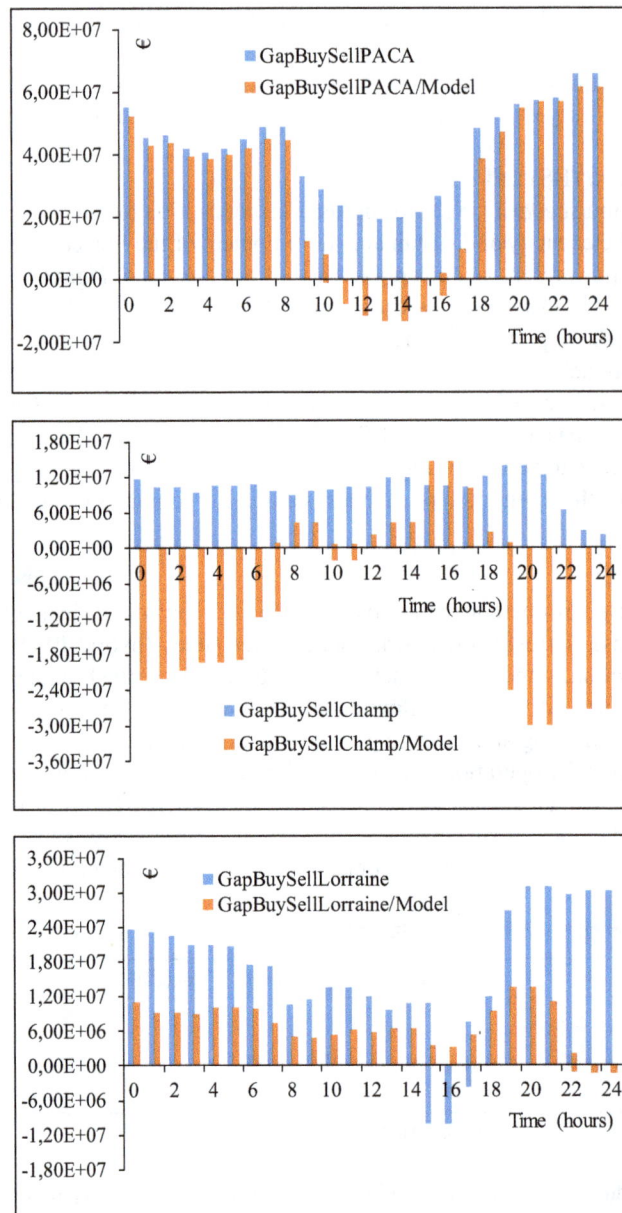

Figure 9. Comparison of Gaps between power purchase and sale for three regions: interconnection scenario (IS)/baseline scenario (BS).

The aim of the work is to demonstrate that it is possible to reduce the dependence on conventional energy by developing local hybrid sources such as "PV-Wind" and pool them through the national network. A detailed economic analysis has been performed based on two scenarii: A baseline scenario (BS) and an interconnection scenario (IS) favouring the interconnection. Numerical Simulations are carried out for 3 French regions: PACA, Champagne-Ardenne and Lorraine. The results clearly show that the interconnection of regions present significant environmental advantages through the use of renewable energy and can reduce considerably financial investment on storage means. They also show pertinently economic advantages over classic model with 42%, 53.53% and 156% of benefices for PACA, Lorraine and Champagne-Ardenne respectively.

Acknowledgements

The authors would like to thank the support of the ECAM-EPMI, Graduate School of engineering.

References

[1] IPCC (2014) Climate Change 2014: Synthesis Report. Contribution of Working Groups I, II and III to the Fifth Assessment Report of the Intergovernmental Panel on Climate Change IPCC [Core Writing Team, Pachauri, R.K. and Meyer, L.A. (Eds.)], Geneva.

[2] Amghar, B., Ikram, E.A., Mladjao, M.A.M. and Moumen, D.A. (2015) A New Hybrid Control Method Of Power Electronics Converters For Wind Turbine Systems. *WIT Transactions on Information and Communication Technologies*, **60**, 667-684.

[3] Palizban, O., Kauhaniemi, K. and Guerrero, J.M. (2014) Microgrids in Active Network Management—Part I: Hierarchical Control, Energy Storage, Virtual Power Plants, and Market Participation. *Renewable and Sustainable Energy Reviews*, **36**, 440-451. http://dx.doi.org/10.1016/j.rser.2014.04.048

[4] Minchala-Avila, L.I., Garza-Castanon, L.E., Vargas-Martinez, A. and Zhang, Y. (2015) A Review of Optimal Control Techniques Applied to the Energy Management and Control of Microgrids. *The 6th International Conference on Sustainable Energy Information Technology (SEIT* 2015), *Procedia Computer Science*, **52**, 780-787. http://dx.doi.org/10.1016/j.procs.2015.05.133

[5] Tazvinga, H., Xia, X. and Zhu, B. (2014) Optimal Energy Management Strategy for Distributed Energy Resources. *The 6th International Conference on Applied Energy (ICAE* 2014), *Energy Procedia*, **61**, 1331-1334. http://dx.doi.org/10.1016/j.egypro.2014.11.1093

[6] Kriett, P.O. and Salani, M. (2012) Optimal Control of a Residential Microgrid. *Energy*, **42**, 321-330. http://dx.doi.org/10.1016/j.energy.2012.03.049

[7] Sortomme, E. and El-Sharkawi, M.A. (2009) Optimal Power Flow for a System of Microgrids with Controllable Loads and Battery Storage. *IEEE/PES Power Systems Conference and Exposition*, Seattle, 15-18 March 2009, 1-5. http://dx.doi.org/10.1109/psce.2009.4840050

[8] Riffonneau, Y., Bacha, S., Barruel, F. and Ploix, S. (2011) Optimal Power Flow Management for Grid Connected PV Systems with Batteries. *IEEE Transactions on Sustainable Energy*, **2**, 309-320. http://dx.doi.org/10.1109/TSTE.2011.2114901

[9] Lagorse, J., Paire, D. and Miraoui, A. (2010) A Multi-Agent System for Energy Management of Distributed Power Sources. *Renewable Energy*, **35**, 174-182. http://dx.doi.org/10.1016/j.renene.2009.02.029

[10] Crossley, D. (2011) Demand-Side Participation in the Australian National Electricity Market. A Brief Annotated History: The Regulatory Assistance Project. http://www.efa.com.au/Library/David/Published%20Reports/2011/RAP_Crossley_DSParticipationAustralianNatlElectricityMkt.pdf

[11] European Commission (2014) Communication from the Commission to the European Parliament, the Council, the European Economic and Social Committee and the Committee of the Regions, "A Policy Framework for Climate and Energy in the Period from 2020 to 2030".

[12] French National Assembly (2015) Bill concerning the Energy Transition to Green Growth.

[13] Ruther, R., Martins, D.C. and Bazzo, E. (2000) Hybrid Diesel/Photovoltaic Systems without Storage for Isolated Mini-Grids in Northern Brazil. *Proceedings of the 28th IEEE Conference of Photovoltaic Specialists Conference*, Anchorage, 15-22 September 2000, 1567-1570. http://dx.doi.org/10.1109/pvsc.2000.916196

[14] Yamegueu, D., Azoumah, Y., Py, X. and Zongo, N. (2011) Experimental Study of Electricity Generation by Solar PV/Diesel Hybrid Systems without Battery Storage for Off-Grid Areas. *Renewable Energy*, **36**, 1780-1787. http://dx.doi.org/10.1016/j.renene.2010.11.011

[15] Ai, B., Yang, H., Shen, H. and Liao, X. (2003) Computer-Aided Design of PV/Wind Hybrid System. *Renewable Energy*, **28**, 1491-1512. http://dx.doi.org/10.1016/S0960-1481(03)00011-9

[16] Diaf, S., Diaf, B., Belhamel, M., Haddadi, M. and Louche, A. (2007) A Methodology for Optimal Sizing of Autonomous Hybrid PV/Wind System. *Energy Policy*, **35**, 5708-5718. http://dx.doi.org/10.1016/j.enpol.2007.06.020

[17] Evans, D.L., Facinelli, W.A. and Koehler, L.P. (1980) Simulation and Simplified Design Studies of Photovoltaic Systems. Sandia National Laboratories, Arizona State. http://dx.doi.org/10.2172/5084440

[18] Nema, P., Nema, R.K. and Rangnekar, S. (2009) A Current and Future State of Art Development of Hybrid Energy System Using Wind and PV-Solar: A Review. *Renewable and Sustainable Energy Reviews*, **13**, 2096-2103. http://dx.doi.org/10.1016/j.rser.2008.10.006

[19] Thapar, V., Agnihotri, G. and Sethi, V.K. (2011) Critical Analysis of Methods for Mathematical Modelling of Wind Turbines. *Renewable Energy*, **36**, 3166-3177. http://dx.doi.org/10.1016/j.renene.2011.03.016

[20] Evans, D.L. (1981) Simplified Method for Predicting Photovoltaic Array Output. *Solar Energy*, **27**, 555-560. http://dx.doi.org/10.1016/0038-092X(81)90051-7

[21] Koutroulis, E., Kolokotsa, D., Potirakis, A. and Kalaitzakis, K. (2006) Methodology for Optimal Sizing of Stand-Alone Photovoltaic/Wind-Generator Systems Using Genetic Algorithms. *Solar Energy*, **80**, 1072-1088. http://dx.doi.org/10.1016/j.solener.2005.11.002

[22] Zhou, W., Lou, C., Li, Z., Lu, L. and Yang, H. (2010) Current Status of Research on Optimum Sizing of Stand-Alone Hybrid Solare/Wind Power Generation Systems. *Applied Energy*, **87**, 380-389. http://dx.doi.org/10.1016/j.apenergy.2009.08.012

[23] Gurkaynak, Y. and Khaligh, A. (2009) Control and Power Management of a Grid Connected Residential Photovoltaic System with Plug-In Hybrid Electric Vehicle (PHEV) Load. *Proceedings of the 24th IEEE Applied Power Electronics Conference and Exposition*, Washington DC, 15-19 February 2009, 2086-2091. http://dx.doi.org/10.1109/apec.2009.4802962

[24] Ruiz-Romero, S., Colmenar-Santos, A., Mur-Pérez, F. and López-Rey, Á. (2014) Integration of Distributed Generation in the Power Distribution Network: The Need for Smart Grid Control Systems, Communication and Equipment for a Smart City—Use Cases. *Renewable and Sustainable Energy Reviews*, **38**, 223-234. http://dx.doi.org/10.1016/j.rser.2014.05.082

[25] Lu, D., Fakham, H., Zhou, T. and François, B. (2010) Application of Petri Nets for the Energy Management of a Photovoltaic Based Power Station including Storage Units. *Renewable Energy*, **35**, 1117-1124. http://dx.doi.org/10.1016/j.renene.2009.12.017

[26] Wang, B.C., Sechilariu, M. and Locment, F. (2013) Power Flow Petri Net Modelling for Building Integrated Multi-Source Power System with Smart Grid Interaction. *Mathematics and Computers in Simulation*, **91**, 119-133. http://dx.doi.org/10.1016/j.matcom.2013.01.006

[27] Mladjao, M.A.M., El Abbassi, I., El Ganaoui, M. and Darcherif, A.-M. (2014) Modélisation et Optimisation de systèmes multi sources/Multi charges pour la ville durable. *Proceedings of the 3ème Congrès scientifique et technique du bâtiment durable*, Paris, 19-20 March 2014, 210-220.

[28] Zhu, F., Zheng, Y., Guo, X. and Wang, S. (2005) Environmental Impacts and Benefits of Regional Power Grid Inter-connections for China. *Energy Policy*, **33**, 1797-1805. http://dx.doi.org/10.1016/j.enpol.2004.02.018

[29] Turvey, R. (2006) Interconnector Economics. *Energy Policy*, **34**, 1457-1472. http://dx.doi.org/10.1016/j.enpol.2004.11.009

[30] Minciardi, R. and Sacile, R. (2012) Optimal Control in a Cooperative Network of Smart Power Grids. *Systems Journal*, **6**, 126-133. http://dx.doi.org/10.1109/JSYST.2011.2163016

[31] Hooshmand, A., Malki, H.A. and Mohammadpour, J. (2012) Power Flow Management of Microgrid Networks Using Model Predictive Control. *Computers & Mathematics with Applications*, **64**, 869-876. http://dx.doi.org/10.1016/j.camwa.2012.01.028

[32] Hammad, E.M., Farraj, A.K. and Kundur, D. (2015) Cooperative Microgrid Networks for Remote and Rural Areas. *Proceedings of the 28th IEEE Canadian Conference on Electrical and Computer Engineering (CCECE)*, Halifax, Nova Scotia, 3-6 May 2015, 1572-1577. http://dx.doi.org/10.1109/ccece.2015.7129515

[33] Hammad, E.M., Farraj, A.K. and Kundur, D. (2015) Grid-Independent Cooperative Microgrid Networks with High Renewable Penetration. *Proceedings of the IEEE PES Innovative Smart Grid Technologies Conference (ISGT)*, Washington DC, 17-20 February 2015, 1-5. http://dx.doi.org/10.1109/isgt.2015.7131865

[34] Ouammi, A., Dagdougui, H. and Sacile, R. (2015) Optimal Control of Power Flows and Energy Local Storages in a Network of Microgrids Modeled as a System of Systems. *IEEE Transactions on Control Systems Technology*, **23**, 128-138. http://dx.doi.org/10.1109/TCST.2014.2314474

[35] Wang, Y., Mao, S. and Nelms, R.M. (2015) On Hierarchical Power Scheduling for the Macrogrid and Cooperative Microgrids. *IEEE Transactions on Industrial Informatics*, **11**, 1574-1584.

[36] The Institution of Engineering and Technology (2014) What Is a Smart Grid? http://www.theiet.org/factfiles/energy/smart-grids-page.cfm

[37] Office of Deputy Prime Minister Creating sustainable Communities (2006) Low or Zero Carbon Sources: Strategic Guide. London.

[38] Hocaoglu, F.O., Gerek, O.N. and Kurban, M. (2009) A Novel Hybrid (Wind-Photovoltaic) System Sizing Procedure. *Solar Energy*, **83**, 2019-2028. http://dx.doi.org/10.1016/j.solener.2009.07.010

[39] Borowy, B.S. and Salameh, Z.M. (1996) Methodology for Optimally Sizing the Combination of a Battery Bank and PV Array in a Wind/PV Hybrid System. *IEEE Transactions on Energy Conversion*, **11**, 367-375. http://dx.doi.org/10.1109/60.507648

[40] Kellogg, W.D., Nehrir, M.H., Venkataramanan, G. and Gerez, V. (1998) Generation Unit Sizing and Cost Analysis for Stand-Alone Wind, Photovoltaic, and Hybrid Wind/PV Systems. *IEEE Transactions on Energy Conversion*, **13**, 70-75. http://dx.doi.org/10.1109/60.658206

[41] Ashok, S. (2007) Optimised Model for Community-Based Hybrid Energy System. *Renewable Energy*, **32**, 1155-1164.
http://dx.doi.org/10.1016/j.renene.2006.04.008

[42] Diaf, S., Notton, G., Belhamel, M., Haddadi, M. and Louche, A. (2008) Design and Techno-Economical Optimization for Hybrid PV/Wind System under Various Meteorological Conditions. *Applied Energy*, **85**, 968-987.
http://dx.doi.org/10.1016/j.apenergy.2008.02.012

[43] Short, W., Daniel, J.P. and Holt, T. (1995) A Manual for the Economic Evaluation of Energy Efficiency and Renewable Energy Technologies. National Renewable Energy Laboratory, Colorado.
http://dx.doi.org/10.2172/35391

[44] Athanasia, A.L. and Anastassios, D.P. (2000) The Economics of Photovoltaic Stand-Alone Residential Households: A Case Study for Various European and Mediterranean Locations. *Solar Energy Materials & Solar Cells*, **62**, 411-427.
http://dx.doi.org/10.1016/S0927-0248(00)00005-2

[45] Perera, A.T.D., Attalage, R.A., Perera, K.K.C.K. and Dassanayake, V.P.C. (2013) Designing Standalone Hybrid Energy Systems Minimizing Initial Investment, Life Cycle Cost and Pollutant Emission. *Energy*, **54**, 220-230.
http://dx.doi.org/10.1016/j.energy.2013.03.028

Appendix A

Table A.1. Transitions and its firing conditions for PN1.

Transitions[a]	Firing conditions	Configuration reached
$t_{1,2,3,4,5,6,7 \to 0}$	$\Delta P_{i=1,2,3}(t) < 0$	$\theta = 0$ (000)
$t_{0 \to 1}$ **et** $t_{x \to 1}$	$\Delta P_{i=1,2}(t) < 0$ **et** $\Delta P_{i=3}(t) \geq 0$	$\theta = 1$ (001)
$t_{0 \to 2}$ **et** $t_{x \to 2}$	$\Delta P_{i=1,3}(t) < 0$ **et** $\Delta P_{i=2}(t) \geq 0$	$\theta = 2$ (010)
$t_{0 \to 3}$ **et** $t_{x \to 3}$	$\Delta P_{i=1}(t) < 0$ **et** $\Delta P_{i=2,3}(t) \geq 0$	$\theta = 3$ (011)
$t_{0 \to 4}$ **et** $t_{x \to 4}$	$\Delta P_{i=2,3}(t) < 0$ **et** $\Delta P_{i=1}(t) \geq 0$	$\theta = 4$ (100)
$t_{0 \to 5}$ **et** $t_{x \to 5}$	$\Delta P_{i=2}(t) < 0$ **et** $\Delta P_{i=1,3}(t) \geq 0$	$\theta = 5$ (101)
$t_{0 \to 6}$ **et** $t_{x \to 6}$	$\Delta P_{i=3}(t) < 0$ **et** $\Delta P_{i=1,2}(t) \geq 0$	$\theta = 6$ (110)
$t_{0 \to 7}$ **et** $t_{x \to 7}$	$\Delta P_{i=1,2,3}(t) \geq 0$	$\theta = 7$ (111)

a) x = 1, 2, 3… 7: previous configuration.

Table A.2. Transitions and its firing conditions for PN2.

Transitions	Conditions	Flow exchanged[a]
$t_{R \to U1}$	$\theta = 0,1,2,3$ et $\sum_{i=1}^{3} \Delta P_i(t) < 0$	(A.0), (A.6), (A.5), (A.4)
$t_{U1 \to R}$	$\theta = 4,5,6,7$ et $\sum_{i=1}^{3} \Delta P_i(t) > 0$	(A.4), (A.5), (A.6), (A.7)
$t_{R \to U2}$	$\theta = 0,1,4,5$ et $\sum_{i=1}^{3} \Delta P_i(t) < 0$	(A.0), (A.6), (A.3), (A.2)
$t_{U2 \to R}$	$\theta = 2,3,6,7$ et $\sum_{i=1}^{3} \Delta P_i(t) > 0$	(A.2), (A.3), (A.6), (A.7)
$t_{R \to U3}$	$\theta = 0,2,4,6$ et $\sum_{i=1}^{3} \Delta P_i(t) < 0$	(A.0), (A.5), (A.3), (A.1)
$t_{U3 \to R}$	$\theta = 1,3,5,7$ et $\sum_{i=1}^{3} \Delta P_i(t) > 0$	(A.1), (A.3), (A.5), (A.0)

a) See Appendix for equations.

Appendix B[2]

$$
\left.
\begin{aligned}
& flowU2U1 = 0 \\
& flowU3U1 = 0 \\
& flowU3U2 = 0 \\
& flowRU2 = \Delta P_2(t) \\
& flowRU3 = \Delta P_3(t) \\
& flowRU1 = \Delta P_1(t)
\end{aligned}
\right\}
\qquad \text{Equation. (A.0)}
$$

---

[2]Whether X and Y, two different users or a user and national network, flowXY > 0 means Y sells |flowXY| to X and flowXY < 0 means Y buys |flowXY| from X.

$$\left.\begin{aligned}
&flowU2U1 = 0\\
&flowU3U1 = \Delta P_1(t)\\
&flowU3U2 = \Delta P_2(t)\\
&flowRU2 = 0\\
&flowRU3 = \Delta P_3(t) + \left[\Delta P_1(t) + \Delta P_2(t)\right]\\
&flowRU1 = 0
\end{aligned}\right\} \qquad \text{Equation. (A.1)}$$

$$\left.\begin{aligned}
&flowU2U1 = \Delta P_1(t)\\
&flowU3U1 = 0\\
&flowU3U2 = -\Delta P_3(t)\\
&flowRU2 = \Delta P_2(t) + \left[\Delta P_1(t) + \Delta P_3(t)\right]\\
&flowRU3 = 0\\
&flowRU1 = 0
\end{aligned}\right\} \qquad \text{Equation. (A.2)}$$

$$\left.\begin{aligned}
&flowU2U1 = \left[\left(\frac{\Delta P_2(t)}{\Delta P_2(t) + \Delta P_3(t)}\right)\Delta P_1(t)\right]\\
&flowU3U1 = \left[\left(\frac{\Delta P_3(t)}{\Delta P_2(t) + \Delta P_3(t)}\right)\Delta P_1(t)\right]\\
&flowU3U2 = 0\\
&flowRU2 = \Delta P_2(t) + flowU2U1\\
&flowRU3 = \Delta P_3(t) + flowU3U1\\
&flowRU1 = 0
\end{aligned}\right\} \qquad \text{Equation. (A.3)}$$

$$\left.\begin{aligned}
&flowU2U1 = -\Delta P_2(t)\\
&flowU3U1 = -\Delta P_3(t)\\
&flowU3U2 = 0\\
&flowRU2 = 0\\
&flowRU3 = 0\\
&flowRU1 = \Delta P_2(t) + \left[\Delta P_3(t) + \Delta P_2(t)\right]
\end{aligned}\right\} \qquad \text{Equation. (A.4)}$$

$$\left.\begin{aligned}
&flowU2U1 = -\left[\left(\frac{\Delta P_1(t)}{\Delta P_1(t) + \Delta P_3(t)}\right)\Delta P_2(t)\right]\\
&flowU3U1 = 0\\
&flowU3U2 = \left[\left(\frac{\Delta P_3(t)}{\Delta P_1(t) + \Delta P_3(t)}\right)\Delta P_2(t)\right]\\
&flowRU2 = 0\\
&flowRU3 = \Delta P_3(t) + flowU3U2\\
&flowRU1 = \Delta P_1(t) - flowU2U1
\end{aligned}\right\} \qquad \text{Equation. (A.5)}$$

$$
\left.
\begin{aligned}
&flowU2U1 = 0 \\
&flowU3U1 = -\left[\left(\frac{\Delta P_1(t)}{\Delta P_1(t) + \Delta P_2(t)}\right)\Delta P_3(t)\right] \\
&flowU3U2 = -\left[\left(\frac{\Delta P_2(t)}{\Delta P_1(t) + \Delta P_2(t)}\right)\Delta P_3(t)\right] \\
&flowRU2 = \Delta P_2(t) - flowU3U2 \\
&flowRU3 = 0 \\
&flowRU1 = \Delta P_1(t) - flowU3U1
\end{aligned}
\right\} \qquad \text{Equation. (A.6)}
$$

13

An Approach to Energy Saving and Cost of Energy Reduction Using an Improved Efficient Technology

Abubakar Kabir Aliyu[1]*, Abba Lawan Bukar[2], Jamilu Garba Ringim[3], Abubakar Musa[4]

[1]Faculty of Electrical Engineering, Centre of Electrical Energy System, Johor Bahru, Malaysia
[2]Department of Electrical Engineering, Faculty of Engineering, University of Maiduguri, Maiduguri, Nigeria
[3]Federal Airport Authority of Nigeria, Katsina, Nigeria
[4]Department of Electrical Engineering, Faculty of Engineering, Ahmadu Bello University, Zaria, Nigeria
Email: *muhammadkabir87@gmail.com

Abstract

The electricity consumption in commercial places like universities has tremendously increased recently. Modern and advanced energy efficient appliances are highly needed to substitute the conventional ones. Energy saving is of great important instead of its wastage, as utilizing the energy efficiently reduces the cost of energy. Energy consumption varies for commercial building due to several factors such as electrical appliance usage, electrical appliance type, management, etc. Due to the advancement in technology, there are new emergence appliances that are of high efficiency and have less energy consumption. A case study is conducted on selected five tutorial rooms, level 4 buildings in the Faculty of Electrical Engineering 19 A, Universiti Teknologi Malaysia. The paper proposes new emergence equipments with high efficiency and less power consumption to replace the existing ones. A survey is conducted on the number of electrical appliances used for each of the tutorial rooms, time table for each tutorial room and the Tenaga Nasional Berhad pricing and tariff are taken into consideration in the analysis of the energy consumption and the cost of energy. This paper aims at reducing the amount of energy consumption by replacing the existing electrical equipments with high efficient electrical equipments; it also tends to reduce the cost of energy paid to the utility. By observing the results, it shows that the proposed efficient electrical equipments are more efficient, less power consumption and less cost compared to the existing electrical equipments.

Keywords

Energy Efficiency, Payback Period, Energy Saving, Energy Auditing, Energy Efficiency Measures

---

1. Introduction

The per capital electricity consumption of the developing countries has been considerably increased due to the increase in population and high demand for energy. Energy growth in Malaysia has been increasing in residential, commercial, industrial and transport [1]. An increase in energy consumption has a great implication on the environment and global warming, therefore the policy to use the energy efficiency should be taken into account [2].

However, an efficient Management of Electrical energy Regulatory was introduced on 15 December, 2008 for the purpose of promoting energy efficiency in Malaysia. Thus, they make it compulsory for large commercial and industrial electrical consumers to manage their equipments so as to develop and implement EEMs to reduce energy losses, cost of energy and enforce efficient utilization of electrical energy [3].

Due to the advancement of technology, researchers and engineers are always working to see that they produce an apparatus or equipments which are very efficient in terms of energy. Energy efficiency can be defined as using less energy to produce the same amount of services or useful output, for example, residential sector, commercial sector and industrial sector. Energy efficiency in terms of mathematical expression can be defined as the ratio of the useful output of a process and the energy input into a process, and it is expressed in percentage [4].

$$\text{Energy efficiency} = \frac{\text{Output}}{\text{Input}}$$

The factors that contribute to high energy usage can be grouped into three.

Firstly, electricity consumption of the equipments itself. That is the purchase of fairly used equipment and non-energy efficient equipments.

Secondly, the number of equipments used for a particular place. As it is known that the number of equipments is directly proportion to the energy consumption. The used of many numbers of equipments such as fans, lights, air conditioner etc. than the design requirements.

Thirdly, the duration usage of the equipments. Long duration use of electrical equipment is directly proportion to the energy consumption [5].

Buildings in the university are usually characterized by high amount of energy consumption. A large portion of energy is being channeled to lecture halls since learning and teaching are the main activities in the campus. As this building consumes a large portion of energy, the energy that being waste should be identified and also should find a way of achieving energy efficiency is very important, that is using less energy to provide the same amount [6]. Some of the advantages of efficient use of energy is shown in **Figure 1** [7].

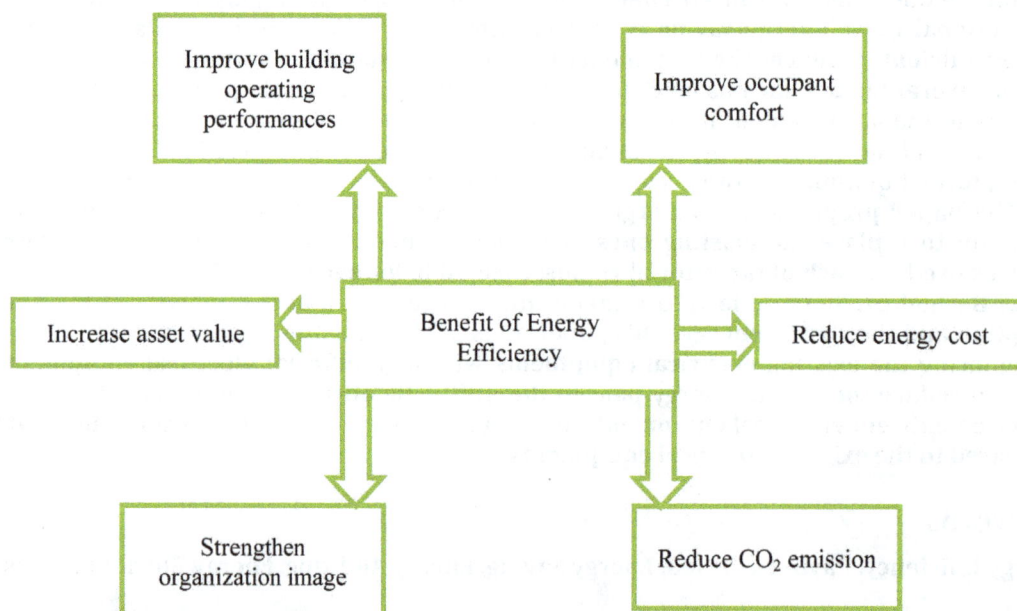

Figure 1. Advantage of energy effifciency.

This paper aims at reducing the amount of energy consumption by replacing the existing electrical with high efficient electrical appliances; these also tend to reduce the cost of energy paid to the utility. It also presents the results and analysis of energy audit of five tutorial rooms, level 4 buildings in the Faculty of Electrical Engineering 19 A, University Teknologi Malaysia (UTM).

2. Methodology

In this section, the method and the steps carried out is briefly described based on the case study. In order to understand fully the concept, energy consumption, price of equipments, cost of energy saving, bill of saving and the payback period are clearly explained and analyzed.

2.1. Energy Management and Energy Auditing

Due to an increase in energy consumption worldwide, energy management and energy auditing are considered to be a global challenge [8].

Energy management is termed as the strategy of adjusting and optimizing energy, using systems and procedures so as to reduce energy requirements per unit of output while holding constant or reducing total cost of producing the output of the systems while the users leave permanent access to the energy they need. The main objectives of energy managements are: resources conservation, climate protection and cost savings [9].

An energy audit is a fundamental of energy management services which employs methods of energy analysis to evaluate the energy usage and develop energy efficiency measures EEMs in the building. The energy audit for this case study disclosed electrical energy usage for lighting, air conditioner and projector. However an improved energy efficient equipments or energy efficiency measures EEMs are used aimed at reducing the cost of energy [3].

2.2. Electricity Consumption

The electricity consumption of the existing appliances and the proposed appliances was computed by multiplication of the number of equipments N, power rating W and the operating hours OH. The mathematical expression is given below:

$$\text{Electricity Consumption} \left(\text{KW} \right) = \frac{N \times W \times OH}{1000}$$

2.3. Cost of Electricity Consumption

Cost of electricity consumption is the multiplication of electricity consumption with the price of electricity per kW·h [dsm5].

$$\text{Cost of electricity consumption} = \text{Electricity Consumption} \times \text{price of electricity}$$

2.4. Energy Saving

Energy saving is the differences between the energy consumption of existing appliances and the proposed appliances.

$$\text{Energy Saving} = \text{Electricity Consumption}_{\text{Existing}} - \text{Electricity Consumption}_{\text{proposed}}$$

2.5. Bill Saving

Bill saving was computed by multiplying energy saving with the electricity tariff. The mathematical expression is given as:

$$\text{Bill Saving} = \text{Energy Saving} \times \text{Electricity Tariff}$$

2.6. Payback Period

The payback period is defined as the time (usually expressed in years) required for the cumulative operational savings of an option (or equipment) to equal the investment cost of that option [2].

$$\text{Simple payback period}\left(\text{years}\right) = \frac{\text{incremental cost}\left(\text{RM}\right)}{\text{Annual energy saving}\left(\text{RM}/\text{year}\right)}$$

3. Study Area

Tutorial Room 1 to 5 located in block P19 at the Faculty of Electrical Engineering (FKE), UTM was selected as the case study area. The classrooms selected are used as a lecture theatre for graduate studies. Electrical wattage and quantity of each of the equipments were studied and recorded. We are able to count and record all the equipments from tutorial Room 1 to 5. Since the tutorial rooms have the same electrical equipments, therefore **Table 1** shows the quantity of electrical equipments of one of the tutorial rooms, which is TR1.

3.1. Energy Audit of the Existing Equipments.

All the tutorial rooms have three things in common; which are; the air conditioning system, the projector and the lightning. The air conditioning system has the highest power consumption of 91.82%, followed by the projector system 7.39% and then the lightening bulbs 0.79%. The power consumption for each load is summarized in **Table 2**.

It should be noted that the level of energy consumption by each of the tutorial rooms differs, as each tutorial room has its own lecture time table. In **Table 3**, analysis of energy consumption characteristics for each of the selected tutorial rooms is carried out.

Although beside the used of lightning, projector and air conditioning, which are mainly provided for the purpose of lectures, sometime student comes with their handset chargers, laptop chargers and variety of electrical loads. These miscellaneous electrical loads are not included in this case study. **Figure 2** shows the percentage of power consumption of the existing equipments. Air conditioner has the highest power consumption having 98.2% followed by projector with 7.39% and lighting with 0.79%.

Table 1. Quantity of electrical equipments in tutorial Room 1.

Room	No. of Lightning	No. of Projector	No. of Air Conditioning System
TR 1	40	1	6

*For the five tutorial rooms × 5.

Table 2. Power consumption of existing equipments.

S/N	Equipment	Model	Power Consumption (W)
1	Air Conditioning System	DAIKIN FCU-4-13-52A	3730
2	Projector	Hitachi CP-EX250	300
3	Lightning Lamps	Philips T8	32

*For the five tutorial rooms × 5.

Table 3. Analysis of energy consumption characteristics of existing equipments.

Room	Hour Class/Week	Energy Consumed by Lightning kW·hr/Week	Energy Consumed by Projector kW·hr/Week	Energy Consumed by ACON kW·hr/Week	Total Energy Consumption per Week kW·hr/Week
Tutorial Room 1	6	7.680	1.8	134.280	143.760
Tutorial Room 2	36	46.080	10.8	805.680	862.560
Tutorial Room 3	15	19.2	4.5	335.700	359.400
Tutorial Room 4	24	30.720	7.2	537.120	575.040
Tutorial Room 5	5	6.4	1.5	111.9	119.8
TOTAL					2060.56

*Energy consumption per year × 52.

Power Consumption of the Existing Equipments

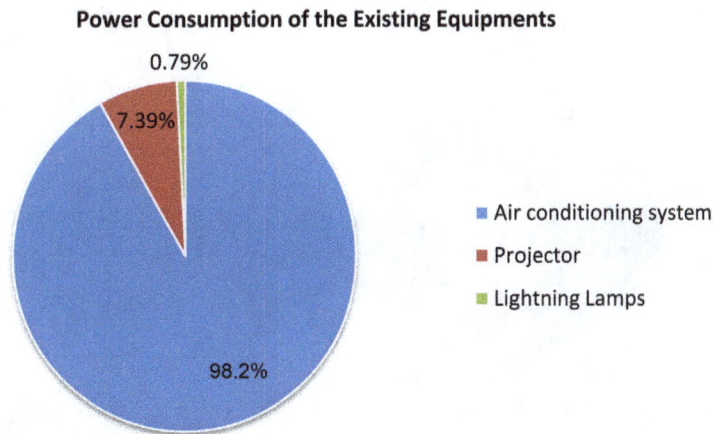

Figure 2. Piechart of power consumption of the proposed equipments.

3.2. Proposed Efficient Energy Equipments and the Analysis of Energy Consumption

In this section, the energy efficient equipments are proposed to replace the existing equipments. **Table 4** shows the less power consumption equipments that have high energy efficient than the existent ones.

In selecting the lightning bulb, A T5 lightning bulb with high lumens of about 3200 Lumens and with a power consumption of 21 W was selected to replace the existing one. The existing T8 lamps have lumens of about 2800 lumens. Therefore, using this T5 lamp reduces the number of lamps in each tutorial room from 40 lamps to 35 lamps. A projector with an 80 W energy consumption was also selected to replace the existing one of 300 W and air conditioner of 3 hp was selected to replace the existing one of 3730 W. **Table 5** shows the of energy consumption for the proposed equipments. **Figure 3** shows the percentage of power consumption of the proposed equipments, air conditioner has the highest wattage consumption with 95.7% followed by projector with 3.4% then the lighting lamp with 0.9%.

3.3. Analysis for Energy Saving, Cost of Energy and Payback Period

$$\text{Energy Savings kW} \cdot \text{hr/Week}$$

$$= \text{Existing} - \text{Proposed} = (2060.56) - (1233.928) = 826.602 \text{ kW} \cdot \text{hr/week}$$

Therefore, energy saving per year = 42,983.304 kW·hr/year

Using the data tariff obtained from the TNB website and as shown in **Table 6**, the cost energy consumption per kW·hr is 36.5 sen/kW·hr.

1) Total energy consumption of the existing equipments per week and per year is 2060.56 kW·hr/week, and 107,149.12 kW·hr/year respectively.

Therefore;

$$\text{Total cost of energy} = 107149.12 \times 36.5 = 3910942.88 \text{ sen/year} = \text{RM } 39109.4288/\text{year}$$

2) Total energy consumption of the proposed equipments per week and per year is 1233.928 kW·hr/week, and 64,164.256 kWhr/year respectively.

$$\text{Total cost of energy} = 64164.256 \cdot 36.5 = 2341995.344 \text{ sen/year} = \text{RM } 23419.95344/\text{year}$$

Total cost of energy saving RM 15,688.91/year.

3) The Payback period for the Air conditioner can be calculated as follows:

$$\text{Payback period}(\text{years}) = \frac{\text{incremental cost}(\text{RM})}{\text{Annual energy saving}(\text{RM/year})}$$

$$\text{Payback period}(\text{years}) = \frac{1488}{14612.17056} = 0.1$$

Power Consumption of the Proposed Equipments

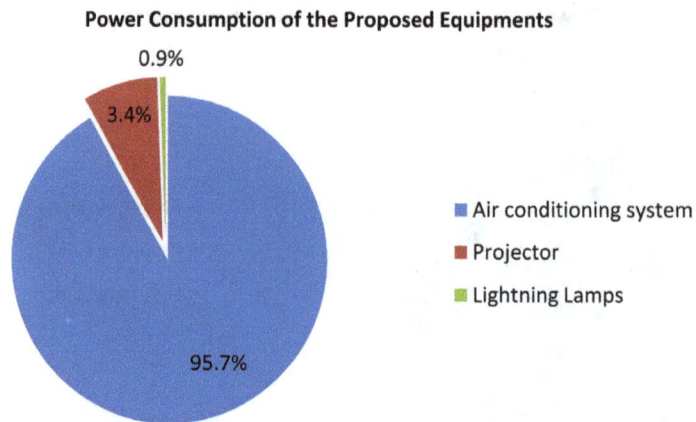

Figure 3. Piechart of power consumption of the existing equipment

Table 4. Proposed equipments to be replaced.

S/N	Equipment	Model	Power Consumption (W)
1	Air Conditioning System	DAIKIN FCQ35C7VEB	2238
2	Projector	BEN Q BEMW853UST	80
3	Lightning Lamps	Philips T5	21

*For the five tutorial rooms × 5.

Table 5. Analysis of energy consumption characteristics for the proposed equipments.

Room	Hour Class/Week	Energy Consumed by Lightning kW·hr/Week	Energy Consumed by Projector kW·hr/Week	Energy Consumed by ACON kW·hr/Week	Total Energy Consumption per Week kW·hr/Week
Tutorial Room 1	6	5.0400	0.480	80.5680	86.0880
Tutorial Room 2	36	30.240	2.880	483.408	516.528
Tutorial Room 3	15	12.600	1.200	201.420	215.220
Tutorial Room 4	24	20.160	1.920	322.272	344.352
Tutorial Room 5	5	4.2000	0.400	67.1400	71.7400
TOTAL					1233.928

*Energy consumption per year × 52.

Therefore the payback period of air conditioner is one (1) year. The same procedure applies to lighting and projector.

Table 7 shows the quantity numbers, incremental price (RM) which is the price difference between the proposed and existing equipment, annual energy saving (kWh), Energy bill saving (RM/years) and the payback period (years) of each equipments.

4. Conclusion

The analysis demonstrates how a meaningful amount of energy can be saved and minimized cost of energy in Tutorial Room 1 to 5 of block 19 in FKE. Although in the previous years, the university has tried tirelessly to reduce its energy consumption and cost of energy paid. Based on this case study, in order to achieve an optimal energy performance, energy audit is a good method to reduce the energy wastes and to improve the energy effi-

Table 6. TNB pricing and tariff of electricity consumption [10].

S/N	Tariff Category	Current Rate (1st June 2011)	New Rates (1st Jan. 2014 to Date)
	Tariff B—Low voltage commercial tariff for overall monthly consumption between 0 - 200 kWh/month		
	For all kWh	39.3 sen/kWh	
	The minimum monthly charge is RM 7.20 For the overall monthly consumption more than 200 kWh/month		
1.	For all kWh (from 1 kWh onwards)	43.0 sen/kWh	
	The minimum monthly charge is RM7.20		
	New structure effective 1st January 2014		
	For the first 200 kWh (1 - 200 kWh) per month		43.5 sen/kWh
	For the next kWh (201 kWh onwards) per month		50.9 sen/kWh
	The minimum monthly charge is RM7.20		
	Tariff C1—Medium voltage general commercial tariff		
2.	For each kilowatt of maximum demand per month	25.9 RM/kW	30.3 RM/kW
	For all kWh	31.2 sen/kWh	36.5 sen/kWh
	The minimum monthly charge is RM600.00		
	Tariff C2—Medium voltage peak/off-peak commercial tariff		
3.	For each kilowatt of maximum demand per month during the peak period	38.6 RM/kW	45.1 RM/kW
	For all kWh during the peak period	31.2 sen/kWh	36.5 sen/kWh
	For all kWh during the off-peak period	19.2 sen/kWh	22.4 sen/kWh
	The minimum monthly charge is RM600.00		

Table 7. Energy efficiency measures for air conditioner, projector and lighting lamps.

Equipment's	Number of Quantity	Incremental Price (RM)	Annual Energy Saving in kWh	Energy Bill Saving (RM/years)	Payback Period (years)
Air conditioner	30	1488	40033.344	14612.17056	0.1
Projector	5	2757.1	983.84	359.1016	7.678
Lighting lamps	80	25	1967.68	718.2032	0.035

ciency of the equipments considered for the case study, which are: lighting lamps, air conditioner and projector. Energy management and energy audit help in identifying several energy saving measures so as to improve the energy efficiency and reduce the cost of energy. From the analysis, it is calculated that if the proposed equipments are replaced, then the energy saved per year is 42,983.304 kW/hr and the cost of energy saving will be RM 15,688.91/year. If the university management can implement the proposed equipments, the energy consumption and the cost of energy will definitely reduce as seen in the analysis. The analysis in this paper is based on five tutorial rooms if compared to the university energy consumption is not up to 1%. Although the proposed equipments have high efficiency and less power consumption, their price is very expensive compared to the existing one. A payback period is calculated to show the time taken to recover its initial outlay from the saving of the cost of energy paid to the TNB.

References

[1] Al-Mofleh, A., Taib, S., Mujeebu, M.A. and Salah, W. (2009) Analysis of Sectoral Energy Conservation in Malaysia. *Energy*, **34**, 733-739. http://dx.doi.org/10.1016/j.energy.2008.10.005

[2] Mahlia, T.M.I., Razak, H.A. and Nursahida, M.A. (2011) Life Cycle Cost Analysis and Payback Period of Lighting Retrofit at the University of Malaya. *Renewable & Sustainable Energy Reviews*, **15**, 1125-1132. http://dx.doi.org/10.1016/j.rser.2010.10.014

[3] Singh, H., Seera, M. and Mohamad Idin, M.A. (2012) Electrical Energy Audit in a Malaysian University—A Case Study. 2012 *IEEE International Conference on Power and Energy*, Kota Kinabalu Sabah, Malaysia, 2-5 December

2012, 616-619. http://dx.doi.org/10.1109/pecon.2012.6450288

[4]   Patterson, M.G. (1996) What Is Energy Efficiency? *Energy Policy*, **24**, 377-390.
       http://dx.doi.org/10.1016/0301-4215(96)00017-1

[5]   (2012) S. Production and E. Products. Energy Efficiency in Malaysia Sustainable Production and Consumption.

[6]   Saidur, R., Rahim, N.A., Masjuki, H.H., Mekhilef, S., Ping, H.W. and Jamaluddin, M.F. (2009) End-Use Energy
       Analysis in the Malaysian Industrial Sector. *Energy*, **34**, 153-158. http://dx.doi.org/10.1016/j.energy.2008.11.004

[7]   Schipper, L. and Meyer, S. (1992) Energy Efficiency and Human Activity: Past Trends, Future Prospects. Press Syn-
       dicate of the University of Cambridge, New York.

[8]   Study, A.C. and Zaria, A.B.U. (2015) Energy Auditing and Management. *Journal of Multidisciplinary Engineering
       Science and Technology*, **2**, 1807-1813.

[9]   Sameeullah, M., Kumar, J., Lal, K. and Chander, J. (2014) Energy Audit : A Case Study of Hostel Building. *Interna-
       tional Journal of Research in Management, Science & Technology*, **2**, 36-42.

[10]   Nasional, T. (2015) Pricing and Tariff. http://www.tnb.com.my/residential/pricing-and-tariff.html

14

Realizing Low Carbon Emission in the University Campus towards Energy Sustainability

Isiaka Adeyemi Abdul-Azeez[1], Chin Siong Ho[2]

[1]Department of Urban & Regional Planning, Modibbo Adama University of Technology, Yola, Nigeria
[2]Faculty of the Built Environment, Universiti Teknologi Malaysia, Johor Bahru, Malaysia
Email: azeezabu@yahoo.com

Abstract

Energy consumption increases with intensity of human activities. People consume energy for movement and other activities and the more fossil-fuel based energy used, the more carbon dioxide (CO_2) emission. Since carbon dioxide is the major element of the greenhouse gases (GHG), this phenomenon has a serious implication for global warming and consequent climate change—a scenario that calls for sustainable development. This research considers the emission of CO_2 from energy use within the campus of Universiti Teknologi Malaysia. Two major sources of energy consumption were identified, namely: electricity and transport. The emission for electricity was estimated based on electricity meter reading and the conversion rate in accordance with the standardized conversion factors for fuel mix of the purchased electric energy as given by PTM (Pusat Tenaga Malaysia), while the associated CO_2 emission for transport was estimated based on the number of miles driven (VMT—Vehicle Miles Travel) within the campus, emissions produced per litre of gasoline, and fuel economy of vehicles plying the campus in line with the Code of Federal Regulations USEPA and consistent with the Inter-governmental Panel on Climate Change (IPCC) guidelines. It was observed that high CO_2 emission resulted from electricity energy consumption, and the highest emission in the transport sector was produced by commuting vehicles while emission from service delivery for cooling, lighting and other equipment was similar to national average.

Keywords

Energy Sustainability, Carbon Dioxide Emissions, Sustainable Campus, Low Carbon Development

1. Introduction

Emission of carbon dioxide continues to increase as energy consumptions grow, partly due to the age-long human habit of burning and the current technological practices that favor the use of fossil fuels as major sources of energy. The amount of carbon produced by the earth in the atmosphere has been of concern in recent time and the realization of low carbon emission will require an understanding of the source, type and quantity of energy use and the extent of emission from such sources. This involves the measurement and determination of extent of emission from the intensity of use. Generally, sustainability includes focusing on raising awareness to improve the overall image as well as bolstering the environmental prestige by encouraging the participation in the development of the strategies or policies for sustainable development, and also by providing incentives to influence and motivate people and institutions to be more active and focused. The assessment of energy sustainability through the emission of carbon dioxide is a key step for university campus sustainability.

Human activity and practices require energy for lighting, cooling and other domestic purposes as well as for movements and manufacturing and to sustain life. However, some sources of energy usually place stress on the environment and result in the emissions of carbon dioxide and other greenhouse gases (GHG). Most carbon emission is as a result of combustion from fossil fuel based energy use. Therefore, the more energy is consumed, the more stress is placed on the environment [1]. This is usually in the form of greenhouse gas emission which impacts negatively on the global environment. The continuous emission of CO_2 into the atmosphere owing to global energy consumption has irreversible effects mostly catastrophic to Earth's global system and requiring attention in the form of low carbon development.

Low carbon development entails safeguarding the environment without slowing down socioeconomic development; this will require technological solutions, to keep pace with the rate of economic growth and to change current unsustainable patterns of consumption and production, in the society. It is essential that every country voluntarily takes part in dealing with the global environmental problem [2] by establishing leadership in both thought and action, to direct sustainability as well as show willingness to define and adopt quantitative sustainability goals specifically on energy to enable more responsibility in operations. This can also be helpful in providing the best options aimed at achieving the goals of energy sustainability in the university campus.

The problem of carbon emission is more pronounced specifically in universities with large population and large spatial size, whose design requires the use of automobile to travel from one place to another within the campus. Similarly, the teaching and learning service delivery, as well as the residential and administrative activities also involves high energy demand for lighting, cooling, and running appliances, while, the movement of vehicles within the campus consumes high amount of fossil fuel energy, whose consumption also results in the emission of carbon dioxide.

Similarly, the electricity consumption in operating machines and transportation fuels of the university campuses, results in high emission of carbon dioxide [3], having serious implication on environmental quality. Also the products of direct and indirect activities such as classrooms, laboratories, offices and the consumption of food and drinks generate negative environmental impacts [4]. The combine activities of the global university population constitute significant energy use; hence, universities offer great potential for sustainability globally [5]. Therefore, focusing on achieving reduction of carbon emission from energy use in university campuses by encouraging low carbon emission through the involvement of the universities and achieving energy sustainability within the campus by the reduction of carbon dioxide emission may benefit global energy sustainability and remedy the current problems of global warming.

The operational approaches to meet sustainability goals in the universities are diverse and the practices are very broad and include improved environmental performances that may not necessarily be equivalent to sustainability. Sustainability is linked with setting quantitative targets in areas such as energy use, water use, use of land, purchases of product and emissions to air, water and land and achieving sustainability in the university is a process of setting goals to determine the extent of the aspects of the university required to be sustained [6].

Therefore, sustainability in university campus infers adopting intellectually defendable target for meeting the transition to sustainability and then developing the approaches and time scale designed to reach the target [7]. Also, focusing on the assessment procedure of sustainability where quantitative or qualitative value measures of the paradigm are developed for particular situations will assist to meet legitimate sustainability targets especially in the universities and colleges.

Hence, each university must determine its goals for itself. This infers taking inventories and setting targets and finally planning the program of implementation of the actions necessary to achieve the targets and then repeating the process all over. However, sustainability may not be easy to achieve in a quantitative manner unless it measures a criteria that is common among universities.

The ability to measure the extent, levels and impact of energy consumption in campuses will facilitate uniformity of practice of sustainability approach through common factors. This will limit global warming and reduce threat to global environment as well as enable the sharing of experience among the university campuses. However, it is necessary to determine the existing levels of emission in order to apply suitable strategies and mitigation measures to stop or reduce the use of the stuff that creates greenhouse gases.

Planning and implementation of policies to reduce carbon emission should be locally decided through the determination and measurement of emission sources because global warming impacts are better decided locally. Therefore, the inventory of carbon dioxide from transport and electricity focuses on energy consumption within UTM. Using the Malaysian University Campus Emission Tool (MUCET), the emissions are reported in percentages of metric tons of carbon dioxide (tCO_2) equivalent towards creating carbon reduction program to achieve more energy sustainable university campus. This will provide an understanding of the pattern and quantity of carbon dioxide emission within the campus, so that the university authority can plan emission reduction based on informed decisions.

2. Importance of Energy Sustainability in University Campus

Energy connects everything to everything else more universally and more quantifiably than any element [8]. It is evident today that carbon emission from fossil fuel based energy consumption is a common global sustainability issue, [9]-[11]. Energy is central to sustainability and there cannot be sustainable university development without sustainable energy development [12], therefore, energy sustainability in university sustainability drive is of great benefit to ameliorating global warming.

Colleges and universities have been at the forefront in addressing sustainability and global warming issues through innovative energy use, energy conservation practices and clean power technologies [13]. The American College and University Presidents' Climate Commitment (ACUPCC) is making effort to achieve sustainability among universities, through an initiative challenging institutions to quantify, reduce and ultimately eliminate their greenhouse gas emissions.

Similarly, other universities are grouping together to form partnerships to pursue the goals of sustainability. For instance the New Jersey (USA) Higher Education Partnership for Sustainability—(NJHEPS) in 2008, is also committed to reducing the greenhouse gas emissions and promoting positive changes in the environment of member universities.

In view of the benefits and importance of energy sustainability to human development and economic growth, the trends in university campus sustainability show the absence of consensus and lack universal direction regarding the approach and the development of sustainability in the sector.

It is evident that sustainability in the university campus presents a confusing scenario in terms of direction and universality. This is because the concept of sustainability presents diverging interpretations [14]. Researches in campus sustainability show concern for environmental issues along sustainable transportation in the university communities [15]-[17], as well as sustainability in waste management programs [18], while innovations in the green design movement including GIS-based evaluation of greenery [19], greening campus restaurant [20] and greenway corridor [21], green transport, green buildings, and green energy among others appear to be the trend. The assessment of the carbon dioxide emissions associated with on-campus electrical, natural gas and oil consumption is more common in the university in recent times [22].

In view of the benefits and importance of energy sustainability to human development, the trends in university campus sustainability show the absence of consensus and lack universal direction regarding the approach and the development of sustainability in the sector. It is evident that sustainability in the university campus presents a confusing scenario in terms of direction and universality.

The knowledge and quantification of energy indicators is important in the analysis and planning towards achieving sustainable energy development [23], as this will enable systematic monitoring of progress made towards the implementation of energy-related targets so as to reduce emission of carbon from energy source and attain effective management of the energy resources in the university campus. Finally, policies on achieving low carbon emission in university campus through implementation of climate action plan [24] are more common today.

As a way to establish the quantities of carbon emission and set targets to mitigate the condition, attempts are made to develop the indices with which campus sustainability could be measured; example is the assessment of carbon dioxide emissions associated with on-campus electrical, natural gas and oil consumption [22]. More universities are developing carbon emission inventories [25]-[29].

The rationale for the investigation of the sources and nature of emission from energy use in UTM is in line with the energy consumption pattern as shown in **Figure 1**, where primary energy is utilized in the form of final energy for various activities, in buildings and for movements. There is evidence of high carbon emission related to energy use in universities. It is necessary therefore to create the awareness among Malaysian universities on the need for carbon reduction strategies to mitigate global warming potentials of the university. The research therefore examine sustainability policies and energy consumption of UTM and the types and sources of energy used on campus to support university core activities such as Teaching and Learning, Administration and support services, ICT, Hostels and movements within the campus of UTM as well as the carbon emission from such energy consumption.

3. Sustainability Policies and Energy Consumption of Universiti Teknologi Malaysia (UTM)

The existing campus sustainability practice in Universiti Teknologi Malaysia (UTM) include policies defined across three major aspects and policy areas, namely, the socio-cultural policy, economic policy and eco-system policy of the university. The socio cultural dimensions consists of two main objectives which is to develop open society with openness and reduce barrier, as well as to develop civil society, through respect high integrity and ethical values so as to promote community spirit that is responsive to local and global context through harmonious and conducive environment.

The economic policy of UTM's campus sustainability aim to achieve cost effectiveness by adopting green building and infrastructure design, promote economic viability, optimization of university assets and to achieve efficiency in operational management of resources and facilities as well as to ensure the smooth implementation of the policies. The eco-system Policy of the UTM focuses on the implementation of low carbon campus initiatives which aim to enhance campus lifestyle through the reduction of energy and water consumption and also to reduce Pollution.

UTM is a pioneer university of technology located within the Iskandar region south of Malaysia. It has a total of 12 faculties and occupies a total area of about 1157 hectares of built up land as well as green area. The gross

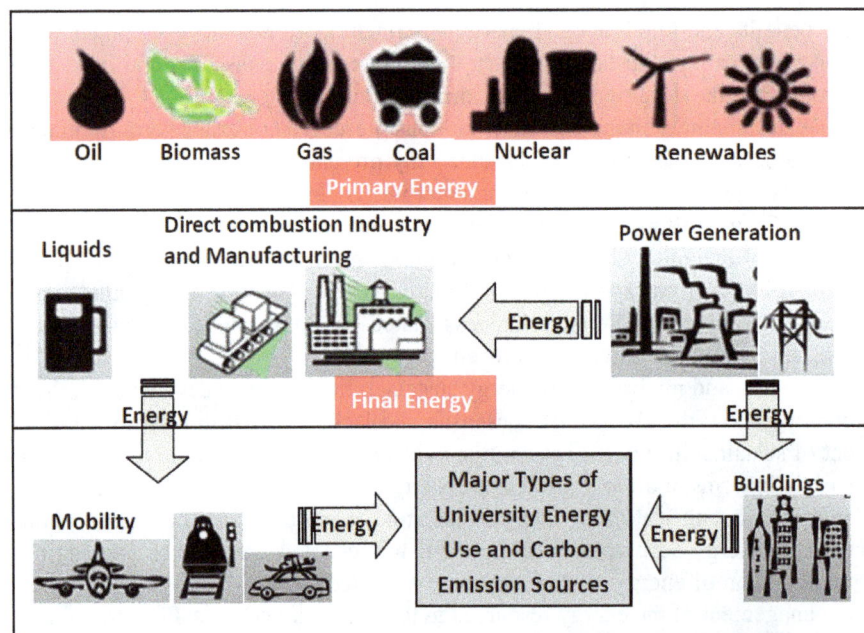

Figure 1. Energy consumption pattern (modified after David Hone, Shell International Ltd).

built up area of UTM is about 813352.21 meter square, consisting of the Zones of Academic Faculty (30%), students Hostels (56%), Administrative and support services (9%) and ICT Facilities area (5%).

The university consists mostly of high rise buildings and structures, fitted with elevators that run for 24 hours, thereby consuming enormous quantity of electricity to run and operate various appliances for its specialized and educational service delivery, specifically teaching and learning activity and for the lighting and cooling, running of laboratory and office equipment for administrative as well as ICT purposes among others. This generates substantial amount of carbon emission from electricity use as well as from the combustion of fuel for internal vehicular movements of the shuttle buses, the university vehicular fleet and the commuting staff and students' vehicles. **Figure 2** is a flow chart of energy pattern of the university, and a representation of the energy consumption pattern in most university campuses, upon which the study is focused.

The study of energy practices in Universiti Teknologi Malaysia is a current planning issue that identifies about 55,317,730 KwHr as total annual electricity consumption, responsible for 70% of the total carbon emission in the campus. Therefore assessing the impact of university operations [10] on the climate and the initiatives of setting targets to reduce emission of CO_2 in the university is a measure of campus sustainability [6] and a step towards reducing contribution of the university campus to global warming—an initiatives directed towards global sustainability.

4. Design Methodology and Approach

Data on electric energy consumption for domestic purpose were collected to identify the sectors with greater emission within the campus. The annual greenhouse gas emissions associated with vehicles were calculated based the number of miles driven (VMT) within the campus, fuel economy of vehicles, the carbon dioxide (CO_2) emissions produced per litre of gasoline, in line with the Code of Federal Regulations USEPA and consistent with the Inter-governmental Panel on Climate Change (IPCC) guidelines. The associated carbon dioxide emission of the total annual consumption of on-campus electrical energy use and the total amount of fuel consumption within the campus based on types of vehicles plying the campus was estimated and calculated using the Malaysian University Campus Emission Tool (MUCET).

The study of carbon emission is specialized and relates to activities that involve combustion of fossil fuel either in the form of liquid energy or generated electricity. For the purpose of accounting for carbon dioxide emission in UTM, the Malaysian University Carbon Emission Tool (MUCET) was employed to determine carbon emission from the university operations. The tool offer opportunity for the determination of the extent of emission and sectors of high emission, which could assist decision making towards reducing carbon emissions and creating energy sustainable campus.

MUCET calculated and provided information on the carbon emission from specific energy use sub systems of the university campus based on the conversion rate in accordance with the Pusat Tenaga Malaysia (PTM), standardized conversion factors for fuel mix for the purchased electricity, while the US Environmental Protection Agency (EPA) accepted method of calculating estimates for the annual GHG emissions associated with vehicle emissions in line with the IPCC standard was used to determine the emission from vehicles transport use within the campus. The tool was tailored towards university campus energy consumption patterns and relied on the structure of the university's service delivery pattern according to the structure of the university's energy flow (**Figure 2**). MUCET is also directed towards the determination of unit and total CO_2 emissions of the campus aggregate input of energy use.

Since the rationale is to determine the extent and intensity of carbon emission, this study concentrated on two main sources of fossil fuel based energy identified within the campus, which is electricity and transport energy use. The data mainly involved:

1) Emissions from electricity based on combustion of fossil fuels for electricity and the fuel mix of external electricity generation (purchased electricity) of university campus;

2) Emissions from energy use in transportation (*i.e.* fuel combustion from vehicular movements of goods and services within the campus).

The study considers a total of five categories of energy service demand sectors namely; energy use for activities of the teaching and learning services, the residential accommodation, administrative and support services, the Information and communication technology and the transport sectors. The carbon emissions were assessed based on energy consumption pattern of the university, and the electricity and transport consumption for the university were categorized as follows:

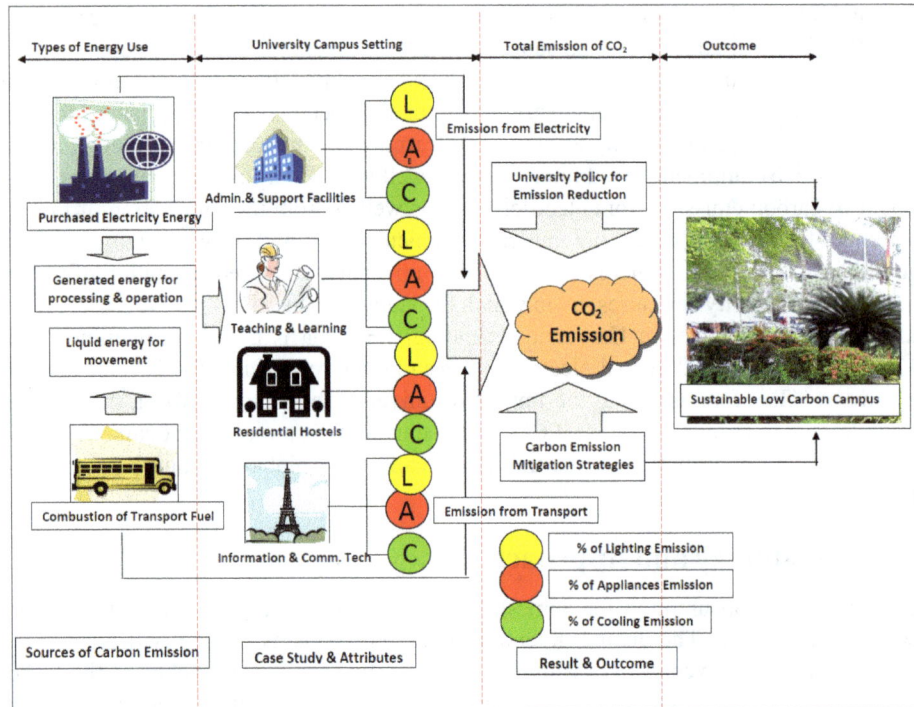

Figure 2. Flow chart of energy consumption and carbon emission of UTM, Malaysia.

Consumption for all the faculties of the university (*i.e.* teaching and learning sector):

1) Consumption for students hostels (residential accommodation);

2) Consumption for central administration and support services; and

3) Consumption for Information Communication Technology (ICT).

While the data on transport was classified as the CO_2 emission from:

1) Fuel consumption of commuting staff and students vehicles;

2) Fuel consumption by on-campus shuttle buses;

3) Fuel consumption for university vehicle fleet.

For the purpose of this research, six (6) indicators responsible for carbon emission in the university were identified from the literature [30] [31]. Based on these indicators, the method used in this study affirms emissions estimates and allows policy prescriptions to reduce CO_2 emissions in the university campus for the future. The indicators are given below as energy use for:

1) Lighting, cooling/space comfort and appliances;

2) Various service demands;

3) ICT and other operations;

4) University shuttle bus;

5) University vehicle fleet;

6) Internal circulation or commuting staff and students vehicles.

Based on these indicators, total emission for energy use [12] is computed as:

$$E_{total} = E_{electric} + E_{transport} \tag{1}$$

where,

$$1) \quad \text{Electric energy, } E_{electric} = E_{TL} + E_{ASS} + E_{HOSTEL} + E_{ICT} \tag{2}$$

$$2) \quad \text{Transport energy, } E_{transport} = E_{SB} + E_{FLEET} + E_{COMMUTING} \tag{3}$$

E_{TL} = Energy for lighting, cooling/space comfort and appliances for Teaching and Learning Sector;

E_{ASS} = Energy for Administration and Support Service;

E_{HOSTEL} = Energy for Hostel Accommodation;

E_{ICT} = Energy for Information & Communication Technology;

E_{SB} = Transport Energy use for Shuttle Bus;

E_{FLEET} = Transport Energy use for University Vehicle Fleet;

$E_{COMMUTING}$ = Transport Energy use for Commuting Staff and Students.

Therefore, the study relies on secondary data based electricity meter readings from Assets and Facilities department of UTM and also on data for fuel consumption of the university shuttle buses and the university vehicle fleet that were obtained from the transport section of the university as well as the fuel consumption for commuting staff and students vehicles, estimated based on average round trip derived from personal interview of staff, students and faculty members.

The data on electricity consumption in UTM was based on 100% survey of total electricity consumption for the year 2011. This was obtained from the recordings of the meter installations at the respective buildings. Twelve (12) academic faculties of UTM were observed and the Annual Area Energy Use Index (AEUI) expressed as $kW \cdot h/m^2$ [32] was presented as total for teaching and learning sector. Population size and energy efficiency are among important factors that affect carbon emission [33], based on the emission pattern for the university buildings, **Table 1** presents the units of population and activity space to enable the measurement of energy intensity and efficiency of energy use and to demonstrate electricity consumption per gross area. While carbon emissions are reported in metric tons of carbon dioxide (tCO_2) equivalent from the conversion of kilowatt hour kWh of electricity and litres of fuel into CO_2 equivalent using the EPA standards based on IPCC guidelines [34] and on fuel mix of electricity for Malaysia Peninsula [35].

Finally, the carbon dioxide of each alternative sector was estimated by entering observed values of fuel and electricity in the baseline of the prototype calculator (MUCET), creating a measure of carbon dioxide emission for each of the sectors which allows a comparison of emission from all the sectors in order to assist university authority in making informed decision. Therefore, savings from the impact of the strategies of campus goal in terms of target setting or alternative policies can be established as percentage reductions of CO_2 in sectors or the percentage of reduction from total campus emission.

5. Result and Findings

The measurement of the carbon emission from energy related sources in the campus is a potential to model transition to a low-carbon future and to facilitate the practice of energy sustainability in a manner easier to understand by university administration so as to set target and guidelines to achieve energy sustainability.

In this regard, it is essential to understand the characteristics of the study area. The basic characteristics of UTM based on the indicators of carbon emission from the university energy consumption are divided into three; basic data, energy consumption and computed findings. This study found that the UTM's activities resulted in approximately 46,000 million tons of carbon dioxide ($MtCO_2$) emissions annually. A total emission of 34,119 $MtCO_2$ was estimated from the consumption of 49,882,746 kWh of electricity in the buildings (**Table 1**). In summary emission per person is given as 1.89 $MtCO_2$, and 56.5 $KtCO_2$ per unit square meter for the gross built up area. These values could be useful in setting targets for emission reduction.

Electricity Energy accounted for the largest component of UTM's carbon emission, at roughly 74%. Transportation makes up the second-largest component of CO_2 emission with a total of 11,872 $MtCO_2$, accounting for 26% of UTM's total annual carbon emission. UTM is largely known as a commuter campus with total commuting vehicles of about 14,540 vehicles plying the campus daily, and annual fuel consumption of 48,707 and 2,591,711 litres of diesel and petrol engine vehicles respectively (**Table 2**). Most of this is due to fuel consumption from staff and students commuting to school by means of private vehicles, which accounts for 75% of the transport emission (**Figure 3**). A relatively small proportion (approximately 14%) of the students and staff live in residences within the campus. Majority of the commuting vehicles are private cars (77%) and about 62% of the commuting cars are owned by faculty members and staff of the university, while 38% belong to students, visitors and private individuals operating some businesses within the campus.

The study also observed that teaching and learning activities and hostel accommodations consumed 43% and 30% of the total university electricity energy use respectively; while Central Administration and Support Services consume about 14% and the ICT consumes 13% of the total electricity supply. The teaching and learning faculties have annual carbon dioxide emission of 14,448 $MtCO_2$ and 10,197 $MtCO_2$ for the residential hostels (**Table 3**). The two sectors have a combined emission of 73%, or about three quarters (3/4) of the total emission

Table 1. Basic characteristics of UTM energy consumption.

Category	Value	Remark
Gross Floor Area (Built Up)	813,352 m^2	
Total University Area	1145 hectares	
Staff Population	4894	
Students Population	19,433	Basic data
Total University Population	24,327	
Total Electricity Purchased (TNB)	55,317,730 kWh	
Total Building Electricity Consumption	49,882,746 kWh	
Miscellaneous Electricity Consumption	5,434,984 kWh	
Estimated Fuel Consumption Diesel	704,268 litres	Campus energy consumption
Estimated Fuel Consumption Petrol	3,116,238 litres	
Electricity Consumption/Capita	2274 kWh	
Electric Carbon Emission	34,119,770 KgCO$_2$	
Transport Carbon Emission	11,872 MtCO$_2$	
Total Carbon Emission	46,000 MtCO$_2$	Computed findings
Carbon Emission/Floor Area	47 Kg of CO$_2$	
Electric Consumption/Floor Area	68 kWh/m^2	

Table 2. Annual fuel consumption and carbon emission of commuting vehicles.

Vehicle type	Total vehicle per day	Annual fuel consumption (litres)		Emission per day (Kg CO$_2$)	Carbon emission	
		Diesel	Petrol		Kg CO$_2$	%
Car	11,159	-	2,212,219	19128	5,088,104	63%
Small van or mini van	442	-	132,538	2287	3,048,381	34%
Medium van	226	-	67,845	684	182,051	2%
Lorries/heavy duty truck	41	14,184	-	144	38,771	0.5%
Bus	115	34,523	-	350	93,212	1%
Motorcycle	2557	-	179,109	1548	411,951	5%
Total	14,540	48,707 litres	2,591,711 litres	24,141 (24 MtCO$_2$)	8,862,407 (8.86 MtCO$_2$)	

Table 3. Percentage of carbon emission by service demand sectors.

Category of uses	Carbon emission (MtCO$_2$)	Percentage of emission	Remark
Faculty	14,448	31%	
Students' hostels	10,219	22%	
Central admin. & support services	4871	11%	Total emission for electricity = 74%
ICT facilities	4591	10%	
Transport	11,872	26%	
Total	46,000 MtCO$_2$	100%	

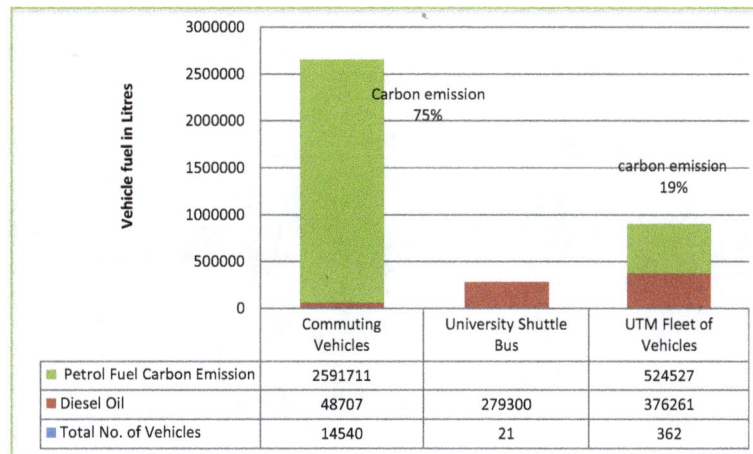

Figure 3. Summary of UTM's carbon emission for transport sector.

	Commuting Vehicles	University Shuttle Bus	UTM Fleet of Vehicles
Petrol Fuel Carbon Emission	2591711		524527
Diesel Oil	48707	279300	376261
Total No. of Vehicles	14540	21	362

of the university from electricity energy use as shown in **Figure 4**. Furthermore, the reason for relatively low consumption of electricity for the administrative and support activity is that the sector is operational for maximum periods of 9 - 10 hours per working day. The ICT facilities include the computer centre; the library and also the Information Technology (IT) support services for the teaching and learning as well as Central Administrative activities. It has an emission of 13% of the electricity CO_2 emissions (**Figure 4**) and 10% of total carbon emission of the entire service sectors as presented in **Table 3**. This is large in view of its physical size of 5% of the total built up area.

The high annual energy use index (AEUI) and carbon emission for ICT sector could be as a result of the electricity consumption for the population served, for instance, the library entertains about 1,083,677 visitors annually and emit about 4 kg of CO_2 per visitors. Similarly, the computer centre (CICT) serves the entire university population, supports all university activities and is continuously operational for 24 hours of 7 days, hence, the high emission intensity.

The carbon dioxide emission of electric energy use based on demand for the cooling, lighting and other electrical appliances is as given in **Figure 5**. The total carbon emission for all the categories of electricity energy consumption is about 34,036 $MtCO_2$. The total electricity demand for cooling or air-conditioning is 24,711 MWh (Megawatt hour) with CO_2 emission of 16,907 $MtCO_2$ and about 50% of the total emission from the university electric energy consumption. The emission for lighting purpose is about 16% while equipment and other appliances is 34% of the total. This figure compare favorably with the national averages as shown in **Figure 5**, and indicates strong similarity between UTM and National emission for these categories of uses.

The percentages of emissions from service demand sectors are presented in **Figure 6**. The Teaching and Learning Faculty tops the list with 31%, followed by the transport sector with 26% of total campus emission. The students hostel accommodation has about 22% of the emission, while central administration and support services has 11% and the ICT sectors emits 10% of the total electricity emission.

Carbon emission is proportional to energy demand, therefore, the emission from service demand sectors offer the opportunity to plan the reduction of carbon emission based on service demand sectors. By this, it will be possible to view percentage emission and determine the high energy demand sectors/high emission sectors on campus which can be targeted by carbon reduction strategies towards realizing low carbon emission.

The knowledge of the sources and extent of carbon emission will facilitate the reduction of the university's contribution to global warming as a means of promoting sustainability and will enable administrators and university leaders to understand, quantify, and manage the emissions as well as make informed decisions towards reducing the global warming impact of the campus through CO_2 emissions reduction.

6. Conclusions

Planning energy sustainability should be a continuous process that can be practiced through the adoption of a goal, research analysis, planning and effectuation of policies. Achieving low carbon and sustainable university campus requires the enshrinement of policy in the form of statements, strategies and plans to direct the physical

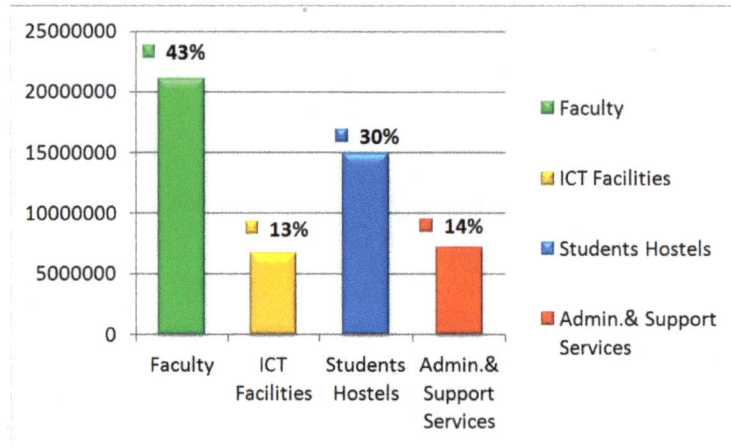

Figure 4. Carbon emission of service activities based on electricity consumption. Source: field survey 2011.

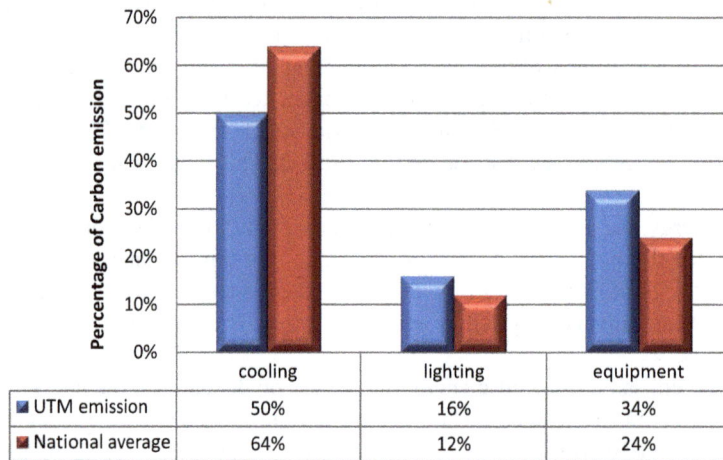

	cooling	lighting	equipment
■ UTM emission	50%	16%	34%
■ National average	64%	12%	24%

Figure 5. Carbon emission by category of uses. Source: modified after Aun 2004.

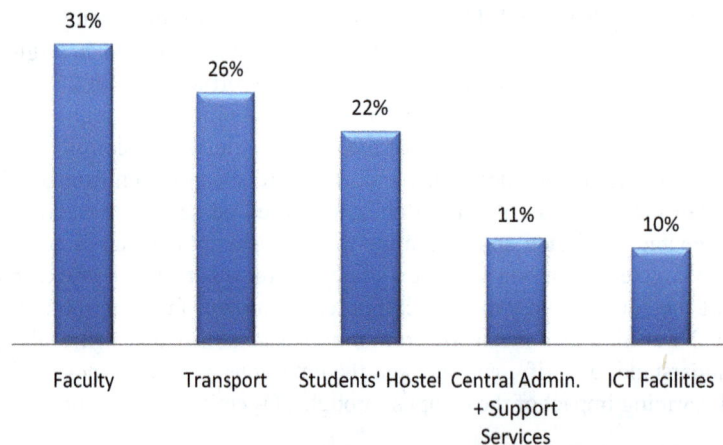

Figure 6. Percentage of annual overall university carbon emission in UTM.

operations of the university through education, research and practices that can be evaluated by a control system that would ensure optimization of energy use. However, the measures put in place to ensure implementation are equally important and capable of facilitating change in many institutions. Intensifying efforts to reduce the universities' contribution of greenhouse gases (GHG) through inventory of carbon emission and an assessment tool that is easier to use as well as the search for alternative energy sources would enhance the achievement of energy sustainability.

The ability to measure the extent and levels of energy use in the campus will facilitate uniformity of sustainability approach where common factors among universities are observed, *i.e.*, energy use. Therefore, for the purpose of achieving more sustainable environment, it is desirable to practice sustainability through effective measurement of CO_2 emission from energy consumption. Sustainability could be more popular among universities, by creating a consensus approach and encouraging collaboration and partnership among universities as well as the establishment of effective implementation and monitoring process of energy sustainability among the university campuses. This could limit global warming and reduce threat to global environment as well as enable the sharing of experience among the university campuses.

However, it is necessary to determine the existing levels of emission in order to apply suitable strategies and mitigation measures. For instance, strategies to stop the use of the stuff that creates greenhouse gases as much as possible as well as other measures of behavioral change such as turning off lights when not in use as well as using the fan instead of an air conditioner on a warm day could save energy in campuses and assist to achieve energy sustainability. Similarly, commitment towards exploring alternative, environmentally friendly forms of energy generation and reduced reliance on imported energy generation and also encouraging on-campus environmentally-friendly electric energy generation can enhance carbon emission reductions.

Furthermore, emission from electricity could be reduced by using more electricity efficient facilities and changing consumption behavior, while the on-campus transport emission could be improved with the adoption of policies for better on-campus transportation facilities and options. Similarly, encouraging the use of more environmentally-friendly forms of transportation as well as the use of Cleaner-burning, renewable fuels, such as biodiesel and encouraging cycling will also achieve low carbon emission in the campus.

Reducing the aggregate emission of universities worldwide is a pathway to achieving the objective of Greenpeace International of holding temperature rise to below 2 degrees Celsius [36]. Therefore, completing an assessment of energy use to determine university's contribution to global warming and conducting surveys of energy consumption within the campus is ideal for the achievement of energy sustainability. Also working with administrators and staff to implement recommendations for low carbon emission is critical to realizing low carbon emission and making improvements in university campus.

Finally, providing an energy impact assessment for the administration and developing carbon saving action plans that cut across energy, transport, buildings, housing, waste, and other issues in the university campus will ensure that rapid development occurs in an environmentally responsible and sensitive manner. However, clear responsibilities for actions of universities to set targets to reduce global CO_2 emissions to below 4Gt/a in 2050 could be achieved through outlined strategies developed to assist reduction of $MtCO_2$ in the priority areas of the university campuses.

References

[1]   Abdalla, K. (2007) Energy Indicators for Sustainable Development: Country Studies on Brazil, Cuba, Lithuania, Mexico, Russian Federation, Slovakia and Thailand, United Nations Department of Economic and Social Affairs.

[2]   Park, K., *et al.* (2003) Quantitative Assessment of Environmental Impacts on Life Cycle of Highways. *Journal of Construction Engineering and Management*, **129**, 25-31. www.ascelibrary.org
http://dx.doi.org/10.1061/(ASCE)0733-9364(2003)129:1(25)

[3]   Alshuwaikhat, H.M. and Abubakar, I. (2008) An Integrated Approach to Achieving Campus Sustainability: Assessment of the Current Campus Environmental Management Practices. *Journal of Cleaner Production*, **16**, 1777-1785.
http://dx.doi.org/10.1016/j.jclepro.2007.12.002

[4]   Lukman, R. (2009) Towards Greening a University Campus: The Case of the University of Maribor, Slovenia. *Resources, Conservation and Recycling*, **53**, 639-644. http://dx.doi.org/10.1016/j.resconrec.2009.04.014

[5]   Rappaport, A. (2008) Campus Greening: Behind the Headlines. *Environment: Science and Policy for Sustainable Development*, **50**, 6-17. http://dx.doi.org/10.3200/ENVT.50.1.6-17

[6]   Graedel, T.E. (2002) Quantitative Sustainability in a College or University Setting. *International Journal of Sustainability in Higher Education*, **3**, 346-358. http://dx.doi.org/10.1108/14676370210442382

[7]   Bosshard, A. (2000) A Methodology and Terminology of Sustainability Assessment and Its Perspectives for Rural

Planning. *Agriculture, Ecosystems and Environment*, **77**, 29-41. http://dx.doi.org/10.1016/S0167-8809(99)00090-0

[8] Jiusto, J.S. (2003) Spatial Indeterminacy and Power Sector Carbon Emissions Accounting. Ph.D. Thesis, Faculty of Clark University, Worcester.

[9] Pope, J., Annandale, D. and Morrison-Saunders, A. (2004) Conceptualising Sustainability Assessment. *Environmental Impact Assessment Review*, **24**, 595-616. http://dx.doi.org/10.1016/j.eiar.2004.03.001

[10] Hardy, D. (2008) Cities That Don't Cost the Earth Published by Jon Land for TCPA in Housing and Also in Communities, Local Government.

[11] Arrow, K.J. (2007) Global Climate Change: A Challenge to Policy. The Berkeley Electronic Press, Berkeley. www.bepress.com/ev

[12] Abdul-Azeez, I.A. (2012) The Development and Application of Malaysian University Carbon Emission Tool (MUCET) towards Creating Sustainable Campus. Ph.D. Thesis, UTM, Johor.

[13] Eagan, D.J., *et al.* (2008) Higher Education in a Warming World. The Business Case for Climate Leadership on Campus National Wildlife Federation's Campus Ecology. www.nwf.org/CampusEcology/BusinessCase

[14] Lourdel, N., Gondran, N., Laforest, V. and Brodhag, C. (2005) Introduction of Sustainable Development in Engineer's Curricula Problematic and Evaluation Methods. *International Journal of Sustainability in Higher Education*, **6**, 254-264.

[15] Balsas, C.J.L. (2003) Sustainable Transportation Planning on College Campuses. *Transport Policy*, **10**, 35-49. http://dx.doi.org/10.1016/S0967-070X(02)00028-8

[16] Toor, W. and Havlick, S. (2004) Transportation and Sustainable Campus Communities: Issues, Examples, Solutions. Island Press, Washington DC.

[17] Dorsey, B. (2005) Mass Transit Trends and the Role of Unlimited Access in Transportation Demand Management. *Journal of Transport Geography*, **13**, 235-246. http://dx.doi.org/10.1016/j.jtrangeo.2004.07.004

[18] Mason, I.G., Brooking, A.K., Oberender, A., Harford, J.M. and Horsley, P.G. (2003) Implementation of a Zero Waste Program at a University Campus. *Resources, Conservation and Recycling*, **38**, 257-269. http://dx.doi.org/10.1016/S0921-3449(02)00147-7

[19] Wong, N.H. and Jusuf, S.K. (2008) GIS-Based Greenery Evaluation on Campus Master Plan. *Landscape and Urban Planning*, **84**, 166-182. http://dx.doi.org/10.1016/j.landurbplan.2007.07.005

[20] Nilsson, J., Bjuggren, C. and Frostell, B. (1998) Greening of a Campus Restaurant at Stockholm University: Sustainable Development Audits by Means of the SDR Methodology. *Journal of Environmental Management*, **52**, 307-315.

[21] Arendt, R. (2004) Linked Landscapes Creating Greenway Corridors through Conservation Subdivision Design Strategies in the Northeastern and Central United States. *Landscape and Urban Planning*, **68**, 241-269. http://dx.doi.org/10.1016/S0169-2046(03)00157-9

[22] Riddell, W., Bhatia, K.K., Parisi, M., Foote, J. and Imperatore, J. (2009) Assessing Carbon Dioxide Emissions from Energy Use at a University. *International Journal of Sustainability in Higher Education*, **10**, 266-278. http://dx.doi.org/10.1108/14676370910972576

[23] International Atomic Energy Agency (IAEA) (2007) Energy Indicators for Sustainable Development: Country Studies on Brazil, Cuba, Lithuania, Mexico, Russian Federation, Slovakia and Thailand. United Nations Department of Economic and Social Affairs, New York.

[24] Spirovski, D., Abazi, A., Iljazi, I., Ismaili, M., Cassulo, G. and Venturin, A. (2012) Realization of a Low Emission University Campus through the Implementation of Climate Action Plan. *Procedia—Social and Behavioral Sciences*, **4**, 4695-4702. http://dx.doi.org/10.1016/j.sbspro.2012.06.321

[25] Beatty, B., *et al.* (2002) Building Environmental Sustainability at Bowling Green State University (Executive Summary). http://www.bgsu.edu/departments/envh/ES-summary.pdf

[26] Isham, J.T., *et al.* (2003) Carbon Neutrality at Middlebury College: A Compilation of Potential Objectives and Strategies to Minimize Campus Climate Impact. Draft Prepared for the Carbon Reduction Initiative at Middlebury College.

[27] Stewart, C. (2005) Ecological Footprint Progress Report. University of Toronto Mississauga, Mississauga. http://geog.utm.utoronto.ca/ecofootprint/doc/efprogressreport2005.pdf

[28] Elderkin, R. (2007) A First Step toward a Climate Neutral Pomona College: Greenhouse Gas Emissions Inventory and Recommendations for Mitigating Emissions. *Pomona Campus Climate Challenge*, 27 April 2007.

[29] McNeilly, L. (2008) UC Berkeley Campus Sustainability. University of California, Berkeley.

[30] Willson, R. and Brown, K. (2008) Carbon Neutrality at the Local Level: Achievable Goal or Fantasy? *Journal of the American Planning Association*, **74**, 497-504. http://dx.doi.org/10.1080/01944360802380431

[31] Filippin, C. (2000) Benchmarking the Energy Efficiency and Greenhouse Gases Emission of School Buildings in Ar-

gentina. *Building and Environment*, **35**, 407-414. http://dx.doi.org/10.1016/S0360-1323(99)00035-9

[32] Chan Seong Aun, A. (2004) Energy Efficiency: Designing Low Energy Buildings Using Energy 10. Pertubuhan Arkitek Malaysia (PAM), CPD Seminar, 7th August 2004.

[33] Sathiendrakumar, R. (2003) Greenhouse Emission Reduction and Sustainable Development. *International Journal of Social Economics*, **30**, 1233-1248. http://dx.doi.org/10.1108/03068290310500643

[34] United State Environmental Protection Agency (2006) Greenhouse Gas Emissions from the US Transportation Sector, 1990-2003. USEPA, Washington DC. www.epa.gov/otaq/climate.htm

[35] WED, IEA/ESD, Energy Statistics Division (ESD) of the International Energy Agency (IEA) (2011) IEA Energy Statistics. www.iea.org/statist/index.htm

[36] Greenpeace International, Implementation of All the Elements of Decision 1/CP.17, (b) Matters Related to Paragraphs 7 and 8 (ADP), 2013.

15

Analysis of DVFS Techniques for Improving the GPU Energy Efficiency

Ashish Mishra, Nilay Khare

Department of Computer Science and Engineering, Maulana Azad National Institute of Technology, Bhopal, India
Email: ashishmishra81@gmail.com

Abstract

Dynamic Voltage Frequency Scaling (DVFS) techniques are used to improve energy efficiency of GPUs. Literature survey and thorough analysis of various schemes on DVFS techniques during the last decade are presented in this paper. Detailed analysis of the schemes is included with respect to comparison of various DVFS techniques over the years. To endow with knowledge of various power management techniques that utilize DVFS during the last decade is the main objective of this paper. During the study, we find that DVFS not only work solely but also in coordination with other power optimization techniques like load balancing and task mapping where performance and energy efficiency are affected by varying the platform and benchmark. Thorough analysis of various schemes on DVFS techniques is presented in this paper such that further research in the field of DVFS can be enhanced.

Keywords

GPGPUs, DVFS, Task Mapping, Energy Efficiency

1. Introduction

As we move from mega scale to petascale era, the requirements of data processing and computation are growing exponentially. In order to accomplish this high computation demand, researchers have moved from serial computation platforms to high performance computation (HPC) platforms such as multicore processor, FPGAs and heterogeneous system (GPU supported systems) etc. GPUs, in particular, have been widely used for HPC applications due to their extremely high computational powers. A large number of supercomputer found in TOP500 list use GPU to achieve unprecedented computational power [1].

Today, GPU has become the core part of high performance system having hundreds to thousands of processor cores and much higher peak performance than CPUs. Hence, many HPC applications utilize the power of GPUs. For example, the recently built supercomputer Tianhe-1A has won the second spot on the TOP 500 list [1] and is equipped with Intel Xeon 5670 processors and NVidia's CUDA-enabled Tesla M2050 general purpose GPUs. Having GPU in Tianhe-1A makes supercomputers able to achieve more than two fold energy efficiency than the third place CPU-based Jaguar TOP 500 list. However, the electricity bill of Tianhe-1A is estimated around annual electricity bill of $2.7 million [2]. This high power consumption is another reason that forces researchers to work in a direction to reduce the power consumption of GPUs.

On the other hand, manufacturers increase the number of processing core to gain the high performance which has resulted in raising the power consumption of GPUs. They consume much high power as compared to CPUs and the raised levels of power consumption of GPUs have significant impact on reliability, architecture design, economic feasibility and deployment into widespread range of application domains. In recent years, several research has been accomplished for the reduction of power consumption of homogenous as well as heterogeneous systems and various techniques to reduce the power consumption of both the systems have been proposed. In [3], Sparsh and Jeffery present superior categorization of various power reduction techniques which are categorized as follows:

1. DVFS based techniques;
2. CPU-GPU workload division techniques;
3. Saving energy in GPU components;
4. Dynamic resource allocation techniques;
5. Application specific and programming level techniques.

In this paper, literature survey and thorough analysis of various schemes on DVFS techniques to reduce the energy efficiency of GPUs only are presented. The rest of the paper is organized as follows.

2. Background

2.1. GPU

As shown in **Figure 1**, all GPUs have two important component one is number of parallel streaming processor and second one is memory used by GPUs core. Each streaming processor again has a number of processing elements. Performance of kernel application is vastly depending on the frequency on which these two components operated.

2.2. Need to Improving Energy Efficiency of GPU

Due to application limitation, it is not always possible for an application to map all the available cores. In several applications, memory bandwidth [4] [5] of GPU act as bottleneck to affect the performance of GPU. Due to this bottleneck, the core of GPU remains unutilized. Therefore, a good power management technique is required

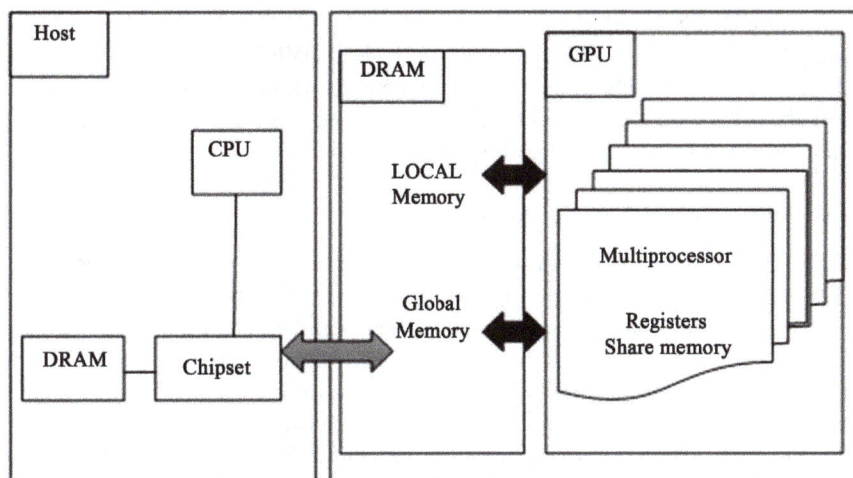

Figure 1. GPU architecture.

to save the power consumed by these unutilized cores. As said by Anderson in [6], a 15 degree increase in temperature is responsible to increase the failure rates of component by a factor of two. This component failure may lead to system malfunction that in turn affect economic of the system as GPU becomes popular accelerator among the super computer and business services. Thus, an efficient energy model is needed to ensure the reliability. Although, GPU's get admired for the performance improvement and have to be energy efficient to ensure the reliably and improve business gain.

3. Dynamic Voltage Scaling Techniques

Dynamic voltage and frequency scaling (DVFS) is a technique widely used for reducing energy consumption of processors by varying the voltage and frequency at run time [7]. The main idea is to reduce frequency or voltage during periods when the processor has a reduced workload. If DVFS is done wisely, energy can be saved without any noticeable effects on the speed at which the processor performs its tasks. Most systems are designed with fixed voltage and frequency settings in order to make the system stable. However, the activity levels of applications are variable, and in many cases, applications have idle periods when no useful task is performed. By reducing the processor voltage and frequency levels at run-time when the application has low-activity or idle periods, energy can be saved.

DVFS can be used to eliminate power-wasting idle times by lowering the processor's voltage and frequency during low workload periods so that the processor will have meaningful work at all times leading reduction in the overall power consumption. The energy consumed by GPU is given by the following equation [3] [8]:

$$E = CV^2 * f \tag{1}$$

where,

E = Energy consumed by GPU Measured in joules (J);
C = Capacitance;
V = Voltage supply to GPU;
f = Clock frequency of GPU.

Thus, the power consumed by a task may be decreased by reducing V or F, or both. However, for tasks that require a fixed 5 amount of work, reducing the frequency may simply take more time to complete the work. As a result, little or no energy will be saved. Therefore, intelligent DFVS techniques are required to improve the energy efficiency of GPUs. Many techniques are used to control power consumption by controlling the frequency, since processor frequency has a strong effect on power consumption and temperature. Dynamic voltage and frequency scaling (DVFS) are the most commonly used techniques in modern processors [9].

This section describes various DVFS techniques explored by the researchers exclusively for GPUs. We found that DVFS not only work solely but also work in coordination with other techniques like workload divisions/ task division techniques to give the best result. This section categorized the DVFS techniques in the following heads:

A. Schemes using core DVFS technique
B. Schemes using DVFS with other GPU optimization (Hybrid DVFS)

In Section 3.1, detailed description of energy saving methods with only DVFS technique is presented. In Section 3.2, those methods which not only used DVFS but also used some other optimization such as task mapping, workload division, load balancing etc. in coordinated manner are discussed.

3.1. Schemes Using Core DVFS Technique

Intelligent use of DVFS technique may reduce energy consumption of GPU's energy demand. It is a challenging task for DVFS to save energy while preserving the performance [8]. The schemes presented in [10]-[14] apply only DVFS technique for conserving energy on various benchmarks. All of them managed to save energy while maintaining the performance aspects. A triple domain DVFS scheme is proposed in [11] for graphical processor where the frequency and voltage of RISC, geometry processor (GP) and rendering engine (RE) are independently managed. This multi-domain power management scheme used three power management unit that is fully integrated on chip to apply DVFS on graphical processing unit (GPU).With this triple domain power scheme in [11], authors managed to save power up to 65%. GPU only consume 52.4 mW at runtime benchmarks test in contrast to 154 mW without power management techniques. In [8], performance and power of various application kernels under varying frequency settings are characterized. In order to conduct research, the authors choose three computationally diverse applications namely:

1. Compute Intensive Application
2. Memory Intensive Application
3. Hybrid Application

The authors identified these three classes of kernel on the basis of two metrics proposed in [15]:

1. Rate of instruction issues
2. Ratio of Global memory transaction to Computation Instruction

On the basis of above ration, following application kernel belonging to above three categories are identified. The kernel categories and application kernels are shown in **Table 1**.

In [8], the number of utilized GPU cores is not varied because of the restrictions of GPUs in "suspending" unused GPU cores so as to utilize lesser power causing inconsistent results. At the end of study [8], the authors concluded that performance and power consumption of GPU are largely determined by two characteristics: the rate of issuing instructions and the ration of global memory transactions to computation instructions. The vital goal of research work in [8] is to investigate the power and performance, and power consumption of typical GPU application kernel under different memory and core frequency.

In [12], the authors improve throughput of GPUs by adjusting the number of operating cores and voltage/frequency of cores and/or on chip interconnects/cached for different application under the power constraint environment. Even they further improve throughput by dynamically scaling the number of core and voltage/frequency of both core and on-chip caches at runtime. The objective of [12] is to improve the throughput only by keeping power consumption constant and the results are shown in **Table 2** on the basis of experiments conducted on GPGPU-SIM [16] simulator.

On average, [12] achieved 20% improvement in power constraint environment. In addition to core/memory frequency, the method proposed in [12] also varies the number of active core. Instead of adding one more parameter, the proposed method does not have robust runtime mechanism to deal with all scenarios. However, the benefits of implementing DVFS on GPU without describing the detailed process of the runtime system [17] are shown. The scheme do not focus on memory side DVFS at all as DVFS can be useful for embedded GPUs.

A power management approach is presented in [10] that takes a unified view of the CPU-GPU DVFS to reduce the power consumption of latest 3D mobile games on android platform as compared to independent CPU-GPU DVFS based power management approach. The main objective of scheme is to provide expected frame per second for games while reducing the power consumption. Besides Asphalt 7, high end android games like Anomaly 2, Call of Duty, Need for Speed Most Wanted, Final Strike, Real Football 2013 and AVP are used. In [10], the authors examined that increasing the GPU frequency has no impact on the frame per second. Instead, they employ the concept of CPU COST and GPU COST [18] which is given by:

Table 1. Kernel categories and application kernels.

S. No	Kernel Category	Kernel
1	Compute intensive	Dense matrix multiplication
2	Memory intensive	Dense matrix transpose
3	Hybrid	Fast Fourier transform

Table 2. Throughput improvement under various uses case and power constraints.

Method Adopted	Throughput Improvement	Power Constraint
Appropriately choosing the number of operating cores and their voltage/frequency for a given application.	29%	No Power Constraint
Changing the number of operating cores and the voltage/frequency of on-chip interconnects/caches for a given application	13%	No Power Constraint
Vary the number of operating cores and the voltages/frequencies of both cores and on-chip interconnects/caches.	10%	Power Constraint
Vary the number of operating cores and the voltages/frequencies of both cores and on-chip interconnects/caches every 20 μs within the power constraint	38%	Power Constraint

$$\text{Cpu cost} = \text{Cpu utilization} * \text{frequency} \qquad (2)$$

$$\text{Gpu cost} = \text{Gpu utilization} * \text{frequency} \qquad (3)$$

Proposed integrated approach is able to reduce power consumption of 3-D games by up to 26% for comparable frame per second range.

A broad study of GPU DVFS conducted on 37 benchmark kernel is presented in [13]. The scheme not only increase performance by 4% but also conserves energy by 19.28% and it is shown that frequency scaling is effective approach to save the energy. By scaling down the core frequency, run time energy can also be saved. The scheme is compared with performance matrices obtained from default setting.

Matrices $\hat{R}$ and $R_{\max}$ are used to evaluate energy conservation in [13].

$$\hat{R} = 1 - E_{\min} / \hat{E} \qquad (4)$$

$\hat{E}$ = Energy consumption at default GPU Configuration.

$E_{\min}$ and $E_{\max}$ are the minimum and maximum energy consumption under different voltage/frequency setting for a given application.

$$R_{\max} = 1 - E_{\min} / E_{\max} \qquad (5)$$

Core scaling and memory scaling does not work well for every application kernel and some application kernel gives best result at default frequency settings.

Ge *et al.* [14] applied frequency scaling on both CPU and GPU with three typical parallel applications. They found that scaling GPU frequency higher would not consume more energy [14]. To investigate the impact of DVFS, they use following four classes of the performance metrics:

1. Performance
2. Power
3. Energy
4. Energy Efficiency

Experiments are performed on Tesla K20 series GPU form the family of Keplers architecture that support power management and power accounting features. The scheme presented in [14] not only concentrate on GPU energy but also focus on system lever energy and concluded that GPU DVFS affect system energy less as compared to CPU DVFS.

3.2. Schemes Using DVFS with Other GPU Optimization (Hybrid DVFS)

DVFS shown in the previous section suffers from major energy/performance trade-off issues. If not intelligently selected, it may affect performance/energy or both. Therefore, researchers combine DVFS with some other optimization techniques to further improve the performance as well as energy efficiency of running kernel. Over the years, considerable work has been proposed in [9] [17] [19]-[22] to improve the energy efficiency by adding some more optimization approaches to DVFS.

Liu *et al.* [19] proposed power aware time sensitive mapping technique for heterogeneous system that are able to meet application timing requirement while reducing power consumption of applying DVFS on both CPU and GPU. The scheme is executed in three phases as shown in **Figure 2**.

Assignment phase is responsible to assign application to processor like GPU/CPU. Thereafter, Load Balancing phase will manage the Load among CPU and GPU. Finally, DVFS phase scale the frequency as per requirement while meeting all the deadlines. Assignment phase calculate the heterogeneous ratio for each of application to take the assignment decision. Heterogeneous ratio is given by H_i

$$H_i = \max \left\{ \frac{e_i^c}{e_i^g}, \frac{e_i^g}{e_i^c} \right\} \qquad (6)$$

where e_i^c is the worst case execution time of ith workload on CPU under maximum voltage and e_i^g is worst case execution time of ith workload on GPU under maximum voltage.

If $\dfrac{e_i^c}{e_i^g} > 1$, applications are more suitable to run on GPU than on CPU otherwise CPU will be more suitable.

Phase-1	Assignment Phase
Phase-2	Load Balancing
Phase-3	DVFS for GPU & CPU

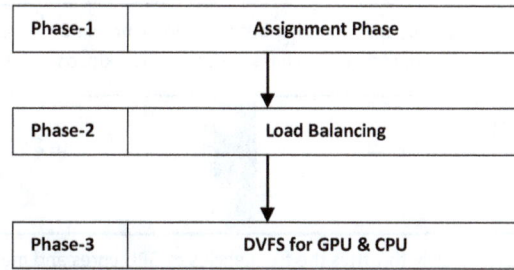

Figure 2. Different phases of equalizer.

With the proposed method in [19], Liu *et al.* managed to save energy more than 20%. Although they develop algorithm for time sensitive applications like data analysis, stock trading, real time scoring of bank transaction, live video processing etc., but does not experiment with them. Behavior of proposed method should be investigate with memory/core bounded applications.

There has been lot of work done on saving energy consumption of either CPU or GPU but, the work in isolated manner cannot achieve maximum performance. Ma *et al.* [20] proposed an energy-management framework known as GreenGPU for GPU-CPU heterogeneous architecture. The framework presented 2-Tier design framework for saving energy as shown in **Figure 3**. As an example, the workload share of CPU and GPU may be 15% and 85% respectively. Tier-1 ensure load balancing which avoid the energy-waste due to idling.

Tier-2 adjusted the frequency of GPU cores and memory is adjusted along with the frequency and voltage of the CPU to achieve largest possible energy savings with marginal performance degradation. However, [20] use DVFS and workload division individually, and so their method cannot set optimal parameters of DVFS and task mapping [22], otherwise existing marginal performance degradation can be improved. An efficient power capping technique through coordinating task mapping and DVFS in a CPU-GPU heterogeneous system is proposed in [22]. The proposed empirical model predict execution time and power consumption of heterogeneous system in order to avoid power violation and load imbalance between the CPU and GPU. The scheme was using benchmark application form Rodinia and form BLAS library under the power constrained of 200 w, 220 w, 240 w, 260 w, 280 w. Their proposed Power model is represented by

$$p_{node} = p_{idle}\left(f_{cpu}, f_{gpu}\right) + p_{cpu}\left(f_{cpu}, f_{gpu}, r_{cpu}\right) + p_{gpu}\left(f_{cpu}, f_{gpu}, r_{gpu}\right) \tag{7}$$

where p_{idle} is represent idle power consumption which is dependent on frequency of CPU (f_{cpu}) and GPU (f_{gpu}) but independent of task mapping function. On the other hand, power consumption of CPU (p_{cpu}) and GPU (p_{gpu}) is dependent on the percentage of task mapping along with the frequencies. The proposed power capping techniques can achieve more than 93% of performance as compared to the ideal one. In [17], Sethia *et al.* proposed a runtime system known as Equalizer that provides adaptive approach so that the hardware will best match the needs of running kernel. Equalizer was designed to work on two modes:
1. Energy efficiency mode
2. High performance mode

Working of Equalizer can easily understand with the help of **Figure 4**. In energy efficient mode, equalizer throttled the frequency of underutilized resources. No performance degradation reported, in fact it saves 15% energy by improving the performance by 5%. On the other hand, in high performance mode only bottleneck resource is boosted by scaling up the frequency to provide higher performance.

On the cost of 6% extra energy consumption, this mode achieves 22% performance improvement. To achieve this improvement in both the modes, equalizer tunes three major architectural parameters: No of Concurrent Thread, Core Frequency, and Memory Frequency according to the mode selected. As per the requirement, Equalizer tunes these three parameters. Wang and Nagarajan [9] proposed a feedback controlling algorithm, known as Proportional integral derivative dynamic frequency scaling (PIDDFS) to scale the core and memory frequencies for GPU architecture. PIDDFS minimized the energy consumption for the memory intensive applications. This technique basically targeted to memory bound application and can be able to save more power if application is memory bounded. The reason is: when memory access intensity is higher, low frequency period maintained by PIDDFS will be more and power saving during that time will be more. The technique was simulated

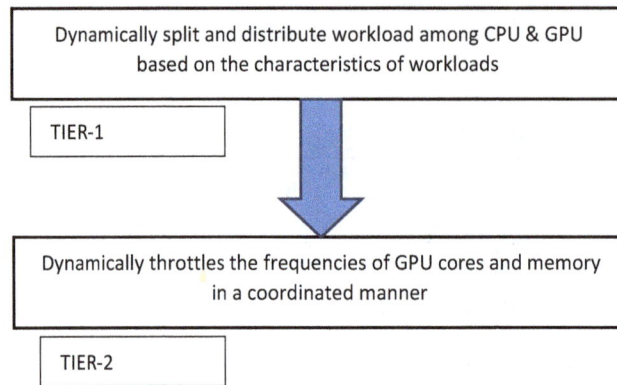

Figure 3. Green GPU two tier design architecture [20].

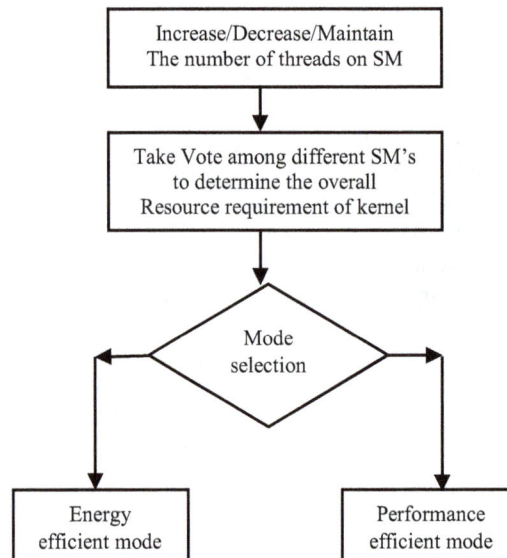

Figure 4. Flow chart of equalizer [17].

with GPGPU-SIM [16] in coordination with power model GPUWattch [23]. Simulation results show that application gain 23% on average power saving with performance improvement of average 4% for all benchmarks. Although proposed method is less complicated and has cross platform adaptability but give best result only for the memory bound applications. The authors compared PID based approach with only CPU DVFS rather than GPU DVFS.

Since the authors in this field usually applied DVFS on single GPU, Ren *et al.* [21] proposed a method that apply DVFS on CPU and load balancing on multiple GPUs. In this scheme, they identify that the power consumption behavior of the application is highly dependent on the underlying design of the algorithm. The scheme used the algorithmic level power model to predict the execution time and power related parameters. The method converts instruction mixture information, pipelining structure and out of order processing in SIMD flow, so that it can be measured in optimum accuracy. This is the only document that used the CPU DVFS technique with load balancing in multiple GPU and successfully saved the 4.4% energy. The scheme used CPU DVFS only for improving the executing time and load balancing technique to improve the energy consumption of applications. Sufficient overhead was saved by avoiding the GPU DVFS. The authors created three scenarios depicted in **Table 3** and test their proposed algorithm model.

The scheme demonstrated that intelligent use of GPU parallelization, CPU frequency scaling and power load scheduling methods will improve the performance of application while reducing the energy consumption of processing elements in multiple GPU platforms. Wu *et al.* [24] proposed a machine learning based power estimation model that learn itself to scale application according to different hardware configuration. The ultimate objective of their research is to predict the power and performance of GPUs across a range of hardware configuration. The machine learning algorithm requires training data set which is formed by varying the number of computing core, frequency of core and memory. Although DVFS not directly used but varying the frequency of core and memory are used to get the performance metrics. Total 448 training sets are acquired by varying the range of eight computing unit (4, 8, ..., 32) eight core frequencies (300, 400, ..., 100 MHz) and eight memory frequencies(475, 625, ..., 1375 MHz).

4. Comparative Study

DVFS can be used either in isolated manner or in coordination with some other techniques. As shown in **Table 4**, energy/performance improvement depends on type of application kernel and platform chosen. DVFS can be

Table 3. Measurement result under three power aware CPU-GPU configuration [21].

	Load Balancing	DVFS (CPU ONLY)	Execution Time Improvement	Energy Improvement
CPU + GPU	NO	YES	YES	NO
CPU + MULTIPLE GPU	NO	YES	YES	NO
CPU + MULTIPLE GPU	YES	YES	YES	YES

Table 4. Comparative analysis.

Author	Technique Used	No of Benchmark Used	Benchmark Or Application Kernel	Energy Improvement	Performance Improvement	Platform for Parallel Implementation
Core DVFS						
Lee et al. [11]	DVFS	Not specified	Not specified	65%	Not specified	Not specified
Jiao et al. [8]	DVFS	3	• Dense matrix multiplication • Dense matrix transpose • Fast Fourier transform	4%	Not specified	NVidiaGTX-280
Lee et al. [12]	DVFS	39	• GPGPU-Sim • Rodinia • ERCBench	Power constraint	20%	GPGPU-Sim (Simulate Quadro FX 5800)
Mei et al. [13]	DVFS	37	• CUDA SDK 4.1 • Rodinia	19.28%	4	NVIDIA GeForce GTX 560 Ti
Ge et al. [14] K20c	DVFS	1	• Matrix multiplication • Traveling salesman problem • Finite state machine	Not specified	Not specified	NVIDIA Tesla K20c
Hybrid DVFS						
Liu et al. [19]	DVFS with Load Balancing	4	• AMD OPENCL Sdk • IBM	20%	Performance constraint	AMD Radeon HD 5770
Ma et al. [20]	DVFS with Task Mapping	9	• Rodinia	21.04%	Marginal performance degradation	NVIDIA GeForce 8800 GTX GPU
Komoda et al. [22]	DVFS with Task Mapping	25	• Rodinia • BLAS Library	Power constraint	93%	NVIDIA Tesla K20c
Sethia and Mahlke [17]	DVFS with Vary No of Thread	27	• Rodinia • Parboil	15% (Energy efficiency mode)	20% (Performance mode)	GPGPU-Sim (Simulate GTX480)
Wang & Nagarajan [9]	DVFS with PID	12	• CUDA Sdk	23%	4%	GPGPU-Sim (Simulate GTX480)

designed to improve either performance [19] or energy [12] or both [9] [12] [13] [17] [19]. A variation in implementation platform is observed although NVidia is the favorite for researchers [14]. Presented a deep study on advance GPU K20c by varying the frequency of CPU and GPU and observe the performance/energy efficiency improvement. It is observed that applying CPU DVFS and Load balancing in multiple GPU can also improve the energy and performance efficiency [21].

5. Conclusion

In this paper, survey and analysis of several DVFS techniques aimed at analyzing and improving the energy efficiency of GPUs are presented. The key emphasis is on the need of power management in GPUs and identification of important trends in DVFS which are admirable for future study. In our study, we classify the research on DVFS into schemes using core DVFS technique and schemes using DVFS with other GPU optimization (Hybrid DVFS) and highlight the underlying similarities and differences between them. Energy efficiency and performance variation of applications running on GPU are presented in this paper such that breakthrough invention of designing Green GPUs for further research can be accomplished. In future, DVFS can be pooled with other techniques such that energy saving in an optimized way can be attained and electric bill as well as carbon footprint of IT infrastructure can be reduced.

References

[1]   TOP500 Supercomputing Sites. http://www.top500.org/
[2]   The Green500 List. http://www.green500.org/lists/2010/11/top/list.php

[3]   Mittal, S. and Vetter, J.S. (2014) A Survey of Methods for Analyzing and Improving GPU Energy Efficiency. *ACM Computing Surveys*, **47**, 1-23. http://dx.doi.org/10.1145/2636342

[4]   Hong, S. and Kim, H. (2010) An Integrated GPU Power and Performance Model. *ACM SIGARCH Computer Architecture News*, **38**, 280. http://dx.doi.org/10.1145/1816038.1815998

[5]   Cebri'n, J.M., Guerrero, G.D. and Garcia, J.M. (2012) Energy Efficiency Analysis of GPUs. 201*2 IEEE 26th International Parallel and Distributed Processing Symposium Workshops & PhD Forum*, Shanghai, 21-25 May 2012, 1014-1022. http://dx.doi.org/10.1109/ipdpsw.2012.124

[6]   Dave, A., Dykes, J. and Riedle, E. (2003) More than an Interface-SCSI vs. ATA. *FAST'03 Proceedings of the 2nd USENIX Conference on File and Storage Technologies*, 2003, 245-257.

[7]   Hsu, C.-H. and Kremer, U. (2002) Compiler-Directed Dynamic Voltage Scaling for Memory-Bound Applications. In: Hsu, C.-H. and Kremer, U., *Compiler-Directed Dynamic Voltage Scaling for Memory-Bound Applications*, Technical Report DCS-TR-498, Department of Computer Science, Rutgers University, New Brunswick/Piscataway, Camden and Newark.

[8]   Jiao, Y., Lin, H., Balaji, P. and Feng, W. (2010) Power and Performance Characterization of Computational Kernels on the GPU. *2010 IEEE/ACM International Conference on & In Conference on Cyber, Physical and Social Computing (CPSCom) Green Computing and Communications (GreenCom)*, Hangzhou, 18-20 December 2010, 221-228. http://dx.doi.org/10.1109/greencom-cpscom.2010.143

[9]   Wang, Y. and Ranganathan, N. (2014) A Feedback, Runtime Technique for Scaling the Frequency in GPU Architectures. 2014 *IEEE Computer Society Annual Symposium on VLSI*, Tampapp, 9-11 July 2014, 430-435. http://dx.doi.org/10.1109/isvlsi.2014.34

[10]   Pathania, A., Jiao, Q., Prakash, A. and Mitra, T. (2014) Integrated CPU-GPU Power Management for 3D Mobile Games. *Proceedings of the the 51st Annual Design Automation Conference on Design Automation Conference*, 2014, 1-6. http://dx.doi.org/10.1145/2593069.2593151

[11]   Lee, J., Nam, B.-G. and Yoo, H.-J. (2007) Dynamic Voltage and Frequency Scaling (DVFS) Scheme for Multi-Domains Power Management. 2007 *IEEE Asian Solid-State Circuits Conference*, 12-14 November 2007, Jeju, 360-363.

[12]   Lee, J., Sathisha, V., Schulte, M., Compton, K. and Kim, N.S. (2011) Improving Throughput of Power-Constrained GPUs Using Dynamic Voltage/Frequency and Core Scaling. 2011 *International Conference on Parallel Architectures and Compilation Techniques*, *PACT*, Galveston, 10-14 October 2011, 111-120. http://dx.doi.org/10.1109/pact.2011.17

[13]   Mei, X., Yung, L.S., Zhao, K. and Chu, X. (2013) A Measurement Study of GPU DVFS on Energy Conservation. *Proceedings of the Workshop on Power-Aware Computing and Systems*, *HotPower* '13, Farmington, 3-6 November 2013, Article No. 10. http://dx.doi.org/10.1145/2525526.2525852

[14]   Ge, R., Vogt, R., Majumder, J., Alam, A., Burtscher, M. and Zong, Z. (2013) Effects of Dynamic Voltage and Frequency Scaling on a K20 GPU. 2013 42*nd International Conference on Parallel Processing*, Lyon, 1-4 October 2013,

826-833. http://dx.doi.org/10.1109/ICPP.2013.98

[15] Ryoo, S., Rodrigues, C.I., Baghsorkhi, S.S., Stone, S.S., Kirk, D.B. and Hwu, W.W. (2008) Optimization Principles and Application Performance Evaluation of a Multithreaded GPU Using CUDA. *Proceedings of the* 13*th ACM SIGPLAN Symposium on Principles and Practice of Parallel Programming, PPoPP* '08, Salt Lake City, 20-23 February 2008, 73-82. http://dx.doi.org/10.1145/1345206.1345220

[16] Bakhoda, A., Yuan, G.L., Fung, W.W.L., Wong, H. and Aamodt, T.M. (2009) Analyzing CUDA Workloads Using a Detailed GPU Simulator. 2009 *IEEE International Symposium on Performance Analysis of Systems and Software*, Boston, 26-28 April 2009, 163-174. http://dx.doi.org/10.1109/ISPASS.2009.4919648

[17] Sethia, A. and Mahlke, S. (2014) Equalizer: Dynamic Tuning of GPU Resources for Efficient Execution. 2014 47*th Annual IEEE/ACM International Symposium on Microarchitecture*, Cambridge, 13-17 December 2014, 647-658. http://dx.doi.org/10.1109/MICRO.2014.16

[18] Bai, Y. and Vaidya, P. (2009) Memory Characterization to Analyze and Predict Multimedia Performance and Power in Embedded Systems. 2009 *IEEE International Conference on Acoustics, Speech and Signal Processing*, Taipei, 19-24 April 2009, 1321-1324. http://dx.doi.org/10.1109/ICASSP.2009.4959835

[19] Liu, C., Li, J., Huang, W., Rubio, J., Speight, E. and Lin, X. (2012) Power-Efficient Time-Sensitive Mapping in Heterogeneous Systems. *Proceedings of the* 21*st International Conference on Parallel Architectures and Compilation Techniques, PACT* '12, Minneapolis, 19-23 September 2012, 23-32.

[20] Ma, K., Li, X., Chen, W., Zhang, C. and Wang, X. (2012) GreenGPU: A Holistic Approach to Energy Efficiency in GPU-CPU Heterogeneous Architectures. 2012 41*st International Conference on Parallel Processing*, Pittsburgh, 10-13 September 2012, 48-57. http://dx.doi.org/10.1109/icpp.2012.31

[21] Ren, D.Q., Bracken, E., Polstyanko, S., Lambert, N., Suda, R. and Giannacopulos, D.D. (2012) Power Aware Parallel 3-D Finite Element Mesh Refinement Performance Modeling and Analysis with CUDA/MPI on GPU and Multi-Core Architecture. *IEEE Transactions on Magnetics*, 48, 335-338. http://dx.doi.org/10.1109/TMAG.2011.2177814

[22] Komoda, T., Hayashi, S., Nakada, T., Miwa, S. and Nakamura, H. (2013) Power Capping of CPU-GPU Heterogeneous Systems through Coordinating DVFS and Task Mapping. 2013 *IEEE* 31*st International Conference on Computer Design* (*ICCD*), Asheville, 6-9 October 2013, 349-356. http://dx.doi.org/10.1109/ICCD.2013.6657064

[23] Leng, J., Hetherington, T., Tantawy, A.E., Gilani, S., Kim, N.S., Aamodt, T.M. and Reddi, V.J. (2013) GPUWattch: Enabling Energy Optimizations in GPGPUs. *Proceedings of the* 40*th Annual International Symposium on Computer Architecture—ISCA*'13, New York, 2013, 487.

[24] Wu, G., Greathouse, J.L., Lyashevsky, A., Jayasena, N. and Chiou, D. (2015) GPGPU Performance and Power Estimation Using Machine Learning. 2015 *IEEE* 21*st International Symposium on High Performance Computer Architecture* (*HPCA*), Burlingame, 2015, 564-576.

16

Feasibility Study of Solar Energy Steam Generator for Rural Electrification

Mouaaz Nahas[1], M. Sabry[2,3], Saud Al-Lehyani[2]

[1]Department of Electrical Engineering, College of Engineering and Islamic Architecture, Umm Al-Qura University, Makkah, KSA
[2]Department of Physics, College of Applied Sciences, Umm Al-Qura University, Makkah, KSA
[3]Solar Research Department, National Research Institute of Astronomy and Geophysics, Cairo, Egypt
Email: mmnahas@uqu.edu.sa

Abstract

In Middle East region, where there are plentiful amounts of solar radiation and great desert areas, solar energy can play a potential role in replacing conventional fuel-operated electricity generation methods with a cost-effective, sustainable solution. This paper presents a feasibility study of a low-cost solar energy steam generator for rural areas electrification. The proposed system is based on the use of trough concentrator which converts solar radiation into thermal energy in its focal line (where a receiver pipe is installed with a fluid flowing in its interior). The aim of the paper is to predict the feasibility and potential for steam generation using a stand-alone solar concentrator with a small dimension for domestic and small-scale electricity generation. The study presented here is based on modelling of the system to determine the points at which the system is expected to produce sufficient steam energy at the tube outlet to drive a steam engine for producing electricity. Results are presented in graphical forms to show the operating points and the effect of changing selected input parameters on the behavior of the system in order to set some limits (boundaries) for such parameters. Results show that among the three input design parameters selected, the tube diameter is the most dominant parameter that influences steam energy, then the tube length and finally the flow rate of the water passing through the tube. The results of this paper can provide a useful guideline for future simulation and/or physical implementation of the system.

Keywords

Solar Radiation, Trough Concentrator, Radiation Intensity, Tube Diameter, Tube Length, Flow Rate, Steam Energy

1. Introduction

A continuously rising energy demand along with increasingly limited natural resources is challenging energy suppliers, industry as well as consumers to rethink how energy can be produced and used efficiently. Energy efficiency, smart energy use, and energy savings are keys to meet this challenge in a sustainable way [1] [2].

State power grid systems supply electricity to the majority of the population living in state capitals and industrial centres [3]. It is highly uneconomical to extend the electrical power grid system into the sparsely populated regions. Hence, there are many small remote communities that need an independent source of electrical energy, especially in the Middle East region. These locations represent a significant potential for renewable energy applications. The importance of using renewable energy not only will be confined to meet the demands of remote sites, but also can contribute to the national grid, helping to meet the peak-load demand during the summer months [3].

Renewable energy is becoming the focus of concern to both oil and non-oil producing countries. Nowadays, many countries around the world are keenly interested in taking an active part in the development of new technologies for exploiting and utilizing renewable sources of energy. The main motivation as to why renewable energy is given much attention is because of its contribution to reduce harmful emissions to the environment, especially carbon dioxide. There are rising concerns around the globe over the high oil and gas prices because of growing demand as well as the aspect to reserve oil for the next generation [4]. In countries located within the equatorial "Sun Belt" (where more solar radiation hits the earth than any other part of the globe), there is a massive amount of freely available solar energy which can be exploited. Besides receiving a lot of solar energy, desert countries in the Middle East have other competitive advantages when it comes to the potential of developing solar energy markets and technology. For instance, there can be lots of open lands, and more importantly, lots of sand which might contain a high percentage of silicon, the starting material for silicon solar "photovoltaic" (PV) cells and panels as well as semiconductor chips. Also, such countries have a relatively fast-growing young and educated population, many of whom are looking for good private sector jobs and careers (for further details, see [2]).

Middle East, Arabia and Gulf area present very high solar radiation potential especially Direct Normal Irradiance (DNI), *i.e.* the fraction of solar radiation which is not deviated by clouds, fumes or dust in the atmosphere and that reaches the Earth's surface as a parallel beam [5]. The regional DNI map is shown in **Figure 1**. Clearly, DNI in such region is amongst the highest values in the world. Moreover, in this region, there are fewer restrictions in space available due to desert areas, while some of the rural areas have not been electrified yet or are under electrification with decentralized ways of connection. The large space combined with the abundant solar resources has made this region one of the promising areas for the installation of solar energy plants for providing electricity [6].

Rural electrification is a global challenge in the developing countries especially those whose area is huge and has low population in scattered communities or tribes as is the case in most countries of the mentioned region. The socio-economic development processes revolve around suitable and sustainable power supply. In fact, it is the nucleus of operations and subsequently the engine of growth for all sectors of the economy. It also determines the living standard of the people and stops the immigration to urban areas as well [7].

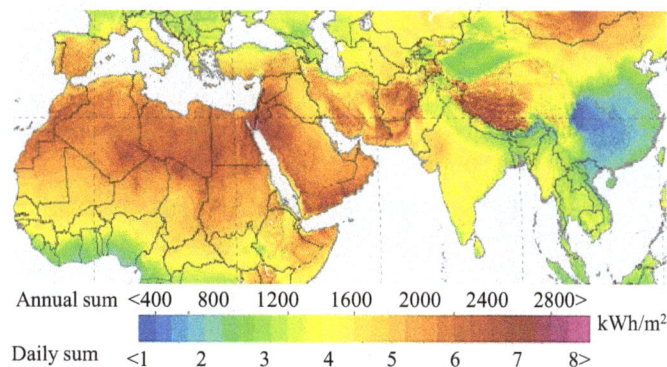

Figure 1. Solar radiation distribution over the Middle East, Arabia and gulf.

Amongst various renewable resources, solar energy could contribute in solving energy-deficiency problems like using electric-powered wells to obtain clean water for domestic use and/or some related activities in such rural communities and tribes.

Solar radiation incident could be concentrated using different imaging or nonimaging solar concentrators like Lenses, Parabolas, Troughs, etc. The only sunlight component that can be concentrated is the "Direct Normal Irradiance" (DNI) component—those rays which come directly from the sun without any scattering by dust or sands suspended in the sky. The other component, the diffuse solar radiation component cannot be concentrated because it occurs due to scattering by suspended particles in the sky. Increasing the percentage of direct solar radiation means that one can use solar concentrators effectively.

Concentrated sunlight has been used to perform useful tasks since long time ago. History mentions that the first one who used concentrated sunlight was Archimedes who used it on the invading Roman fleet and repelled them from Syracuse [8]. In 1866, Auguste Mouchout used a parabolic trough to produce steam for the first solar steam engine.

The first patent for a solar collector was obtained by the Italian Alessandro Battaglia in Genoa, Italy, in 1886. Over the following years, inventors such as John Ericsson and Frank Shuman developed concentrating solar-powered devices for irrigation, refrigeration, and locomotion. In 1913, Shuman finished a 55 HP parabolic solar thermal energy station in Meadi, Egypt for irrigation.

Giovanni Francia (1911-1980) designed and built the first concentrated-solar plant which entered into operation in Sant'Ilario, near Genoa, Italy in 1968. This plant had the architecture of today's concentrated-solar plants with a solar receiver in the centre of a field of solar collectors. The plant was able to produce 1 MW with superheated steam at 100 bar and 500 degrees Celsius [9].

Different types of concentrators produce different peak temperatures and correspondingly varying thermodynamic efficiencies, due to differences in the way that they track the sun and focus light. New innovations in "Concentrated Solar Power" (CSP) technology are leading systems becoming increasingly more cost-effective.

CSP systems use mirrors or lenses to concentrate a large area of DNI onto a small area. Electrical power is produced when the concentrated light is converted into heat, which drives a steam turbine that is connected to an electrical power generator [10].

The main focus of this paper is to study the feasibility of developing an electrical generator system based on the use of efficient solar concentrator. The solar concentrator is mainly used for heating the fluid that will produce the steam (vapour) through the receiving of solar radiation. Our study will involve determining a particular set of concentrator parameters that can be used to design the sought system, keeping in mind that the system should be able to work efficiently in regions located near the equatorial (*i.e.* Middle East) as well as being simple, safe, portable, and cost-effective. Before determining the design parameters of our proposed concentrator-based system, it is important to understand its structure and how it works. This is carried out in the next section.

2. Solar Energy Steam Generator System

This section describes—in brief—the proposed solar energy electrical generator system for which the feasibility study detailed in this paper is carried out. The proposed system idea is based on collecting solar energy by using solar concentrator which concentrates solar radiation on its focus by using a stationary nonimaging trough horizontal concentrator as shown in **Figure 2**.

Concentrators can absorb perpendicular incidence and scattered radiation in the received range causing the work temperature to reach 250°C or even higher. An example of trough solar concentrator is the Compound Parabolic Concentrator (CPC) having a pipe set in the focus as shown in **Figure 3(a)**. Optical concentration ratio of a solar concentrator X_c is defined as [10]:

$$X_c = \frac{\text{Collector aperture width}}{\text{Receiver diameter}} \tag{1}$$

In the system proposed, a copper tube is situated exactly on the concentrator focus, which is heated by means of the concentrated solar radiation falling homogeneously over its external surface. The tube inlet is connected to a liquid reservoir, which passes through the tube till it reaches the tube outlet.

The liquid has to be chosen with a low boiling point such that when passing through the hot tube, its temperature increases till it reaches boiling point and converts to steam with relatively high pressure (hence speed)

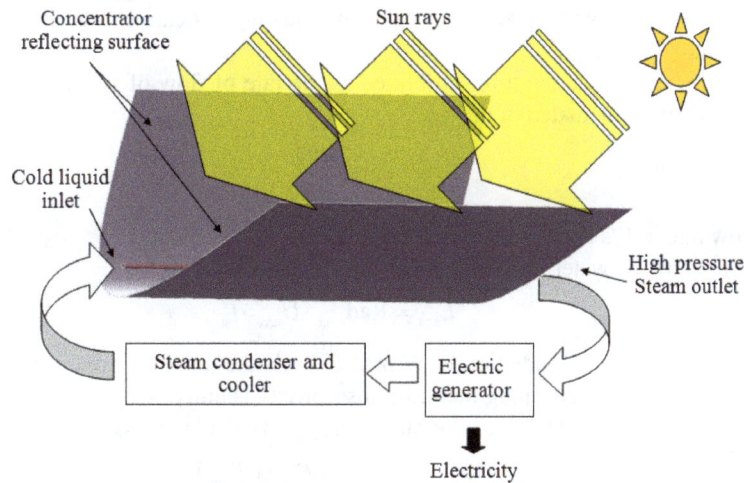

Figure 2. Flow diagram of the proposed system.

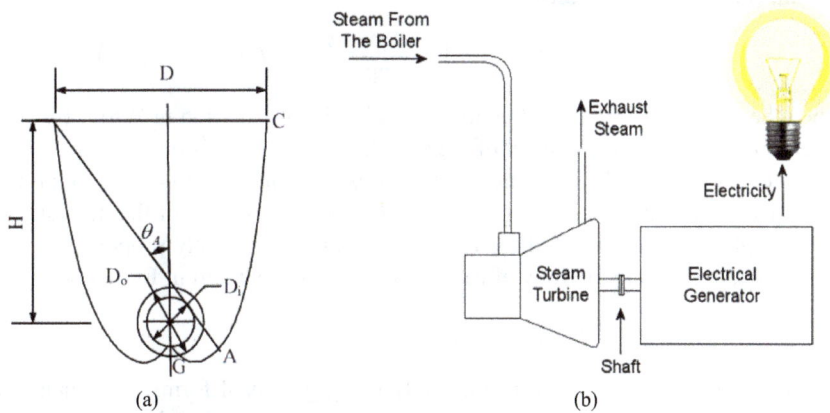

(a)   (b)

Figure 3. (a) Cross sectional view of a CPC; (b) Steam generator model.

before reaching the tube outlet. Generated high pressure steam is then directed to a steam turbine, which rotates generating electricity for use in various applications (see **Figure 3(b)**).

The study presented here is based on mathematical modelling of the system to determine the parameter values at which the system would produce sufficient amount of power (*i.e.* sufficient steam quantity which will rotate the turbine).

3. Methodology and Mathematical Models

The study presented in this paper was carried out using MathCAD to develop mathematical models (equations) for calculating both energy absorbed by the water flowing in the tube and energy of the steam generated at the tube outlet to investigate the generated steam quantity and energy.

The input parameters examined in this study are:

1) Incident solar radiation intensity.

2) Diameter of the tube (0.005, 0.01 and 0.015 m).

3) Length of the tube (1, 2, and 3 m).

4) Flow rate of water inside the tube (15, 10, and 7 kg/hr).

Then, various graphs were generated to define the points at which the system would operate effectively (*i.e.* points at which the system is expected to produce sufficient steam energy at the tube outlet to drive the steam engine). Graphs were also used to demonstrate the effect of changing each parameter on the behaviour of the system in order to set some limits (boundaries) for the input design parameters (more details are provided in Section 4).

In this section, we develop some equations to calculate the output steam energy in terms of the various input system parameters stated above.

The velocity of the water inside the tube is defined as the rate of flow of water over a specific area. Mathematically, the velocity vel is calculated in m/s as:

$$\text{vel} = \frac{4\text{FR}}{\pi d^2} \tag{2}$$

where FR is the flow rate in L/s and d is the diameter in m.

The energy absorbed by the water inside the tube E_{abs} is calculated as:

$$E_{\text{abs}} = \text{Rad}_{\text{conc}} \cdot U_{\text{area}} \cdot t_p \tag{3}$$

where Rad_{conc} is the concentrated radiation in W/m^2, U_{area} is the unit side area in m^2, and t_p is the passage time of the water inside the tube; assuming that tube absorptivity is unity.

The energy of the steam generated at the tube outlet E_{steam} is calculated as:

$$E_{\text{steam}} = E_{\text{abs}} - \left(E_{\text{boil}} + E_{\text{ltnl}}\right) \tag{4}$$

where E_{boil} and E_{ltnt} are the boiling energy and latent energy of water (respectively).

Therefore, steam energy E_{steam} is calculated as:

$$E_{\text{steam}} = \pi^2 d^2 \cdot X_c \cdot \text{Int} \cdot \frac{\text{len}}{4\text{FR}} - \frac{\pi}{4} \cdot \rho \cdot d^2 \left(C_p \Delta T - E_{\text{ltnt}}\right) \tag{5}$$

where d is the tube diameter in m, Int is the intensity of incident radiation in W/m^2, FR is the flow rate in L/s, len is the tube length in m, ρ is the density of water in kg/m^3 and C_p is the specific heat capacity of water in J/kg·K. The collector aperture width is assumed to be a constant value of 1 m throughout this study.

Clearly from Equation (5), the steam energy is proportional to the square of the tube diameter, whereas its relation to the tube length and flow rate is directly proportional and inversely proportional (respectively). An intuitive schematic diagram showing the flow of calculations is demonstrated in **Figure 4**.

4. Results

This section presents the results obtained in this study using graphical forms. The main aim of the presented graphs is to show the effect of various input parameters (*i.e.* tube length, tube diameter and flow rate) on the output of the system, namely the steam energy. The graphs are also used to determine the points at which the system is expected to produce sufficient steam energy at the tube outlet to drive a steam engine for producing electricity. From such graphs, it is possible to set some limits (boundaries) for the input parameters for practical implementation and/or simulation of the system.

We begin by showing the effect of tube diameter and tube length on the velocity and/or passage time of the water traveling in the tube. This is to begin to understand how such parameters will affect the energy of the steam produced at the tube outlet. **Figure 5** shows the effect of tube diameter on the velocity and passage time of the water flowing in the tube for different flow rates. It is clear that as the tube diameter increases the velocity decreases and the passage time increases. Also, increasing flow rate results in increasing velocity and reducing passage time at each tube diameter.

Figure 6 shows the effect of tube diameter on the passage time of the water for different tube lengths. The figure clearly shows how the increase of tube length results in increasing the passage time at each tube diameter.

Figure 7 shows the effect of flow rate on the energy of steam generated for different tube. Clearly when flow rate increases, the energy of the produced steam decreases since the water does not spend enough time to heat up while traveling in the tube. Moreover, for different tube lengths, the steam energy will increase as the tube length increases at a given flow rate. This is simply because the water will travel for longer period of time in the tube and hence absorb more energy while traveling. The figure also shows the flow rate points above which the system will produce steam for the different tube lengths considered. For example, when using 1 m tube, only the three lowest flow rates will produce steam. As the tube length increases the steam will be produced with the higher flow rate values. For example, with 3 m tube, all flow rates considered here are expected to produce steam at the tube outlet. Note that the zero-line shown in the graph presents the threshold level above which the system is expected to produce steam and under which it will not produce any steam. This threshold value de-

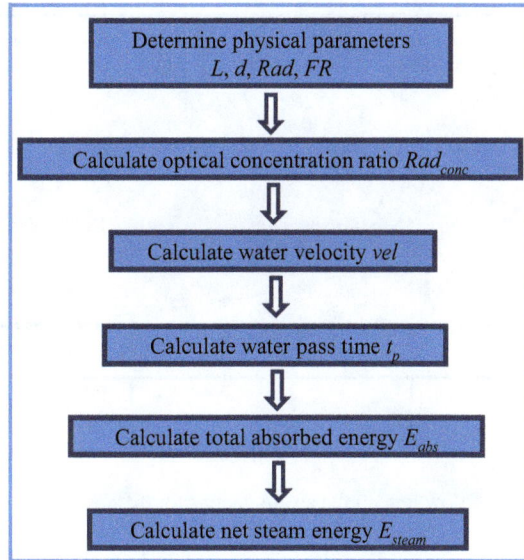

Figure 4. Intuitive schematic diagram.

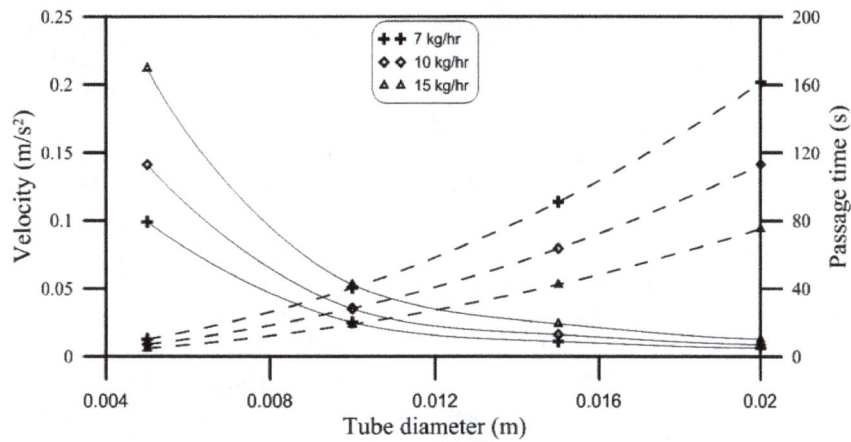

Figure 5. Effect of tube diameter on the velocity (solid) and passage time (dotted) for different flow rates.

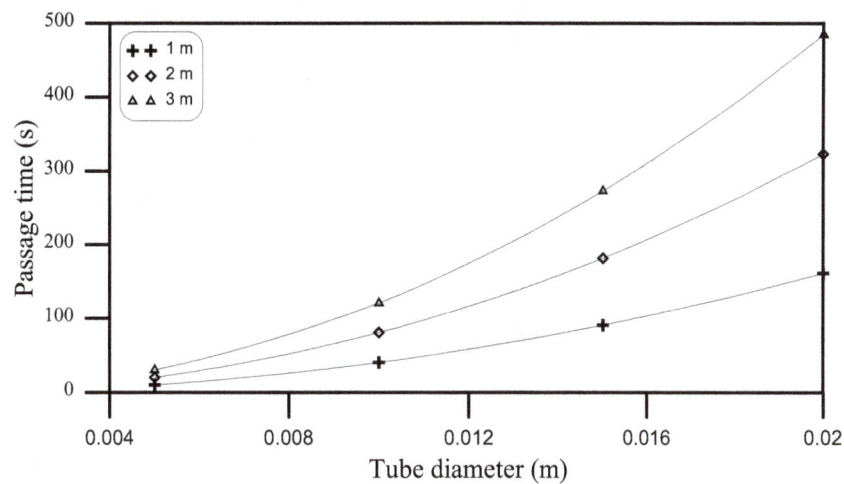

Figure 6. Effect of tube diameter on the passage time for different tube lengths.

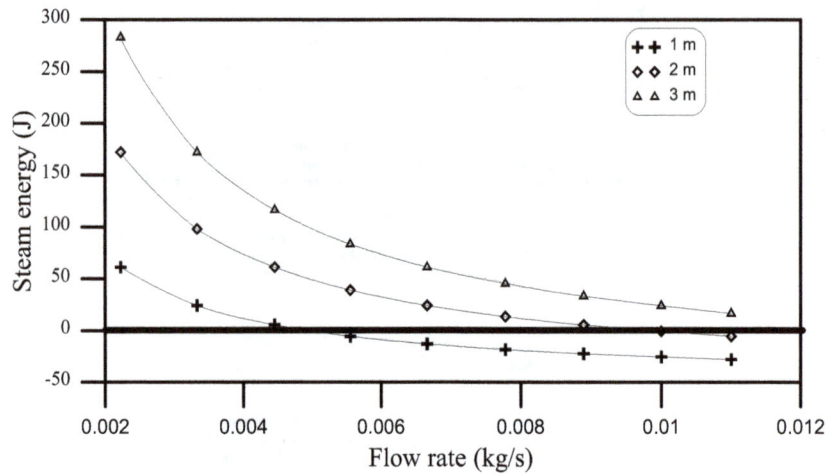

Figure 7. Effect of flow rate on the steam energy for different tube lengths.

pends on concentration ratio X_c, solar radiation intensity Int and rise in the inlet temperature ΔT. At this level, the absorbed energy is equal to the sum of boiling energy and latent energy of water, hence, the steam energy becomes zero; see Equation (4) above.

Figure 8 shows the effect of flow rate on the energy of steam generated for different tube diameters when the tube length is fixed to 1 m. Also here, it is clearly shown that as tube diameter increases, more energy will be produced from the system at a given flow rate. However, when increasing the tube diameter, steam energy grows faster than the case of increasing the tube length (compare with **Figure 7**). This is simply because the steam energy is squarely proportional to the tube diameter while it is directly proportional to the tube length, as in Equation (5). Moreover, it is clear that with all tube diameters considered, only flow rates below 0.004 kg/s will produce steam. Again this depends on the other operating conditions like tube length, incident radiation intensity and inlet temperature.

Figure 9 shows the effect of radiation intensity on the energy of steam generated for different flow rates.

Here, the tube length and diameter are set to 1 and 0.005 m (respectively) with an inlet temperature of 30°C. Obviously, steam energy increases linearly as radiation intensity increases. Moreover, as flow rate decreases, steam can be produced by lower radiation intensity values. For example, using 7 kg/hr flow rate, the system will produce steam at all radiation intensities considered except at the lowest one (which is 100 W/m²). Obviously, this is due to the low velocity and high passage time of the water inside the tube which makes it possible to convert into steam even with low radiation intensities. In contrast, for the 15 kg/hr flow rate (which is relatively high, resulting in high velocity and low passage time), the minimum radiation intensity needed to produce steam is 400 W/m². With lower intensity values, the water inside the tube will not absorb sufficient energy to convert into steam before reaching the tube outlet end.

Figure 10 shows the effect of radiation intensity on the energy of steam generated for different tube diameters. Here, tube length is set to 1 m and water flows with a rate of 15 kg/s. Again, steam energy increases linearly as radiation intensity increases. However, when increasing the tube diameter, steam energy grows faster than the case of increasing the flow rate (compare with **Figure 9**). Recall that the steam energy is squarely proportional to the tube diameter while it is inversely proportional to the flow rate, as in Equation (5). Also from the graph, with all tube diameters considered, the minimum radiation intensity needed to produce steam is 400 W/m² at the abovementioned operating conditions.

To investigate the system's performance under realistic operating conditions, a daily profile of solar radiation intensity has been chosen along with ambient temperature. Then, steam generated from the different combinations of the abovementioned system parameters were calculated and compared.

Figure 11 shows the daily profile of radiation intensity and ambient (which is set equal to the inlet) temperature in a selected day in the concerned region.

Figure 12 to **Figure 14** show the total accumulated steam energy for all input parameters considered in this study over that selected day from 8 am to 6 pm. The aim of these graphs is to investigate the effect weight of each parameter against the other parameters. More particularly, **Figure 12** shows the total accumulated steam energy

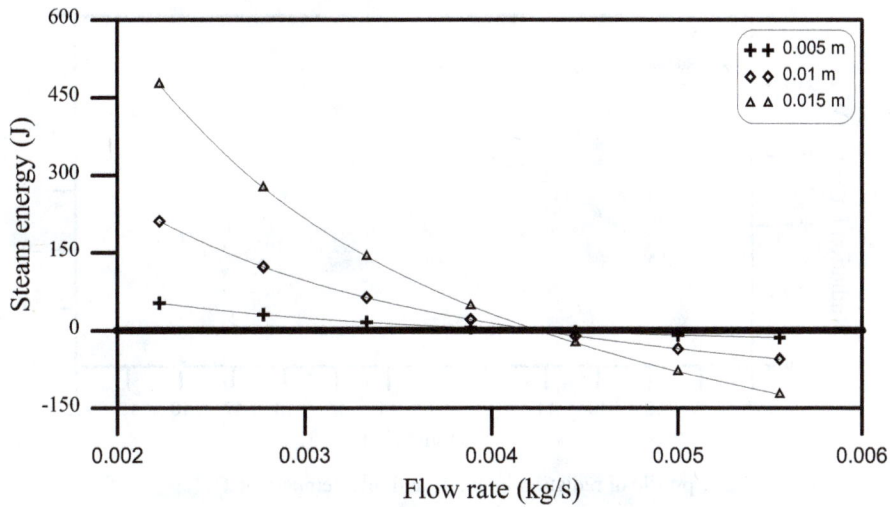

Figure 8. Effect of flow rate on the steam energy for different tube diameters.

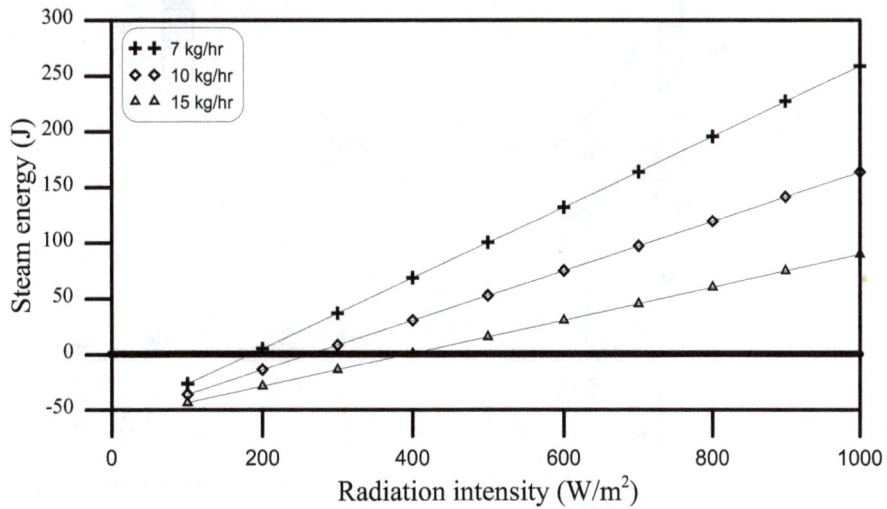

Figure 9. Effect of radiation intensity on the steam energy for different flow rates.

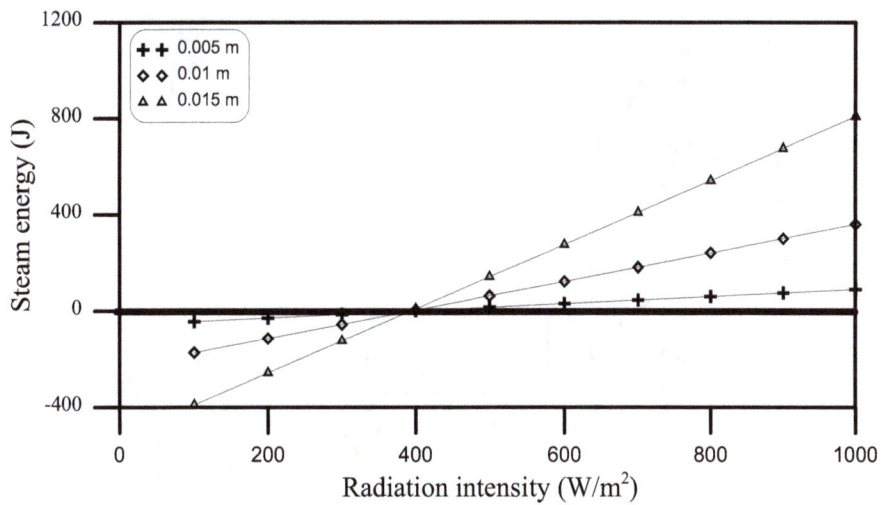

Figure 10. Effect of radiation intensity on the steam energy for different tube diameters.

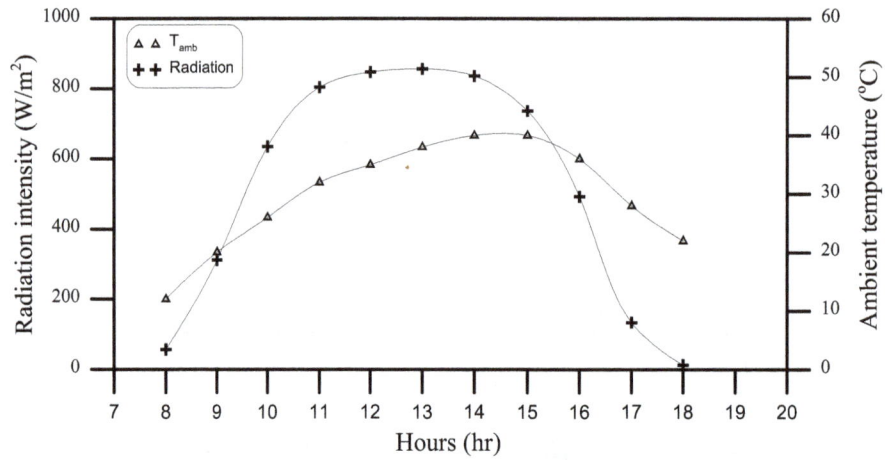

Figure 11. Daily profile of radiation intensity and inlet temperature.

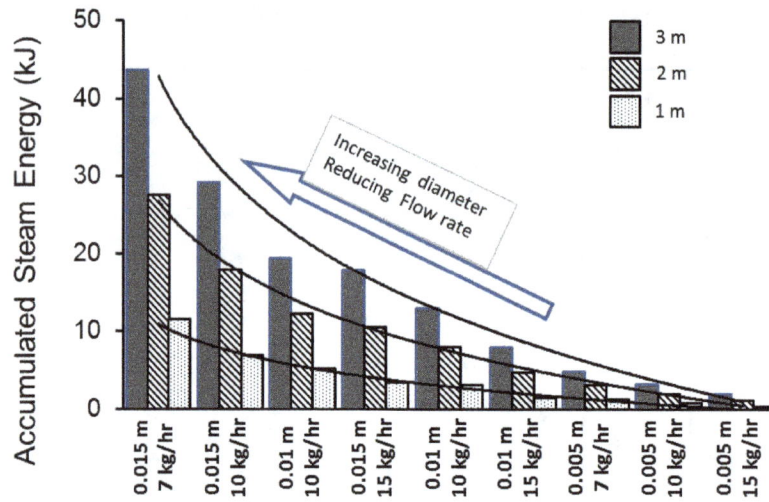

Figure 12. Accumulated steam energy for all flow rates and tube diameters considered when fixing the tube length.

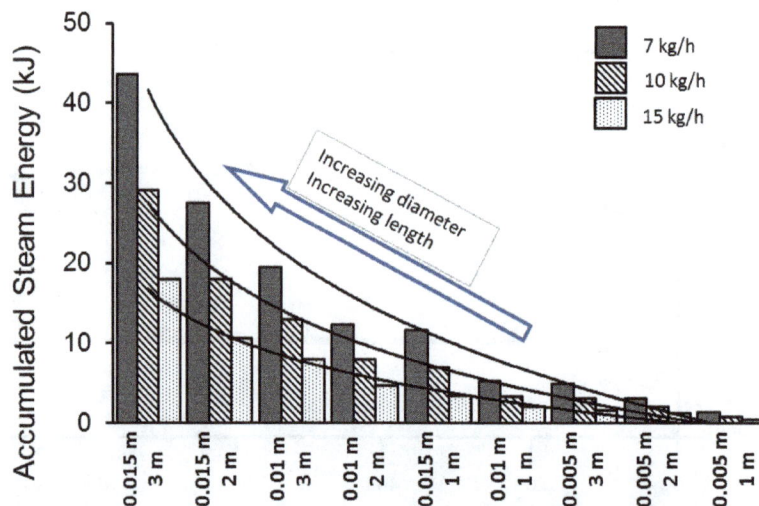

Figure 13. Accumulated steam energy for all tube lengths and tube diameters considered when fixing the flow rate.

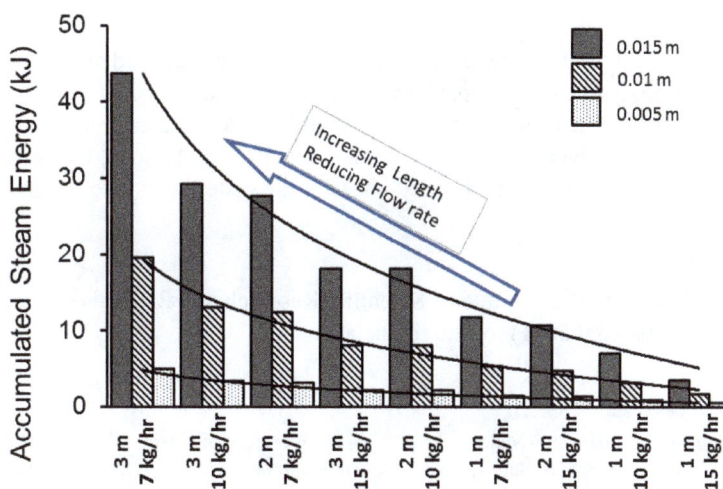

Figure 14. Accumulated steam energy for all tube lengths and flow rates considered when fixing the tube diameter.

for all flow rates and tube diameters considered when fixing the tube length. Similarly, **Figure 13** shows the total accumulated steam energy for all tube lengths and tube diameters considered when fixing the flow rate, and **Figure 14** shows the total accumulated steam energy for all tube lengths and flow rates considered when fixing the tube diameter.

The three graphs clearly show that the best steam quantity over the day is achieved with the largest tube diameter, largest tube length and lowest flow rate (and vice versa). However, by looking at the details, it can be noticed that:

1) For a given tube length, the effect of changing tube diameter overwhelms the effect of changing flow rate. Moreover, the total accumulated steam energy increases linearly with increasing the tube length.

2) For a given flow rate, the effect of changing tube diameter overwhelms the effect of changing tube length. Moreover, the total accumulated steam energy increases linearly with decreasing the flow rate.

3) For a given tube diameter, the effect of changing tube length overwhelms the effect of changing flow rate. However, the total accumulated steam energy increases nonlinearly (squarely) with increasing the tube diameter. It can also be noticed that with the very low tube diameters (such as the case of 0.005 m tube), total accumulated steam energy is not affected much by manipulating the other parameters.

5. Conclusions

The study outlined in this paper intended to investigate the feasibility of designing a small-size, stand-alone solar energy steam-based electric generator to use for domestic and small-scale electricity generation purposes. The proposed system was based on using nonimaging "Compound Parabolic Concentrator" (CPC) in which a copper tube is placed on the concentrator focus and heated up by receiving homogeneous concentrated solar radiation on its external surface. A fluid (chosen here to be water) is injected in the hot tube that will pass through it—while being heated up—till it converts from liquid to steam before reaching the tube outlet end.

The study was based on developing mathematical equations to calculate the energy of the steam produced from the concentrator system in terms of four main parameters: radiation intensity, tube length, tube diameter and water flow rate. The three parameters—tube length, tube diameter and flow rate, were considered to be the main input design parameters of the system. Graphs were then presented mainly to show the effect of changing input design parameters on the quantity of steam generated at the tube outlet. In addition, graphs were also used to determine the values at which the system is expected to produce steam so as to set initial boundaries for the input design parameters for further design processes of the system.

Overall, the results obtained demonstrate that among the three input design parameters, tube diameter is the most dominant parameter that influences steam energy, then the tube length and finally the flow rate. This implies that for achieving better steam quantity the designer shall begin by increasing the tube diameter before increasing the tube length or reducing the flow rate at last. Such results can provide a guideline for simulating and/or implementing the system in practice.

It is worth noting that, in this study, the steam produced from the proposed system was only analyzed quantitatively since qualitative analysis cannot be performed by the approach considered in this study which is based on mathematical calculations. It is therefore suggested to conduct a computer simulation of the system using appropriate Computational Fluid Dynamics (CFD) simulation software. As such, the results presented here can be used effectively in the initial design (or modelling) phase of the system that is to be simulated.

Acknowledgements

The authors would like to thank the Institute of Scientific Research and Revival of Islamic Heritage at Umm Al-Qura University (*Project ID* **43305021**) for the financial support.

References

[1] Rahman, F., Rehman, S. and Abdul-Majeed, M.A. (2012) Overview of Energy Storage Systems for Storing Electricity from Renewable Energy Sources in Saudi Arabia. *Renewable and Sustainable Energy Reviews*, **16**, 274-283. http://dx.doi.org/10.1016/j.rser.2011.07.153

[2] Al-Ammar, E. and Al-Aotaibi, A. (2010) Feasibility Study of Establishing a pv Power Plant to Generate Electricity in Saudi Arabia, from Technical, Geographical, and Economical Viewpoints. *International Conference on Renewable Energies and Power Quality (ICREPQ'*10), Granada, 23-25 March 2010,

[3] Said, S.A.M., El-Amin, I.M. and Al-Shehri, A.M. (2004) Renewable Energy Potentials in Saudi Arabia. *Beirut regional Collaboration Workshop on Energy Efficiency and Renewable Energy Technology*, American University of Beirut, Beirut, 76-82.

[4] Alnatheer, O. (2006) Environmental Benefits of Energy Efficiency and Renewable Energy in Saudi Arabia's Electric Sector. *Energy Policy*, **34**, 2-10. http://dx.doi.org/10.1016/j.enpol.2003.12.004

[5] Fernández-García, A., Zarza, E., Valenzuela, L. and Pérez, M. (2010) Parabolic-Trough Solar Collectors and Their Applications. *Renewable and Sustainable Energy Reviews*, **14**, 1695-1721. http://dx.doi.org/10.1016/j.rser.2010.03.012

[6] Tsikalakis, A., Tomtsi, T., Hatziargyriou, N.D., Poullikkas, A., Malamatenios, C., Giakoumelos, E., Jaouad, O.C., Chenak, A., Fayek, A., Matar, T. and Yasin, A. (2011) Review of Best Practices of Solar Electricity Resources Applications in Selected Middle East and North Africa (MENA) Countries. *Renewable and Sustainable Energy Reviews*, **15**, 2838-2849. http://dx.doi.org/10.1016/j.rser.2011.03.005

[7] Afa, J.T. (2013) Problems of Rural Electrification in Bayelsa State. *American Journal of Scientific and Industrial Research*, **4**, 214-220. http://dx.doi.org/10.5251/ajsir.2013.4.2.214.220

[8] Africa, T.W. (1975) Archimedes through the Looking-Glass. *The Classical World*, **68**, 305-308. http://dx.doi.org/10.2307/4348211

[9] Butti, K. and Perlin, J. (1980) A Golden Thread: 2500 Years of Solar Architecture and Technology, Vol. 514. Cheshire Books, Palo Alto.

[10] Rabl, A. (1976) Comparison of Solar Concentrators. *Solar Energy*, **18**, 93-111. http://dx.doi.org/10.1016/0038-092X(76)90043-8

17

Energy Planning: Brazilian Potential of Generation of Electric Power from Urban Solid Wastes—Under "*Waste Production Liturgy*" Point of View

Neilton Fidelis da Silva[1,2,3], Angela Oliveira da Costa[1], Rachel Martins Henriques[1], Marcio Giannini Pereira[1]*, Marcos Aurelio Freitas Vasconcelos[1,2]

[1]Energy Planning Program (PPE), Coordination of Postgraduate Programs in Engineering at the Federal University of Rio de Janeiro (COPPE/UFRJ), Cidade Universitária, Rio de Janeiro, Brazil
[2]International Virtual Institute of Global Change—IVIG, Centro de Tecnologia, Cidade Universitária, Rio de Janeiro, Brazil
[3]Federal Institute of Education, Science and Technology of Rio Grande do Norte (IFRN), Natal, Brazil
Email: *giannini@cepel.br

).

Abstract

The use of Urban Solid Waste (USW) as sources of energy has acquired rising importance in current discussions of alternative energy supplies, in particular in Brazil. This paper brings to these discussions an examination of the concept of solid wastes, including their historic origins and formation, taking their social, economic and cultural characteristics into account, including point view of *waste production liturgy*. Consequently, a spendthrift society slanted towards the decreasing marginal utility of assets must make efficient use of its USW in order to reduce excessive output. Besides that, this document presents the Brazilian potential of urban solid waste to produce electric power.

Keywords

Urban Solid Wastes, Renewable Energy, Energy Planning, Brazil

1. Introduction

From a systemic standpoint, the definition of solid wastes may be presented as the outcome of poorly balanced

---

*Corresponding author.

flows of certain elements in a specific ecological system, implying the instability of the system itself. However, once the orderly arrangements between the whole and the parts of an ecosystem have been taken into consideration through relations based on complementarity, with all the parts dependent on the life-cycles of the others, additional elements blur the definition of the concept of solid wastes, "as elements produced by the metabolisms of organisms or their life cycles could may be used as nutrients by other organisms, thus perpetrating the life of the system" [1].

Among all the many different ways of identifying an element distinguishing human beings from animals, Marx selects the capacity to produce their means of existence, a skill granted only to humans and the hallmark of this distinction [2]. Marx affirms:

"by producing their means of existence, human beings indirectly produce their material lives" [2].

However, the way in which human beings produce their "material lives" necessarily depends *"on the nature of the means of existence already found, and that they need to reproduce"* [2]. The human development process has always been closely linked to the expansion of mastery over the exploitation and use of the resources available in Nature. The imbalances imposed on the environment in the form of solid wastes are consequently the outcome of the choices made, the technological routes adopted and the speed of production and reproduction of goods.

The process of human occupancy of the land is spurred by a steadily expanding population, followed by an equally steady expansion of settled regions. Solid wastes generated as by-products of human activities soon outstrip the possibilities of dilution, regeneration and reintegration of the elements in the natural cycles of environmental change. As a result, the surrounding environment is unable to achieve satisfactory results in its attempts to absorb discarded elements into its original cycles.

During the XX century, visible alterations of a qualitative and quantitative order were imposed on urban arrangements and functions. These changes boosted the volumes of Urban Solid Wastes (USW) to a significant extent, as well as altering their composition. Prior to the XVIII and XX centuries, the world's population grew sevenfold, soaring from around one billion in the mid-XVIII century to some seven billion at the new millennium.

The problems caused by higher output of USW are not due only to population growth and concentration in major urban centers. The main culprits behind all the chaos caused by the imperative need to dispose of these solid wastes are life styles centered on the liturgy of consumption.

The development model adopted by modern society and the pace of its progress are reflected in an upsurge of the supply of goods and consequently energy consumption—which is in turn the basis of the production system. The consumption structure is shaped by life styles that define family arrangements, income levels and distribution, ownership and use of consumer goods, dissemination of heating/cooling equipment, transportation structure and housing expansion models, among other aspects.

Modern life styles have stepped up and concentrated family demands for goods through widespread use of household appliances, incentives for individual transportation, and endless appeals to squander, with new demands generated all the time, but without extending their marginal utility. Along these lines, Mészáros [3] affirms:

"The notorious 'planned obsolescence' of 'consumer durables' that are mass produced; the replacement, junking or deliberate destruction of goods and services offering an intrinsically higher potential use (collective transportation, for example), substituted by those whose usage rate tend to be far lower or even minimal (such as private automobiles) and that absorb a considerable part of the purchasing power of society; the artificial imposition of a production capacity that is almost completely unusable; the upsurge in waste resulting from the introduction of a new technology; and the deliberate 'elimination' of maintenance skills and services. All this belongs in this category, dominated by these imperatives and underlying determinations to heedlessly lower the practical usage rates".

For Mészáros [3], the decreasing usage rate law is historically endowed with a civilizatory importance:

"the movement that made two pairs of shoes available to workers instead of a single pair may certainly be rated as positive, regardless of the hidden agendas and motivations of the capitalist side."

With this same approach, Marx [2], affirms:

"Despite all 'pious', discourses, he [the capitalist] seeks ways of encouraging [the workers] to consume, trying to endow his products with new charms, triggering new needs through non-stop advertising etc. It is exactly this aspect of the relationship between capital and labor that is at one time essentially civilizatory, and on which all historical justifications are supported, such as the contemporary power of capital".

However, destructive repercussions are inherent to the decreasing use rate law—a trend driven strongly by the formation of the military-industrial complex—*"which appeared on the scene with dramatic emphasis during the XX century, particularly during its last four or five decades. As a result, the old socialist approach of warding off shortages by producing unimaginable abundance also requires radical re-examination, in the light of these same occurrences"* [3].

Within this context, it is noted that the possibilities of stepping up production, as a result of the civilizatory potential based on progress in science and technology, are distorted in terms of the adoption and consolidation of a capitalist practice whose bases are destructive and profligate. Pressured by the necessary expansion of the production segment, natural needs are thus constantly edged out by "historically created needs".

According to Mészáros [3] waste production liturgy can be defined:

"the positive output of the dialectic interaction between production and consumption is far from being assured, as the capitalistic drive to expand production is not necessarily linked at all to human needs as such, but only to the abstract imperative of 'capital achievement'".

It is a known fact that consumption styles vary drastically between the more and less developed nations, with very marked differences also apparent between urban and rural consumption profiles. However, market appeals have extended these boundaries of action, spreading an ideal of rising consumption that is driven by technological progress but hobbled by economic and regional constraints.

In today's world, there are many enthusiasts urging the benefits of USW as an alternative energy source, particularly electricity and process heat. This paper discusses what society rates as solid wastes, assessing their formation and querying their role as energy sources. The concept adopted here is that producing all materials discarded as solid wastes requires amounts of energy far higher than the total energy that could be obtained from its use as a primary source by any technology designed to tap the energy contained in USW. Thus, in a society grounded on spendthrift consumption, shorter utility spans and shrinking margins, the most efficient way of making good use of USW lies in complex efforts reining in this profligate production, compatible with the concepts underpinning the concept of sustainable development[1].

This article is divided into five sections. In Section II, the theme of urban solid wastes and the waste production liturgy are presented. Section III discusses the use of USW to generate energy in Brazil considering the potential and stored. Finally, Section VI offers conclusions and recommendations.

2. Urban Solid Wastes and the Waste Production Liturgy

The human body houses the most complex energy conversion system used by humankind. Through digestion, the chemical energy found in foods is processed into heat, as well as muscle and brain energy. Outside their own bodies, human beings work with two basic forms of energy conversion: organic converters (the use of haulage animals to produce mechanical energy, fuelwood and others) and inorganic converters that use the energy stored in the environment (electrical machinery, internal combustion engines and others).

The hallmarks of energy resources and their uses are reflected in the freedom of movement they offer through extending the reach and strength of human beings. The earliest exteriorization processes expanded the use of muscle power and the heat generated in the human body [4].

For Freud [5], this exteriorization process—consisting of the development of knowledge used in the formation of transformation capacity and control of nature—proved to be a major civilizatory trend, together with the rules and actions regulating the distribution of these created values:

"If we look back sufficiently far into the past, we discover that the earliest acts of civilization involved using

---

[1]As used here, this term refers to the adoption of a type of development that guarantees at least the existence and quality of life of current social occupations and future societies. Consequently, it differs from concepts that deploy sustainable development as a tool underpinning the feasibility of entering a new capitalist expansion phase.

tools, learning to control fire, and building shelters. Among them, controlling fire stands out as an extraordinary and unprecedented accomplishment, as others opened up paths that human beings have been following since then. Through each tool, man recreates his own motor or sensory organs, extending the scope of his functions. Motor capacity places massive forces at his fingertips that he can deploy through his muscles: thanks to ships and aircraft, neither water nor air can hamper his movements".

For Marx [6], by exteriorizing his body, "*man turns something in Nature into an organ of his own activity, an organ added to his own body organs, extending his natural body, despite the Bible*".

During the process of building up a usage model for the resources available in nature, human beings were continually exteriorizing their bodies, replacing hard-to-handle organic converters (such as human and animal traction for transportation and generation mechanical energy, natural biomass for cooking and heating, etc.), by inorganic converters that reflect the creative spirit, backed by scientific and technical progress and opening up sources that were formerly unthinkable.

Consolidating the capitalistic production mode, the Industrial Revolution was a watershed in production relations, particularly through energy systems that were structured in earlier times and had to underpin the entire goods production and re-production framework. Ending the unchallenged supremacy of biological energy sources, the Industrial Revolution paved the way for the triumphant march of hegemonic fossil fuels.

"*The steam engine rearranged the relations between man and energy. The clock, the windmill and the watermill used forces within a context that left them intact; in contrast, the 'combustion engine' consumes the materials from which it draws energy. he new lines will require heavier investments, as well as the use of increasingly vast scientific and technical knowledge. Energy will become an independent and autonomous sector that will play a decisive role in economic regulation*" [7].

The driving force behind the Industrial Revolution is credited to the development of machinery that ratcheted up production scales to well beyond the capacities of individual workers and their tools. This approach to production was guided by increasing output and shorter production times, turning valueless resources into profitable products.

Within this context, the use of coal become more important worldwide, while energy production forged steadily ahead in response to rising demands spurred by the industrialization process.

"*For XXI century capitalism, energy production acquired an unprecedented elasticity through the widespread use of non-renewable fossil fuels and the progress of transportation: from this time onwards, energy supplies tended to stay a step ahead of demands. In fact, huge energy grids share the common characteristic of structuring a new energy market, based on technical systems that include important motor activities and consequently generate new energy needs. Thus, the primacy of production over demand was consolidated, characteristic of the capitalistic energy system*" [7].

Since the XIX century, the importance of re-using consumer goods has been discussed in society. It is known that certain goods (or parts of them) can be used in other products or to make new items. The economic system in effect at that time viewed the trend towards generating waste as no more than a "deviation" of "spirit of capitalism", in relation to idealized "sensible economic principles". Thus, rising production was viewed favorably, as some advantages could be assigned to the presence of machines and capitalistic manufacturing practices. Outstanding among them is the conversion of apparently common and valueless substances into profitable products, in parallel to savings in human time through manufacturing these goods.

The importance of time savings is assessed from a positivist viewpoint, and when analyzed from the viewpoint of capital, where a minimum time is a basic target to be attained, it becomes clear that human beings have become subordinate to this. "*The same trend is revealed as a force that degrades the human being, turning him into a 'time drudge' (Marx), while appearing as a measured objective from the standpoint of capital: the ideal solution for all possible lawful disputes between capital and labor*" [3].

The rising participation of machines in the means of production requires constant updating, in order to keep them as modern as possible. The outcome of this quest for the "new" is that they become obsolete, often before the end of their useful lives. The general trend is that large-scale production spurred by competition turns out goods that are less durable, meaning it may well cheaper to purchase a new item instead of re-using it.

Goods become viewed as old when the natural wear and tear caused by time occurs. Moreover, with no reduction in their usefulness, articles become considered as "obsolete" when improvements occur in their production processes, or because they no longer comply with current consumption standards. These discarded articles then become accessible to segments of society that lacks the purchasing power to acquire them new. In general, this generates a steady stream of fresh demands without extending the marginal utility of the goods.

It is important to note that this increase in productivity is not viewed askance, and is good and desirable, according to certain standards. However, gains in productivity trigger alterations in consumption standards in a throwaway society, making it hard to pinpoint the perfect balance point between production and consumption. Although society should ideally take steps to ensure that most of its resources are channeled to the production of re-usable goods, its resources are frittered away, under pressure from dropping usage rates.

"The dropping usage rate assumes a dominant position in the capitalist structure of the socio-economic metabolism, notwithstanding the fact that astronomical quantities of wastes must be now produced in order to impose some of their most disconcerting manifestations on society" [3].

As human beings have countless needs, there are no limits on what will satisfy them, building up consumer markets that extend well beyond basic requirements. Reflection is required on the nature of these needs, as what was once a "luxury" (everything other than basic needs) is soon not rated even as a basic requirement today. As already shown, producing and scrapping goods create new demands in society, although without creating the corresponding uses. Thus, *"it does not matter how absurdly wasteful the production process may be, provided that its outcome can be imposed profitably on the market"* [3].

The generation of waste is consequently the outcome of a society with high consumption standards and dropping utilization rates, encouraging the ongoing production of goods that are (quasi) discardable. These lavish leftovers then become a burdensome nuisance for this same society (**Figure 1**).

One of the negative outcomes inherent to the development process based on the production of goods with a dropping utility rate is what society calls garbage. Within this context, the production of solid wastes necessarily arises from rising outputs of "luxury" goods. Consequently, solid wastes are the outcomes of basic human needs, compatible with the capacity of the planet to re-absorb them within an evenly-balanced population growth model. On the one hand, proper garbage recycling, treatment or disposal is a responsibility that cannot be ignored by

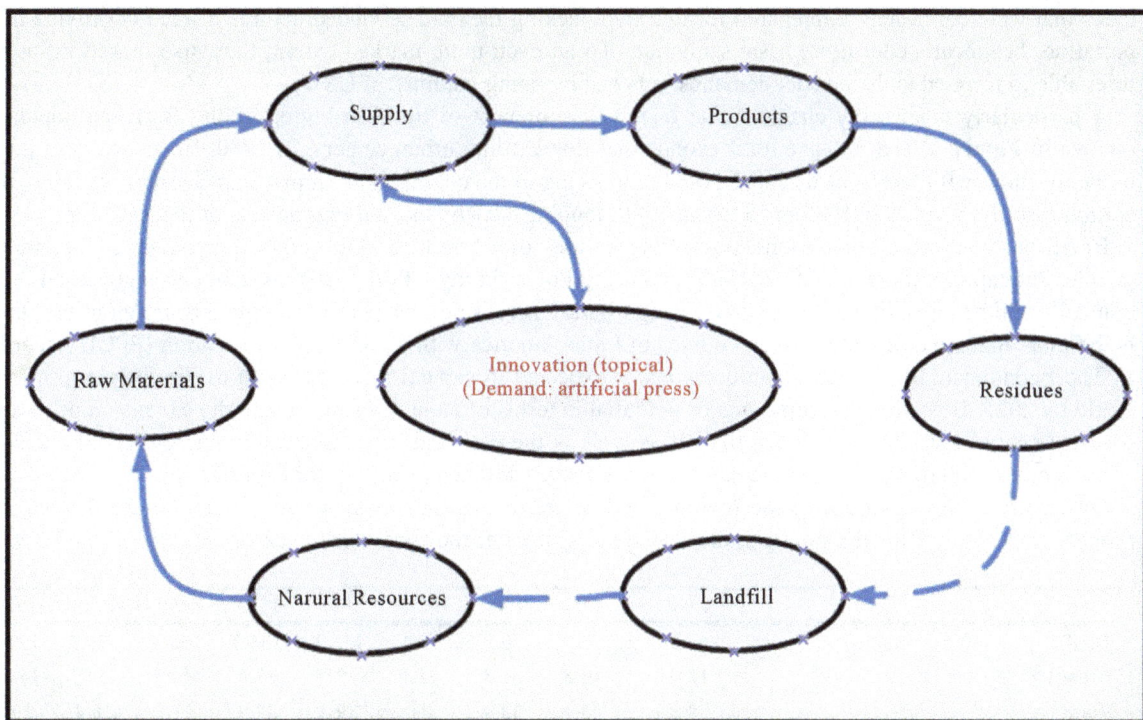

Figure 1. Schematic diagram of cycle process of the solid wastes in society. Source: own elaboration based on [3].

society, while on the other, it must find alternative ways of managing these solid wastes. This necessarily requires exploring ways of generating less USW.

According to its supporters, the use of USW to generate energy should be part of a drive to develop clean and abundant sources. For Illich [8] *"the promotion of clean techniques is almost the promotion of a luxury way of producing goods that meet basic necessities"* [8] in a throwaway society. Along these lines, a contradiction is noted in tagging USW as a clean source of energy, as the incentive for greater participation by this source may trigger efforts to step up USW production (or justify the *status quo* of production and consumption standards), which in turn exceeds the energy recovered by the generation plants.

The increase in USW volumes is intrinsically linked to wider social gaps, as affirmed by Illich [8] *"Believing in the possibility of high levels of clean energy as a solution to all evils is an error of judgment. This means imagining that equal shares in power and energy consumption can grow together."*

For Figueiredo [1] there are many doubts about the subliminal outcomes of adopting the concept of making direct use of USW as a clean source of energy:

"Along these lines, in addition to concerns over rising energy consumption in societies and the widespread effects on the natural environment, this concept prompts an additional concern, represented by the dependence of energy generation on consumption. This might offer an incentive for consumption, particularly energy-intensive fuel products, as solid wastes generated by consumption would contribute to energy generation. This absurd logic would be catastrophic from the environmental standpoint, through encouraging the extraction of natural elements".

3. Using USW to Generate Energy: Brazil as a Case Study

3.1. Urban Solid Wastes in Brazil

Understanding the Brazilian solid waste generation structure is a basic requirement for adopting Government actions designed to shape the development of this approach and/or in other related areas, such as public health campaigns or job and income generations programs, energy use and the mitigation of poverty. This understanding requires an analysis of Brazil's urbanization process.

Similar to the processes noted in many poorer countries, urbanization in Brazil took place within a context of cities that were completely unprepared for this step, lacking the core service structures needed to provide transportation, healthcare, education, basic sanitation. To an even more marked extent, they also lacked infrastructures able to respond to the heavier demands imposed by rising volumes of USW.

A particularly noteworthy characteristic of the development of the distribution of the Brazilian population (shown in **Figure 2**) is a massive rural exodus that flooded into urban centers. In 1940, Brazil was still predominantly rural, with 68.8% of its entire population living in the countryside. In just four decades, its urban population rose from 31.2% in 1940 to 55.9% in 1970, topping 84% by the and first decade of the XXI century.

Brazil has continental dimensions, possessing an area of 8.5 mil-lion km^2, and a population of 191 million people. Per capita income is US\$ 10,414 (Purchasing Power Parity—PPP 2010), the country is governed democratically and has friendly relations with its neigh-bors. There is not perspective of ethnic or religious conflict. It is the most industrialized and diverse country in Latin America with a GDP of 2.017 trillion (PPC). According to The Economist [10], the Brazilian economy is expected to expand from the ninth to the fifth largest in the world by 2025. Brazil also possesses major potential in terms of natural resources and the expansion of an agro-industry geared towards the external market as well as the potential of its renewable energy resource2 (solar, wind, biomass and hydraulic) and pre-salt fossil resources (destined primarily for export) [11].

Disorderly urban expansion made regional and social inequalities even worse, driven by the development process imposed by Brazil's political and economic system at the time. *"The lack of 'social reforms' and the*

Year	1940	1950	1960	1970	1980	1985	1990	2000	2010
Total	41.2	51.9	70.2	93.1	121.3	135.5	150.3	179.5	190.7
Urban	12.9	18.8	31.3	52.1	82	97.6	112.7	134.1	160.9
Rural	28.3	33.1	38.9	41	39.3	37.9	37.6	36.4	29.8

Figure 2. Brazilian population (millions). Source: IBGE [9] and Figueiredo [1].

inability of the Brazilian bourgeoisie to introduce social standards that are compatible with level of economic diversification and growth has given rise to a deeply unequal society" [12]. Thus, the distribution of national wealth is concentrated in the South and Southeast, to the detriment of the States in the North and Northeast. This is mainly why the composition of solid wastes differs among regions, as well as States.

This shows that the South and Southwest are home to most of the Brazilian population, living mainly in urban areas, while also accounting for much of the formation of the nation's wealth. These influences are particularly significant when analyzing the composition of Brazil's USW, as this is intrinsically linked to income and location.

Sectors with higher purchasing power use products that are more industrialized with higher added values. In contrast, the less privileged classes use simpler products with higher organic content. With more packaging and less organic matter, the solid wastes discarded in wealthier areas have higher energy content than the USW produced in poorer areas. It is important to stress that, even in regions with better economic development, such as the South and Southeast, the USW composition differs among areas with varying purchasing power.

Keenly aware of all this diversity, the Technological Research Institute [13] drew up a conservative estimate showing that the average composition of Brazil's USW consists of 65% food wastes, 25% paper, 5% plastic, 2% glass and 3% metals.

The amounts of garbage generated every day in each municipality require careful observation. Only major cities producing vast amounts of garbage have this measurement equipment. According to data issued by the Brazilian Institute for Geography and Statistics (IBGE) in its National Basic Sanitation Survey [14], only 8.4% of Brazilian municipalities actually weigh collected garbage in scales, accounting for 64.7% of the garbage generated in Brazil. The remainder (35.3%) is based on an estimate drawn up by this Institute and based on the number of inhabitants in each municipality.

3.2. Energy Generation Potential

The USW volumes that might be recycled or reused for energy purposes are veiled by uncertainties, due to the difficulties of effectively measuring or even estimating these solid wastes. However, for the purpose of this exercise, the official data issued by the Brazilian Institute for Geography and Statistics [14] will be used, at around 160,000 tons/day. Additionally, selecting various ways of obtaining energy results in different possible values [15].

Among the technologies currently available, incineration is particularly noteworthy. This consists of using the calorie power stored in combustible materials found in garbage by burning it to generate steam. This naturally depends heavily on the calorie power of the USW being processed, with the use of solid wastes with higher calorie power being recommended, such as plastics, papers, etc.

The incinerator design technology currently available allows planned power generation of up to 0.7 MWh/t, [16]. Data issued by the EPA [17] show that incineration may produce up to 0.55 MWh/tons of solid waste.

For Brazil, looking at the amount of solid wastes mentioned above and taking an efficiency rate of 0.7 MWh/t, the amount of energy generated might well reach 45.44 TWh/year, as shown in **Figure 3**.

Produced through anaerobic decomposition of garbage in landfills, biogas offers another alternative way of using the energy stored in USW, possibly through open cycle plants. Once separated from the CO_2, it can be used in gas turbines, performing less well than combined cycle plants, but efficient enough for energy conversion purposes. The **Figure 4** summarizes the energy generation potential of this technical route in Brazil.

These findings indicate ample potential for generating electricity from solid wastes through anaerobic diges-

Item		Unit	Reference
(a) Urban Solid Waste	59.07	M t/ano	(IBGE 2000)
(b) Factor of Energy Produced	550.0	k Wh/t RSU	(EPA 2002)
(c) Factor of Energy Produced	769.2	k Wh/t RSU	(TOLMASQUIM 2003)
Potential Energy Estimated (a) × (b)	**32.49**	**T Wh/ano**	
Potential Energy Estimated (a) × (c)	**45.44**	**T Wh/ano**	

Figure 3. Recovery energy potential by incineration. Source: Henriques [15].

Item		Unit	Reference
(a) Urban Solid Waste	59.07	G kg/year	(IBGE 2000)
(b) Factor of Methane Produced (tCH4/tRSU)	6.5	%	(IPCC 1996)
(c) Density	1.40	m^3/kg	(PERRY 1999)
(d) Conversion Factor	10.76	k Wh/m^3	(MME 2003)
(e) Capacity Factor of Power Station	95	%	(MUYLAERT 2000)
(f) Efficient of Power Station (Open Cycle)	35	%	(GASNET 2004)
(g) Efficient of Power Station (Combined Cycle)	45	%	(GASNET 2004)
Potential Energy Estimated Open Cycle **(a) × (b) × (c) × (d) × (e) × (f)**	19.18	T Wh/ano	
Potential Energy Estimated Combined Cycle **(a) × (b) × (c) × (d) × (e) × (g)**	24.67	T Wh/ano	

Figure 4. Energy recovery potential with landfill gas. Source: Henriques [15].

tion, from 19.18 TWh/year from Garbage Gas at its lowest open cycle yield to incineration at 45.44 TWh/year. This clearly shows the significant energy generation potential of these technologies on the Brazilian energy scenario.

3.3. Energy Stored in Solid Wastes

The energy required to produce a ton of plastic in Brazil hovers around 6.74 MWh/t; 4.98 MWh/t for a ton of paper; 4.83 MWh/t for a ton of glass and 6.84 MWh/t for a ton of metal [18]. Taking into account the gravimetric composition of garbage in Brazil as described above [16], each ton of garbage produced contains 250 kg of paper, 50 kg of plastic, 20 kg of glass and 30 kg of metals, as well as 650 kg of organic matter. The energy required to produce the paper, plastic, glass and metal contained in a ton of garbage dumped in Brazil is presented in **Figure 5**.

With the energy content required to produce a ton of garbage totaling 1.9 MWh, a demand of 110 TWh may be estimated for the total amount of USW dumped each year in Brazil.

As a ton of garbage generates 0.55 MWh [14] and 0.7 MWh [16] through incineration technology, the energy generation potential of the amounts of garbage (59.07 Mt) produced in Brazil each year reaches 32.49 TWh/year for the former possibility and 45.44 TWh/year for the latter. Garbage Gas technology also has the potential to generate energy from USW in Brazil, reaching 19.18 TWh/year (Open Cycle) and 24.67 TWh/year (Combined Cycle).

4. Conclusions

The development of technologies generating energy from USW, specifically through heat conversion and/or gas production, is receiving steadily increasing amounts of attention. However, an entire cluster of concerns surrounds this discussion, related mainly to the sustainability of the planet.

Among the issues under discussion, this paper focuses on criticisms of what is known as the throwaway society, stressing that generating energy from USW can and should be planned in order to avoid dumping high-energy solid wastes in landfills. It also presents the concern that this practice may nourish and encourage a type of production based on meeting "artificial needs" and resulting in social and environmental degradation.

The development model adopted by modern society and the pace of its progress are reflected in an upsurge of the supply of goods and consequently energy consumption—which is in turn the basis of the production system. The consumption structure is shaped by life styles that define family arrangements, income levels and distribution, ownership and use of consumer goods, dissemination of heating/cooling equipment, transportation structure and housing expansion models, among other aspects. It is important to advertise that this development model is not enough so as to reduce the inequality of income around the word, in other words this model to reinforce the liturgy waste production liturgy.

This case study indicates that generating electricity from USW would avoid 41% of the energy expenditures needed to produce this volume of garbage, taking the highest possible USW energy recovery value (using incineration technology at the rates presented by the EPA).

Garbage	Energy Requeride to Produce MWh/t	Participation of Waste per ton Mwh
Metal	6.84	205.2
Vidro	4.83	96.6
Papel	4.98	1245
Plástico	6.74	337

Figure 5. Energy required and participation of waste per ton. Source: Author's calculation sbased on Calderoni [18].

It is important to stress that the energy required to produce the organic fraction of the USW was not considered. Moreover, the calculations for the share held by metal in energy expenditures took the steel value as a reference (6.84 MWh/t). This results in a conservative analysis, as aluminum holds a significant share in the metal fraction and its production is far more energy-intensive (17.6 MWh/t).

According to the 2013 National Energy Balance, the electricity generated in Brazil reached 552 TWh, 3.9% more than in 2012. The Brazil presents an electricity matrix predominantly renewable, and the domestic hydraulic generation accounts for 70.1% of the supply. Adding imports, which are also mainly from renewable sources, it can be stated that 85% of electricity in Brazil comes from renewable sources [19].

When assessing the use of USW as an energy source, it was assumed that all solid wastes contain more potential energy than the energy that could be converted through the conversion technologies that are currently available. This means that energy conversion based on USW falls below the energy required to produce the goods, with this gap widening significantly if including the energy required to ship the finished products.

This paper is not intended to challenge energy generation through USW, but rather attempts to highlight the fact that—in a society structured on the appeal of spendthrift consumption and slanted towards a steady drop in the marginal utility rate of goods—efficient use of USW should be encouraged while curbing profligate output, as an avoided ton of USW greatly exceeds the amount of energy that could be recovered by USW usage technologies.

The rising participation of machines in the means of production requires constant updating, in order to keep them as modern as possible. The outcome of this quest for the "new" is that they become obsolete, often before the end of their useful lives. The general trend is that large-scale production spurred by competition turns out goods that are less durable, meaning it may well cheaper to purchase a new item instead of re-using it. It is important to note that this increase in productivity is not viewed askance, and is good and desirable, according to certain standards. However, gains in productivity trigger alterations in consumption standards in a throwaway society, making it hard to pinpoint the perfect balance point between production and consumption. Although society should ideally take steps to ensure that most of its resources are channeled to the production of re-usable goods, its resources are frittered away, under pressure from dropping usage rates.

In actual fact, the most efficient way of making good use of USW involves moving away from the cult of the one-way pack, redefining the model that continuously generates "new needs" and integrating energy recovery processes into management practices.

References

[1]   Figueiredo, P.J.M. (1994) A Sociedade do Lixo: Os resíduos, a questão e a crise ambiental. Editora Unimep, São Paulo.

[2]   Marx, K. and Engels, F. (2002) A Ideologia Alemã. Editora Martins Fontes, 2ª Edição - 3ª tiragem. São Paulo.

[3]   Mézáros, I. (2002) Para Além do Capital. Editora da Unicamp, primeira edição, Campinas.

[4]   Boa Nova, A.C. (1985) Energia e Classes Sociais no Brasil. Edições Loyola, São Paulo.

[5]   Giannetti, E. (1983) Energia: Seu Conceito Histórico. In: *Energia e a Economia Brasileira: Interações Econômicas e Institucionais no Desenvolvimento do Setor Energético no Brasil*, Livraria Pioneira Editora, São Paulo, 1-31.

[6]   Marx, K. (1975) O Capital. Civilização Brasileira, Livro 1, Volume 1, São Paulo.

[7]   Debeir, J.C., Deléage, J.P. and Hémery, D. (1993). Uma história da energia. UnB, Brasília.

[8]   Illich, I. (1975) Energia e Equidade. Livraria Sá da Costa Editora. Lisboa.

[9]   IBGE (2010) Instituto. Brasileiro de Geografia e Estatística. Rio de Janeiro.

[10]   The Economist (2009) Brazil Takes Off. 14 November 2009.

[11]   Pereira, M.G., Camacho, C.F., Freitas, M.A.V. and Silva, N.F. (2012) The Renewable Energy Market in Brazil: Current Status and Potential. *Renewable and Sustainable Energy Reviews*, **16**, 3786-3802.
        http://dx.doi.org/10.1016/j.rser.2012.03.024

[12]   Costa, L.C. (2000) O Governo FHC e a reforma do Estado Brasileiro. *Pesquisa e Debate*, São Paulo, **11**, 49-79.

[13]   Instituto de Pesquisas Tecnológicas (IPT) (1998) Lixo Municipal - Manual de Gerenciamento Integrado. São Paulo.

[14]   Instituto. Brasileiro de Geografia e Estatística (IBGE) (2000) Pesquisa Nacional de Saneamento Básico. Rio de Janeiro.

[15]   Henriques, R.M. (2004) Análise Comparativa de Tecnologias Para Geração de Energia Elétrica com Resíduos Sólidos Urbanos. Dissertação de Mestrado, Programa de Planejamento Energético—PPE/COPPE-UFRJ, Rio de Janeiro.

[16]   Tolmasquim, M.T., *et al.* (2003). Fontes Renováveis de Energia no Brasil. Interciência, Rio de Janeiro.

[17]   EPA (2002) Solid Waste Management and Green house Gases—A Life-Cycle Assessment of Emissions and Sinks. Environmental Protection Agency. US. EPA.

[18]   Calderoni, S. (1998) Os Bilhões Perdidos no Lixo. Publicações FFLCH/USP, São Paulo.

[19]   MME (2013) Balanço Energético Nacional. Ministério de Minas e Energia - Brasília.

Upcoming Transitions in the Energy Sector and Their Impact on Corporations Strategies

Jose M. "Chema" Martínez-Val Piera

ETSI Minas, Universidad Politécnica de Madrid, Madrid, Spain
Email: chemaval@gmail.com

Abstract

An analysis is presented on a set of enabling technologies which are opening new routes for energy conversion and consumption. This portfolio of innovations is complemented by a new framework in hydrocarbon production. This integration yields an optimization of energy uses that can result in lower greenhouse gases emissions and expand the lifecycle of current available resources. These options are confronted with the need for higher quantities of energy, at affordable costs in order to maintain the economic development. The conclusion is that there are no contradictions among the general objectives in global energy policy and the goals of corporations. Companies can take advantage of their previous expertise to remain competitive, but have to further develop new skills to operate in a new energy sector that is likely to be highly interlinked; evolving for the previous model that had markets segmented by specialty. New goods, such as the electric vehicles or the advanced high temperature high power fuel cells for generating electricity, should pave the way for a more synergetic and efficient energy sector.

Keywords

Energy Enabling Technologies, Integral Energy Efficiency, Hydrocarbon New Uses, Electrochemical, Thermal Hierarchies

1. Introduction

A minimum of $20 trillion investments is estimated for the next 25 years to meet increasing energy demand and to offset the declining reserves hydrocarbons [1] [2]. This is an indicator of the challenges that the sector has to address. Further, local and global contamination is posing a severe problem, which will be intensified as the number of cars increases from a circa 1 billion to 1.7 billion.

Such a global market expansion will concur with a wide portfolio of new technologies that can produce a solution to the otherwise inevitable energy crisis [3]-[5]. This solution will have to go beyond alternative and renewable energy and embrace the economic harvesting of remaining oil and gas resources, including the so called unconventional.

New trend in energy and currency accounting is likely to appear. This will take into account the efficiency of different technologies and the fit among sources, technologies and end-uses. Such scenarios could be characterized by the Integral Energy Efficiency (IEE), a concept presented in this paper. The IEE can be used to find the optimum cost-benefit ratio at a global scale, at a given, worldwide.

If such a deep transition materializes, industrial corporations will have to undergo changes in structure, scope and methodologies. This will be both a threat and an opportunity [6] [7].

Energy policies are primary established at national level. Basic principles apply everywhere: security of supply, environmental quality, and minimum cost. The latter is, however, an incomplete concept if a time frame is not defined during cost minimization. Most of the policies are established only considering a very short term (*i.e.* four years or less).

Spain can be taken as an example, notably for the electricity industry [8]-[13]. The boom of renewable energy sources was the consequence of a generous framework of feed-in tariffs established in the Royal Decree RD 661/2007. Collateral effects of this subside based policy [14] have become a notorious problem to maintain a proper balance between the economy and the profits of the system.

When the energy problem is addressed with a longer run perspective, underlying uncertainties are too broad as to enable the reach of the optimum solution in terms of energy policy. Moreover, collateral effects can appear in policies implementation, and correction measures must be enforced. A similar problem regarding uncertainties appears in the elaboration of a paper on these subjects. Some of the relevant variables are not of physical nature, but rather evanescent and quickly fade or disappear. These variables are related with geopolitics, financial pressures on a currency, environmental trend, ideology and other social developments.

A technical analysis can be aimed at optimizing a given energy problem with defined boundary conditions [15]-[17]. However, the result can become useless because of the interference of the evanescent variables, which usually appear with enormous strength over short periods of time. They are crucial in actual life, but they are almost not admitted in a technical paper. If a paper dares commenting this type of evanescent variables, the paper risks to be deemed unsuitable for standard scientific publications, where the formal procedure has been established according to the scholar tradition.

Let us consider the problem of Global Warming and the risk of increasing the atmospheric greenhouse effect by methane emissions. Methane has a Global Warming Potential that is 70 higher than CO_2. However, accounting of methane emission is much less accurate than that of CO_2. The latter comes from chemical combustion, and CO_2 emission rate is directly calculated from fuel composition and stoichiometric balance. On the contrary, methane can come from natural reservoirs, leaks from gas-pipes, or leaks from extraction and production process. These mechanisms do not have proper instruments for measuring the potential flow of methane.

In 2007, Global Warming [18] was so high in the cultural agenda that the Nobel Prize for Peace was awarded to the former US Vice President, Al Gore, and the Intergovernmental Panel on Climatic Change (IPCC). Seven years later, the IPCC is less emphatic in the declaration that "unambiguously, there is relation between human activity and climatic change". Moreover, President Obama's Administration is favoring fracking [19]-[21], as a way to reach energy independence which represents a milestone in the US international policy. A reliable estimate of methane leaks from fracking-based gas production is not available, which means that the famous "IPCC scenarios" for reaching a certain level of temperature rise must be revisited.

Such abrupt and evanescent changes in energy policy from the biggest economic power in Planet Earth is rather difficult (*i.e.* almost impossible) to model in any scientific analysis [22] and it is, however, second to none in terms of ranking of critical variable.

From the point of view of methodology, the difference between physical variables and evanescent variables is similar to the difference between "optimization process" and "decision-making process". To optimize a system, this must be well known and comprehensively defined so that a minimum and/or a maximum can be found (including local min/max, or a saddle point).

When uncertainties ranges and evanescent variables dominate the description of the system, optimization techniques cannot be applied. In such circumstances different methods must be used, from fuzzy logic to purely stochastic.

In scientific attempt to express all internal relations of a system in terms of integral-differential equations, evanescent variables are usually omitted because they disturb the ideal picture we can get from the energy world, and compromise our goal.

In the presentation of the work pursuing the goal of dissecting the energy world, I focus on the possibilities and potential of the physical mechanisms available to extract, convert or apply energy by the end user. It goes without saying that the path to the goal and the goal itself can be disturbed by an evanescent variable, but an analysis must be carried out on the basis of physical facts and laws [23] [24].

In this context, we propose a new concept to complement the guidance of energy policy making, which is the Integral Energy Efficiency. IEE should lead to maximize the total amount of End Uses of Energy, compatible with a minimum cost at a global scale (relative to the produced benefit).

In this paper, this concept is analyzed considering that some of the enabling technologies currently under research will actually achieve industrial maturity. Hence, opening the energy sector to more degrees of freedom to optimize global efficiency. Those technologies would include deep changes in the transport sector. For instance, the current dominance of petroleum products could be challenged by other hydrocarbons and biological products [25]-[28] or other sources of energy (*i.e.* the electric vehicles).Energy storage would represent a fundamental element in such long term energy scenario. It is therefore mandatory to analyze the role and features of corporations in a much more integrated Energy sector. The classical objective for a corporation is to achieve a niche in a given market. Very likely, a further objective of the corporation will be to consolidate and/or enhance that market quota. However, both the framework and the boundary conditions for energy corporations will not be the same as those experienced in the past. This is why the analysis pays attention to the problem of fitting corporation skills to future energy sector requirements, and accounts for alternative tools to deliver energy policies in the near future.

It seems we are in a crossroad similar to that the energy sector was in mid-seventies of the 20th Century. At that time, new agencies and developers were created (*i.e.* the Energy Research and Development Administration (ERDA) later substituted by the Department of Energy). Technology was one of the key elements for these types of agencies. However, one should underline that some of the most relevant enabling technologies have not being provided by these institutions, but rather by independent initiatives such as the case of fracking [29]-[31].

In order to analyze the world of energy, we will study in Section 2 the current situation, which is fast evolving [32] [33]. The (re)evolution is catalyzed by technology, and the influence of higher efficiency of the energy conversion chain, from raw sources to end uses. Efficiency is the subject of Section 3. Section 4 is devoted to review technology, which is the main driver to change the Energy world. Section 5 deals with the different scenarios identified according to the priorities in the energy policies, which mean that some scope for evanescent variables is needed. Special attention will be paid to the case of unconventional gas (and oil).

The paper ends with a collection of prospects that could be derived from attempts to improve significantly the utilization of available energy sources, which can then open a new age for corporations prepared to meet these opportunities.

2. The Current Energy World: Crossroads or Road-Maps?

There are several studies and series of analysis [1]-[4] devoted to a better knowledge of Energy. The same facts are presented to argue in favor different points of views. The International Energy Agency tries to offer a more eclectic landscape of energy, without pushing it to extreme opinions. The European Union has to integrate different views on topics such as Nuclear Energy, but it is generally well aligned with environmental matters (*i.e.* Directive 20/20/20 [34]). Furthermore, the EU has a taste for technology and development, but lacks establishing strong priorities (*i.e.* the Alternative Fuel Directive).

The US has adopted recently a much liberalized position in the energy market, and has shown a lot of interest in energy information [3] (http://www.eia.gov/forecasts/ieo/). The US has demonstrated very limited interest in developing new energy technologies in recent years. However, several institutions have produced interesting analysis on the true values of energy research for the long run (*i.e.* the IEEE [4] with the document USA-NEPR-2014). This long-run advantage seems to be of secondary importance for the current Obama administration as compared to those related with lowering cost and providing higher energy independence (*i.e.* the unconventional hydrocarbon industry). The 20th Century ended with an international concern about the long term case, and a quest for sustainability was launched to avoid Global Warming [35]. As already quoted, an international body, the IPCC [36], was created to produce forecasts on climatic for the coming decades. This effort was not match to foresee problems in the Energy sector and much less to identify pathways to development markets and technologies.

It is true that an organization as the International Energy Agency is already making forecast about the evolution of the sector, but the IEA only has 29 members, most of them from the European Union, but unfortunately not the EU as such.

We are already aware that Climatic Change deserves attention at the highest level, but Energy is not a minor problem. Energy in turn is dominating the environmental impact at every scale.

Energy is at a manifold crossroads where several concepts open different lines, especially the following ones: sources, investment, technologies, environment, sectors, emerging countries demands, independence, markets, and efficiencies.

A minimum of $20 trillion investments is estimated by the IEA for the next 25 years to meet surging energy demand and to offset the declining reserves and production of the world's major oil fields. These numbers can be challenged; both as a whole and as an aggregation of sectorial figures. However, they are based on projections that are easier to accept. For example, the rise in the number of automobiles in the world goes from circa 1 billion to 1.7 billion.

There is, however, another set of unknowns behind these numbers. For instance, automobiles will not be the same ones as those of present days. Automobiles will be used in a slightly different pattern. People tend to live in macro-urban areas and the use of car mainly in that environment can be substituted by public transportation. As a matter of fact, as seen in **Table 1**, IEA [1] projections on total consumption in transportation for 2030 (and even for 2050) do not change.

Some of the hypotheses underlying in those numbers can be challenged. The forecast for Light Road values in 2050 is based on the following allocation based on energy:

- Oil products, 20
- Natural gas, 1
- Biofuels, 11
- Electricity, 7

This projection considers a very short penetration of Electric Vehicles, and this is even more relevant as it includes H_2 vehicles in this source type.

As presented later, the EV presents very appealing features in fuel consumption and environmental quality. The main limitation on range should not be considered a burden as 85% of the mileage done by private cars corresponds to daily round-trips under 150 km.

This subject will be dealt with later; it is at the very center of a potential revolution in town transportation modes and air quality. It seems to be extremely difficult to calculate the cost of environmental consequences, specifically in terms of health problems as a result of micro-particles aggression. This hazard can only be dramatically reduced by electric cars.

The Future of Energy will be the result of a confrontation among facts from Nature, political principles & programs; and the evolution of both technology and society. **Figure 1** presents a comparison between the current North-American model and the European model on the basis of a 4-corner ring where the energy battle takes place.

Nowadays, the North-American masterpiece is the new aggregation of reserves which appear in the inventory thanks to hydraulic fracturing in horizontal wells, an enabling technology which has changed at depth the oil and gas industry both in the US and at a worldwide scale.

Table 1. Total world final energy consumption in transportation (EJ) for several modes and years (source, IEA).

Mode	Year		
	2011	2030	2050
Heavy road	26	27	29
Light road	51	49	39
Rail	2	3	5
Air	11	12	13
Sea	10	12	13
TOTAL	100	103	99

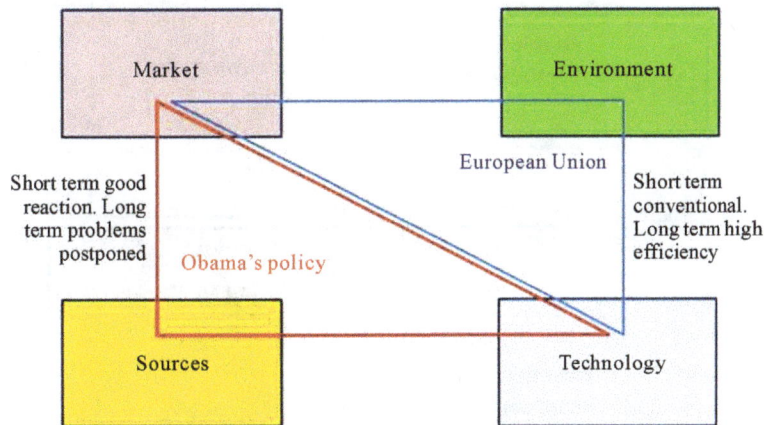

Figure 1. Sketch of the energy policy priorities in the US and the EU.

Breakthroughs in technology can be fundamental pieces in shaping the future. Nuclear Energy was an example in the past; and Shale hydrocarbons are an example right now. The Future of Energy will also be affected by changes in the demand side, but this is still harder to predict. Some of the new products and new activities, as widespread communication and digitalized information, can create new consumption modes (relying on electricity) and can save energy because they reduce the need to move for peoples and documents.

It is worth pointing out that IEA estimates predict a huge increase in energy consumption from the chemicals and petrochemicals industry. Increasing from 40 EJ in 2011 to 80 in 2050. The remaining industrial sectors do not undergo a similar rise. Furthermore, this industry has an important peculiarity: it uses hydrocarbons as primary matter, not only as fuel. This fact points out the importance of preservation of hydrocarbons for the highest added value end use, and it is an advice for looking to petro-chemistry with greater interest.

Such a dual role of fossil fuels is not shared by the rest of electricity energy sources, namely renewables and nuclear. In particular, the success of oil products for powering vehicles comes from one of their important properties; the high energy store capacity in terms of amounts of energy per unit mass and/or unit volume.

However, efficiency of Internal Combustion Engines (ICE) is very small as compared to other energy conversion machines (*i.e.* turbines). That fact was not important when oil was cheap and abundant, but is becoming a problem, from economical to environmental. ICE plus oil products were a very good tandem for powering vehicles, ships and airplanes, but such a good fitting is not enough for the problems of a sector which has to become more efficient and clean.

Figure 2 shows the strong changes produced in our 4-corner ring from the appearance of fracking as a new enabling technology. **Figure 3** has the EV as the technology driver.

Long term energy scenario will likely be more complex in its internal relations than the today's situation. Nowadays, as presented in **Figure 4**, the world of energy is rather specialized and segmented into compartments.

Close to 70% of coal currently goes to electricity generation, and 95% of oil goes to transport. **Figure 5** portrays the long distant future, Electric Vehicles could give much more flexibility to the energy system, and oil could be used for more expensive applications (*i.e.* higher added value). This includes new energy utilizations, as discussed in next section.

Understanding how new scenarios can be built based on new enabling technologies is useful for guiding energy policies and to guide corporations too. This double objective will be treated after an insight into energy enabling technologies.

3. Energy Efficiency Structure

More than 90% of the primary energy goes through the process of combustion; which, as presented in **Figure 6**, has a temperature. This is particularly relevant for thermodynamic cycles, limited in efficiency by Carnot's principle as presented in **Figure 7**.

An alternative to combustion is offered by electrochemistry, left side of **Figure 6**, which is not limited by Carnot's principle. However, it also has limitations [37] because of the generation of entropy. The theoretical

Figure 2. Four pillars of Energy and the sprout of shale gas and oil changing the scenario.

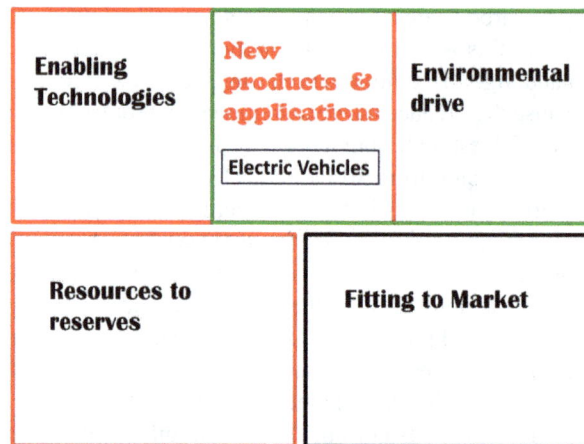

Figure 3. Four pillars of Energy and the sprout of Electric Vehicles, a key element in the future of transportation and environmental quality.

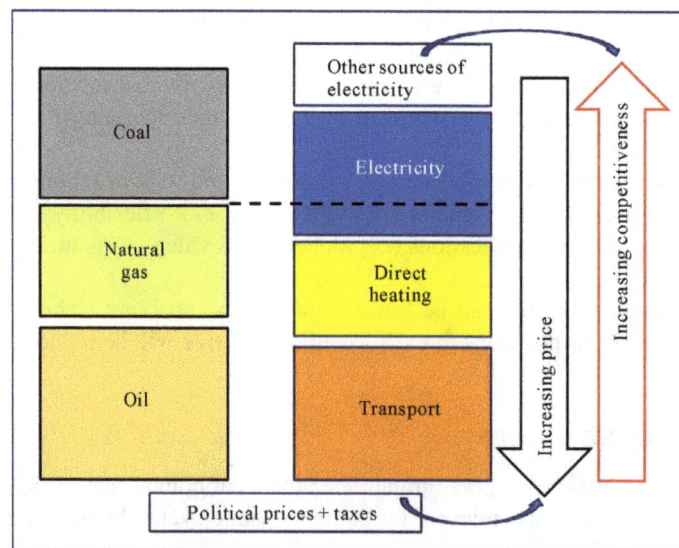

Figure 4. Fossil fuel coupling to end uses. In general, transportation presents lower competitiveness but support higher taxes.

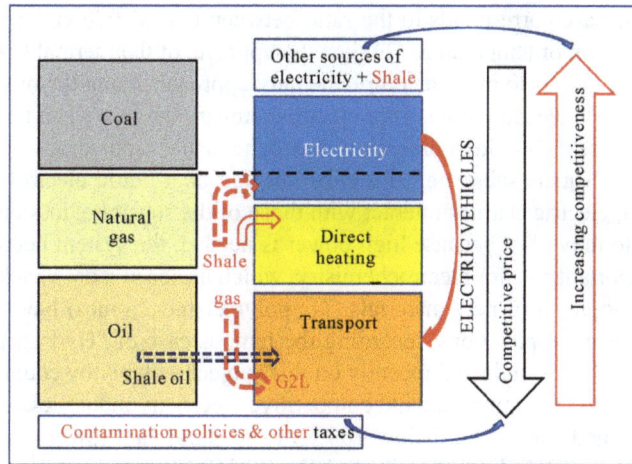

Figure 5. The deployment of Electric Vehicles will convey a deep change in the Energy sector. For instance, coal can contribute to transportation, through the electric grid, so increasing competitiveness in that field. The same can be said about electric renewable sources.

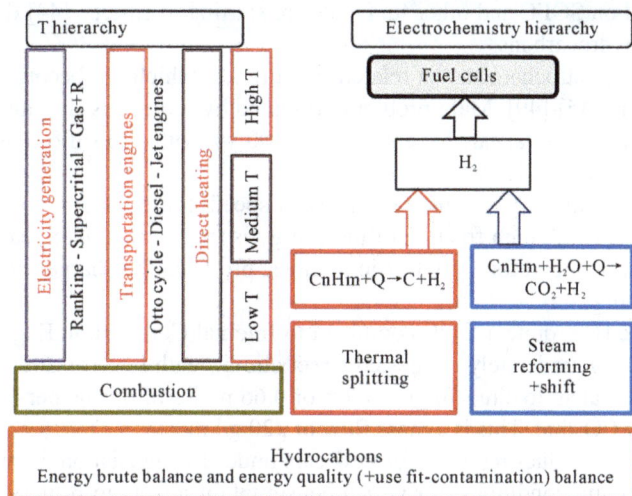

Figure 6. Diagram of the pathways for exploiting the energy of the hydrocarbons. The combustion branch at left, and electrochemistry at right.

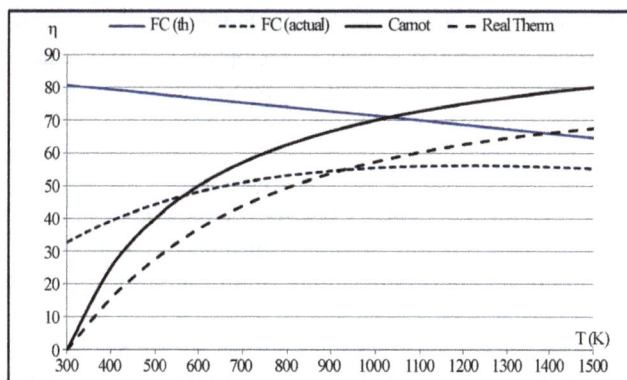

Figure 7. Fuel cell efficiency and thermal energy efficiency versus system temperature.

limit of efficiency in this case corresponds to the ratio between Gibbs' free enthalpy (ΔG), and enthalpy (ΔH), and it is a decreasing function of temperature (T), just the opposite of the thermal branch.

A first reaction to this figure is to make a complementary approach, using the brute force of combustion when high temperatures are achievable, and looking for electrochemistry devices when temperature must be low.

The latter, however, presents a hidden problem: electrochemistry needs that some electric charges (electrons and ions) move over an adequate substrate. This substrate can be a liquid electrolyte, as in lead batteries, or a solid. When moving fast, electric charges interact with those of the substrate, loss energy, and heat the substrate. Hence, if charges have to move fast because high power is needed, the system becomes very hot and efficiency decreases. This is an important rule for electrochemistry, which has been a drawback for its development [38].

Nonetheless, the development of new materials (*i.e.* polymer membranes) have open the way for extending the range of electrochemistry applications, including the famous case of "Hydrogen automobiles". Several studies and assessments have been conducted recently on this subject, notably by countries as Canada [39], Norway [40], and Australia [41] with very important and competitive electricity industries.

Fuel cells working at moderate T (*i.e.* below 100°C) have significant lower efficiencies than the theoretical values (*i.e.* PEFC devices) [38]. Solid Oxide Fuel Cells (SOFC) [42]-[43] need very high operational temperatures, but they can work directly with hydrocarbons. This simplifies the previous phases of the system, with the H2 generator.

They are also less sensitive to poisoning by carbon monoxide and other contaminants. They could be used for electricity generation wherever gas or oil is delivered (*i.e.* underground in the center of a city or any remote location). Proposals based on SOFC and other fuel cells are starting to emerge [44] driven by a better integral use of the energy stored in hydrocarbons.

A different branch of electrochemistry is related to batteries, which are becoming of paramount importance for powering electric cars [45]-[49]. Main problem presented by these cars (in plug-in fully electric type) is the range. For the moment, commercial cars have a range of 200 km with a specific consumption of 12 kWh/100 km (more expensive models can double this range).

Another concern is the time required to recharge (aka speed).Batteries don not cope well with high electric energy rate over a short time. A high fraction of the energy is transformed into heat. Therefore, electric cars appear as machines very well suited for daily traffic in cities, provided the charge can be done at night in a proper place (*i.e.* a private parking site).

Efficiency in this case rises quite a lot as compared to Internal Combustion Engines. Let us take the example of an average petrol car, in moderately congested street circuit, with real consumption of 10 liters per 100 km. This consumption is equivalent to circa 70 grams/km or 0.66 moles of octane per km; that would yield an emission of about 5 moles of CO_2/km. This is a mass flow of 220 g/km.

This is almost double the values required for new cars under EU legislation. However, these "administrative" values refer to ideal running conditions, for which consumption lowers to 6 liters per 100 km (or less), or approximately 130 g/km.

The value of the mechanical energy to drive an average car is 0.12 kWh (*i.e.* 0.432 MJ). The heat load in the combustion of 70 g of gasoline is 700 kcal, equivalent to 2.92 MJ. Therefore, the actual efficiency of the average automobile is about 15%.

The optimum engine performance is almost double the latter figure, circa 28%. Inefficiencies of the engine out of its optimum, accelerations, and idle periods make the optimum performance decline significantly.

An electric motor has the advantage of not having idle periods, as it rests when the car stops. Its performance is very high for most of the regimes of the system, and is exceptionally controlled by power electronics. On average it can support an 88% yield. However, a correction factor due to charge/discharge performance of the battery is required. A conservative estimate can be a correction of 75%. This results in an overall efficiency of 66%. So its consumption (from the electric grid) is 0.18 kWh electric per kilometer.

An additional adjustment is required before the final benchmark. We must add grid and generation losses to the latter figure. Again, a conservative estimate is a 10% of the total loss. This yields a need of 0.2 kWh per km for the electric system.

Octane burned in a combined cycle Brayton-Rankine, currently yields an efficiency of 60% and it is expected that this efficiency is boosted up to 70% in the near future. Taking the latter figure (which best represents the future), we obtain a primary (thermal) energy consumption of 0.28 kWh per km travelled (afterward) by car, which is equivalent to 1 MJ.

This means that we will need to burn 24 grams of octane in an electric power plant to produce the required energy in electric form. This is just over a third of the direct consumption of an ICE gasoline car.

Generation of CO_2 follows a parallel route. It is reduced from 220 to 75 g/km. This accounts for the energy chain after the gasoline is delivered; as neither process account for upstream emission of that delivery point.

It is inevitable to note that CO_2 emissions would be much lower if the batteries were recharged with electricity from nuclear or renewable sources. In this case, full-life cycle emissions are in the range of 30 to 40 g/kWh, or 10 g/km. This is one-twentieth of that emitted by the average car.

As for the cost to the end user, efficiency is not the only advantage. The price in electric domestic market are circa 20 c€/kWh (electricity), hence the cost per km would be 3.6 c€/km, or 3.6 euros per 100 km. With gasoline at 1.5 €/liter, and an average of 10 liters per 100, the cost would increase to € 15 per 100 km. This is a fourfold increase. ON the one side, this value includes the very high tax burden to gasoline, but on the other side it reflects the price of gasoline at either high and low international prices of oil as a commodity in the international market, *i.e.* Brent oil. At 100,000 km the electric car would save more than 11,000 €.This should compensate for the cost of the battery and the ancillary systems.

In summary, the largest environmental cleanup in both global and local pollutants, increased energy efficiency of automobile transportation and the positive economic impact to consumers, all point out that the Electric Car is a very powerful pathway to improve overall energy arena.

Additional economy considerations can be taken into account. If a car owner decides to optimize the investment and running costs for the total requirement of car performances in a given period of his/her life (*i.e.* 5 years) the owner will find that more about 85% of the total distance (number of km) traveled is done in daily roundtrips that are shorter than 100 km.

Owners might dive longer distances some long week-ends throughout the year, and could engaged in few (*i.e.* one or two) trips a year with distances longer than 2.000 km (forth and back). For the latter, an air flight plus a car renting strategy saves the owner a lot of money and a lot of time, and at the same time the total energy balance is also optimized (even for a car with 5 passenger this will not be a worst scenario). An Airbus 320 spends 5.000 liters of kerosene in a trip from London to the Spanish cost in the Mediterranean Sea, and carries around 125 passengers. This means 40 liter per passenger. If driven in a usual European car, the owner will need about 20 hours of effective drive (for 2.100 km), consuming about 200 liters of gasoline.

Similarly, if a depreciation of 0.5 €/km is applied to the car, the trip London-Murcia implies a roundtrip cost of 2.100 €. Adding the cost for fuel, the total cost of the round trip is circa 2.700 €.

Again, the comparison with the flight depends on the number of passengers, but one can find air fares for that flight at less than 250 € (excluding low cost airlines). Therefore, both from a personal and an energy efficiency perspective, the optimum solution is to accommodate the features of the trip to the characteristics of the machine.

It can be argued that the cost for renting a car must be added, but this cost is significantly lower to the associated cost of three driving days for a fifteen-day holidays.

The option of renting an ICE car for long domestic week-end trips is an option that will help optimize the energy consumption over the time-span under consideration (*i.e.* 5 years).

It is true that ordinary people are not used to this type of calculation, but this is rather simple and one can foresee an information campaign to illustrate these calculations. All electric car vendors already have calculation schemes to demonstrate the savings associated to a given type of daily trips and other journeys. And should everything fail an app for smartphone can help deliver this advantage.

Social and cultural changes are in many cases encouraged by technology, with or without government advertisement. There was no governmental campaign to foster mobile phones or the use of Internet.

Cars are much more expensive than a phone or a computer, and the reaction time constant for buying a car is much longer than the period to replace a cellular phone. This means that this "evanescent" variable of public perception and decision making will be not as prompt as in telecommunication. However, many hints point out in that direction.

Electric vehicles can become household goods on this basis faster that they can do so by local government policies to impose them to fight local contamination.

4. Technology Changes in the 21st CENTURY

It is not easy to generate prospects on Energy Technology. Previous experience points out that some promising lines grow little by little while other lines lagging behind for many decades burst suddenly and make an early market breakthrough.

An example of the former is nuclear fusion, which 40 years ago it was considered a matter to be controlled within the forthcoming 30 years. In reality, nuclear fusion has kept in the main road of tokamak reactors but it still considering that industrial maturity will arrive 30 years from now.

An example of the latter is photovoltaics, which was considered a very expensive and inefficient technology for electricity generation, and explode over the last decade. Investment costs have decreased from over 10 to under 2 US$/kW.

The key factor was that PV did not require a significant breakthrough in physics, but rather a development of technology. On the contrary, the control of magnetic fusion reactors was not mastered 40 years ago and there it still holds several unknowns nowadays. It is obvious that many advanced have been made, but research is still at the laboratory scale and with fundamental problems to be addressed.

The situation for renewable energies is just the opposite. These sources of energy rely on well-known physical mechanisms and the main problem has always been to reduce costs (both capital expenditures and operating costs). The best results have been achieved by improving designs and materials.

The learning curves of all renewable energy sources have been successful in the last 10 to 15 years. The first renewable energy to achieve a significant reduction of costs was wind power. It became possible to manufacture a machine (the 3 blade horizontal generator mounted on a tower) with a higher efficiency than the rest of the machines.

PV still presents many potential lines of development, although the best cost reduction step has been associated to cheap semiconductor material (i.e. poly-crystalline silicon).

Nevertheless, the new energy technology port-folio includes advancements in all levels or stages of the energy supply chain and the form of energy and its changes throughout the chain.

In terms of sources of energy, it is obvious that "fracking" represents one of the key milestones of the past years, and efforts should be done to make it cleaner and more effective in terms of the hydrocarbon recovery factor. Otherwise, some areas (i.e. Europe [50] or Quebec (Canada) [51]) will not join the fracking club.

In the energy conversion phases, electrochemistry [37] can give a boost to energy efficiency, particularly in the domain of distributed generation. Fuel cells working on very clean hydrogen [38] need a previous chemical double reactor, with one endothermic process for steam reforming and other exothermic process for shifting the reaction.

For the moment, technological advanced in this technology have been very limited. However, many possibilities can be identified in this technology option such as reactor configuration, materials, and temperature regime.

Two important aspects about fuel cells are the following: first, some fuel cells [42]-[44] run on hydrocarbons, which can easily be stored; and second, fuel cells do not have moving parts, which can an advantage for operation and maintenance.

On the other hand, experience with fuel cells is still limited if compared to the enormous potential of electrochemistry.

Electrochemistry has already been identified as a candidate for new energy technology alternatives. Electrochemistry can be considered as the most important one in terms of economic impact on environmental quality and integral energy efficiency, which is the case of the electric vehicle.

It is eloquent to revise the history of electric cars [52]-[54] which were proposed more than one century ago. The electrical vehicle has had peaks and troughs along these years. Even after the oil crisis, when some analysts and researchers propose electric cars fed by nuclear – generated electricity for coping with the energy problem. Troublesomeness for this proposal is that it took too long to develop the nuclear power park for supplying all the energy that was required, and beyond and above, the very limited range that an electrical car had at that days.

Batteries for electric cars and the cars themselves started to change 25 years ago, once electrochemistry was identified as a powerful tool for energy transformation with high efficiency. **Table 2** is a good example of the efforts needed to push technology ahead, and it is homage to those who believed in Electric Vehicles when they were not yet a promise.

However, upcoming energy technology developments will affect most (if not all) segments in the world of energy. This applies for the domain of sources, its conversion, and its applications. Of course, in the short term there will be a clear priority to make natural gas the center of many fundamental decisions, particularly because of the success of fracking and other new techniques, but also because of the vast amount of resources discovered over the past years.

Table 2. An important relic from the past: assessment on batteries for electric cars as foreseen in 1989. Projections were incomplete, but not out of the case [52].

	Characteristics of selected electric passenger vehicles						
							Projected
	ETV-1	ETV-2	ETV-20	ETV-I	1987 BMW	1990 BMW	VW Jetta
Top speed. mph	†	62	60	60	53	75	78
Urban range. Mi	Up to 75	66	<75‡	†	43-77§	62-124§	118
0 - 30 mph acceleration secs	†	8	†	7	14	7	6
Mi/kuh from battery city	3.41	3.14	†	3.61	†	3.73I	†
Passenger capacity	4	4	2	2	4	4	4
Power train	dc	dc	ac	ac	dc	ac?	ac
Battery	Pb/acid	Pb/acid	Pb/acid	Pb/acid	Na/s	Na/s	Na/s
Approx year of tests	1980	1980	1986?	1985	1987	1987	†
Reference	Kurtz (1981)	AiResearch (1981)	Wyczalek (1987)	Ford & GE (1987)	Regar (1987)	Regar (1987)	Angelis & Sedgwick (1988)I

Top speed is maximum continuous cruising speed. Range and efficiency data for ETV-1. ETV-2. and ETX-I refer to FUDS (Federal Urban Drive Schedule);for Jetta. They refer to ECE (European) urban cycle. †Not available. ‡Range is at constant speed of 25 mph. §Lower range estimate at top speed; higher at 30 mph. IBased on ABB Na/S battery projections of Table 3. with an improved powertrain.

Properly speaking, fracking had been known for decades, and it was pursued in the peculiar "Plowshare Project" [55] which was conceived for extending the use of Nuclear Explosives to peaceful applications, as ultra-large civil works and underground intrusions for getting minerals and above all, getting gas from source and/or trapping rocks. It could be said that "fracking gas" was known but remained unknown for many years.

Energy conversion and customization for final uses will surely be another domain for development and novelties. Notably in the process of becoming more dependent on electricity for many applications, including a substantial part of ground transportation in private vehicles.

Fuel cell, batteries and combustion will have to come closer to enable synergies that we are currently missing. Integral Energy Efficiency it is somehow dependent on those synergies.

Currently, oil products are mainly consumed in Internal Combustion Engines with a theoretical efficiency circa 30%, but with a real efficiency in the range 15 to 20%. This is in full contradiction with the inherent capabilities of a chemical fuel, which could burn at 1500 K and above, so creating a very hot energy source. Of course the main feature appreciated in liquid oil products is their capability to store energy in small volumes. This is just the end of the string respect electricity.

Electric engines have very high efficiencies (from plug to shaft) but electricity cannot be stored. A potential way to avoid this mismatch is to rely on hydrocarbon-fueled high temperature Fuel Cells in order to generate electricity with a very high efficiency that will later be used to recharge batteries. Of course it is not possible to assemble a car with a SOFC component, at is operates at 600°C and has a power in the order of few MW. However, it can be located in a secured facility close to consumers.

This leads to an integration of routes for energy conversion in order to optimize the Integral Energy Efficiency within the framework established by the applications. A sketch on this integration between the electrochemical hierarchy and the thermal hierarchy is shown in **Figure 8**.

Integrating all possible mechanisms to optimize the exploitation of energy does not disturb the energy markets. On the contrary, it aids sorting them out in a better manner. In some instances it is considered that the deployment of the electric vehicle at large scale will be harmful for the oil industry, and this is not the case. It is not so because oil has unique properties, that range from a well stablished global market to its own physical and chemical properties.

In particular it can be considered one of the best, if not the best way, to store energy. **Figure 9** depicts this characteristic in comparison with Natural Gas and major Renewable Energies (Solar ones plus Wind). Oil has the largest range for accessibility (because it can be transported easily across long distances) and also has the widest range of applications (including future generation of electricity with a much higher efficiency, and lower

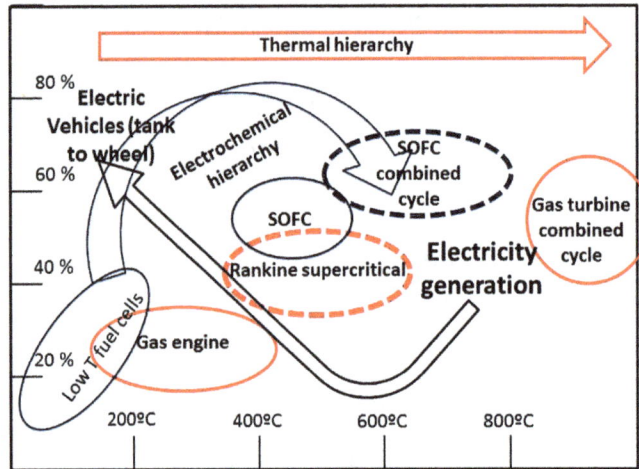

Figure 8. A map of efficiencies and temperatures for different combinations of thermal machines and Fuel cells.

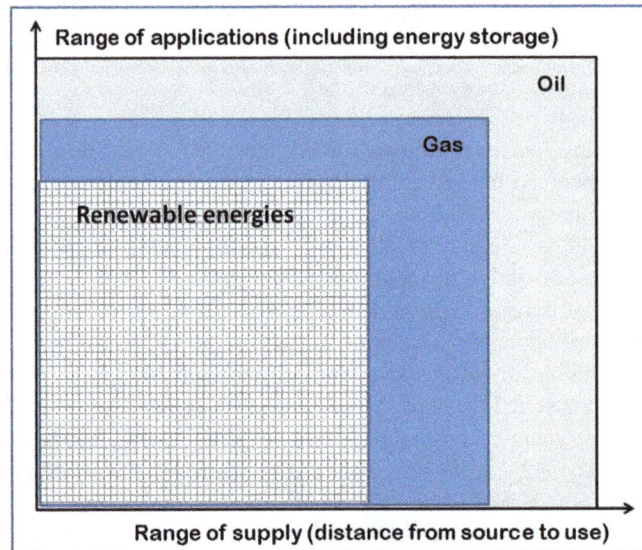

Figure 9. A qualitative graph of the ranges covered by different sources of energy in relation to supply and environment.

CO_2 emissions unit of energy per consumed).

5. Energy Policies and Corporation Strategies

The actual evolution of the Energy sector cannot be outlined with a high level of confidence. Even in centrally planned economies, because of technology leaps, market oscillations and environmental pressure, the task of forecasting evolution of energy is hard. What can be done is to analyze specific scenarios. Relying on coherent assumptions we can identify shortage of materials, or excessive financial support to materialize. Those scenarios can be defined presuming selected evolution in relevant variables, including evanescent ones, which can come from the past in the Energy sector.

This exercise is achievable in the case of automobiles, their engines and the energy infrastructure to power them. Good projections of automobiles have been carried out in the past using specific ratios as a basis for forecasting. This is based on the relationship between the number of automobiles and total population in a specific geographical area (*i.e.* continents or countries), as well as other variables in relation to the number of cars (*i.e.* as work force and gross domestic product). The fraction of total income devoted to vehicle acquisition and fuel is another fundamental ratio. Last but not least, sound judgment is needed to preserve necessary coherence of the study.

The penetration of EV in a given market can be studied as an example of this procedure. This will at least highlight the different requirements on the key variables of the problem. Each scenario includes different patterns in the energy conversion chain (from sources to market) depending in this case of the type of vehicle. A conversion chain is defined by a number of steps where some energy is lost, some contamination is generated and some investments or cost must be expensed.

Some values can be assigned to different steps and patterns to qualify their maturity or readiness, but credibility of the study will largely remain a personal decision. Energy technologies still needing very basic R&D programs as Nuclear Fusion should not be included in the picture.

The International Energy Agency, the US Energy Information Administration and the European Commission (through the Joint Research Center, mainly) have similar models for characterizing energy scenarios. The main difference among them is the role assigned to some of the new technologies.

The main deficiency for this methodology is that it cannot integrate innovations which are not yet in the agenda. The most visible example is happening now in the USA, where the "breakthrough" of fracking and shale resources has changed completely the Energy sector. **Figure 10** shows the impressive evolution of "shale gas" in the USA over the past six years. This revolution deserves a suitable analysis and some further comments.

The most important one is that "unconventional gas" (and oil) is not an interim explosion in the North-American industry but a long-standing trend with important consequences in geopolitics, environmental qualification and economics of Energy. Projections of gas production in the States are given in **Figure 11**, where it can be seen that "unconventional gas" will be dominant from now on.

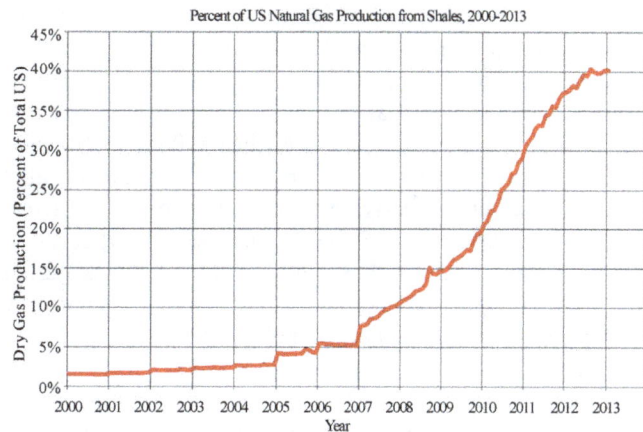

Figure 10. Percentage of shale gas production over total, in the USA.

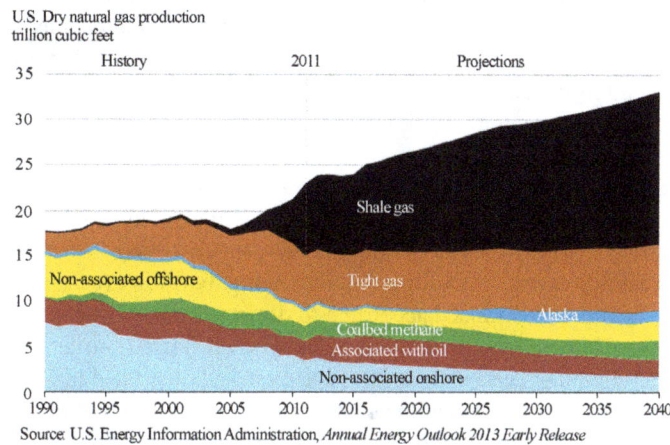

Source: U.S. Energy Information Administration, *Annual Energy Outlook 2013 Early Release*

Figure 11. Share of different types of gas production in the USA.

"Shale" was not in any official agenda, and had not been previously evaluated in environmental terms. It was a "gold rush" created by technology pioneers, sprawling immediately. The full story of "fracking" is still to be told, but we need to highlight that the US DOE was promoting other lines of action. Most of these lines are currently frozen.

The rush has been so fast that potential cost associated with toxic waste or by products are not internalized yet. The main reason for this is a lack of reliable accounting on them. Moreover, shale hydrocarbons have not yet become an international commodity but rather locally traded.

A suitable infrastructure was not available when the gold rush started, and a sort of mismatch went on between drilling and full-scale commercialization. Drilling is not cheap, and completions are expensive. However, the biggest threshold was to develop an infrastructure for exporting goods throughout the country. The core of the infrastructure was for importing crude oil and refined products, and that shift implied a change in mentality and big capital investments.

The development of that network will only take a couple of years (partially aided by the Obama's Administration support and commitment) and will start a new age (one of uncertain time duration) where oil will not be linked only to critical domains.

Important countries as United States and Canada, with huge shale reserves, can convey a lot of stability to the oil market in the mid-term, mainly in terms of supply.

This process to achieve the international market cannot be deployed faster. "Fracking" was born and grew up without asking permission, and without it being present in the energy agenda or planning.

This delay is neither good for the energy system nor goof for shale-hydrocarbons producers. Potential revenues are not realized because they do not have yet access to the global market.

Figure 12 shows the differences in price among three major gas markets. They are paramount, and they have several root causes. Access to cheap reserves is one of them, and this includes a long term of guaranteed supply.

Short term profit has been in this case the driver planning tool. The process was somehow capped; but it has not experienced a strong change, due to a change in priority in President Obama's Administration. At first, this Administration seemed very committed to global environmental issues. For example, it created in 2009 a new entity called "White House Office of Energy and Climate Change Policy". However, it this entity did not quite deliver on its expectations as it was integrated in the "Domestic Policy Council" after two years.

Anyhow, first presidential statements clearly indicated direction in policies related to energy. In the vice-president's memorandum dated December 15, 2009, under the title "Progress report: The transformation to a clean energy economy", the driving force for those policies are defined as "jumpstarting a major transformation of our energy system including unprecedented growth in the generation of renewable sources of energy, enhanced manufacturing capacity for clean energy technology, advanced vehicle and fuel technologies, and a bigger, better, smarter electric grid."

In fact the report is divided in the following sections: Renewable Energy, Vehicles and Fuels of the Future, Grid Modernization, Energy Efficiency, Carbon Capture, Nuclear Power, and Science and Innovation. Neither the word "shale" nor "fracking" appear in the text.

• In current times, "unconventional gas" has already been embodied in Obama's Administration. Efforts have been made to justify it not only by private profit, but according to a philosophy established on the following points: "Unconventional gas" can be made "sustainable" by selecting the regions where it can be exploited and establishing the corresponding regulations.

• It is a very good element to support freedom from Russian energy dependence and freedom from Persian Gulf oil dependence.

• Unconventional gas profits will have to be invested, in a sizeable fraction, into Clean Energy Projects.

Industrial and commercial corporations, as anyone else, live in a certain social and legal environment, including technical regulations. A specific feature of corporations is their strong dependence of being profitably fitted to that environment. Adaption is important for any living being, but it is even more critical in companies. Companies can easily be created or discontinued according to decisions made by investors, who act as watchdogs for corporation profitability. Although each corporation has its own internal structure, coherent to its objective, a generic description of its functional articulation can be depicted as illustrated in **Figure 13**.

Money (and Profit) is usually the criterion to assess the health of a corporation. However, some intangible assets (such as know-how and expertise) should also be evaluated in terms of potential profit for the future.

If money is the outermost shell which is indeed connected to the general financial system, the Physical Assets

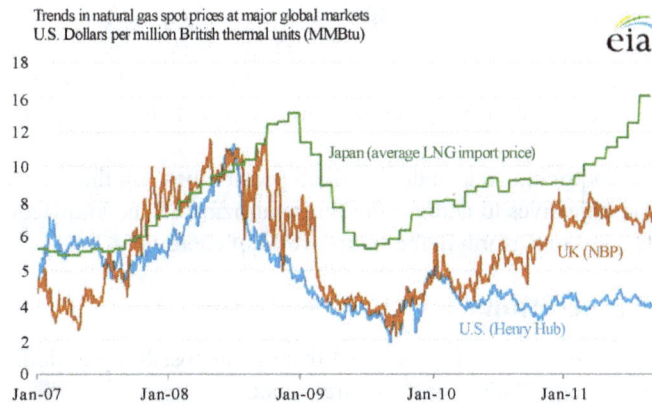

Figure 12. Differences in price among three major gas markets.

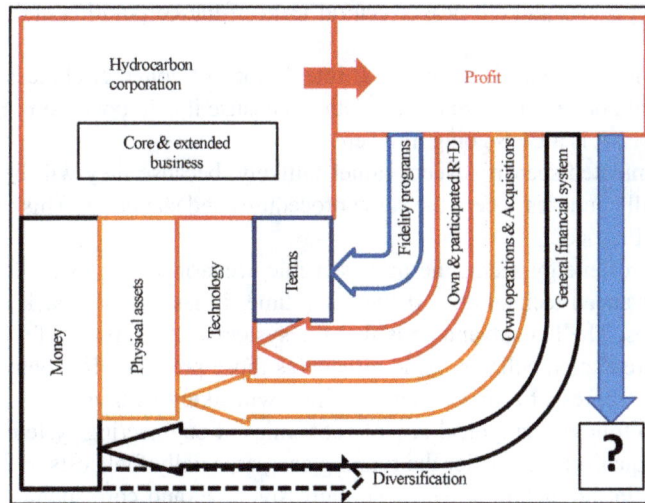

Figure 13. A schematic flowchart of activities in a corporation.

of the company, acquired along its history, constitute the bones of such particular body.

Technology is the third shell of the company; and a clear distinction can be seen between this shell and the Physical Assets: these ones are private property of the corporation, and Technology is in many cases a "cloud" of knowledge, which can be partially protected by patents and other rights, but it is created out of scientific and commercial knowledge, and it can therefore evolve and be improved.

In the innermost shell we find the people, or teams, of the company. They can go out as they can come in, and properly speaking are not assets of the corporation, regardless of confidentiality clauses.

People however, constitute the nervous system and the brain, and this is why the so called "human factor" is so important. This is none the less outside the scope of this paper.

The objective we had when setting the foundations of this work was to analyze the main features of the evolution of Energy macro-policies, and to analyze how Energy corporations should do for being successful in the new Energy scenario where it will have to compete. , which includes a long series of influences characterized booth by physical and evanescent variables.

The answer to this question is very complex, and in some cases will not exist as such, and it will be a collection of considerations and advises coming from the general analysis of the Energy sector.

A Directive from the European Union "European alternative fuels strategy" (COM) 2013, 17) requires Member States to adopt national policy frameworks to development markets for alternative fuels and develop their infrastructure. The Directive sets binding targets for the build-up of alternative fuel infrastructure. An accompanying Impact Assessment (SWD (2013) 5/2) evaluates cost and benefits of different policy options and identifies conditions for a comprehensive coverage of the main alternative fuel options.

The assessment is aimed at being totally neutral in both economics and technology terms. This boundary condition limits drastically the scope of the assessment. The Directive in opens all possible ways (different from the classical ones) and does not attempts to define a priority list of actions based on a given number of criteria's. The Directive seeks insight of long-term potential, environmental impact, security and flexibility of energy supply.

It does not mean that a corporation has to deal with all possibilities. On the contrary, a corporation must identify the ways, means, and objectives to address the future utilizing all the knowledge and judgment acquired in the past to succeed in its quest to perform in the future world of energy.

6. Summary and Expectations

From any known analysis, it must be understood that there is no possibility to define and develop a unified and integrally coherent Global Energy Policy for the entire planet.

Even in the world covered by the International Energy Agency, a unique policy seems out of reach. Nevertheless, corporations can be oriented by analytical approaches to the scenarios envisioned by the prospective work done by several agencies and the proper central team of the corporation and external specialized consultants.

Scenarios identified in this analysis as potential integrations of coherent choices are depicted in **Figure 14**. Business as usual should not be considered as an option, because it is impossible to ignore the strong changes in technologies and markets. It is a BUS going nowhere.

Environmentally dominated scenarios seem rather unlikely, because they will be very expensive and would only be meaningful if all countries accept the same procedures and standards. This is very unlikely based on the aftermath of the Kyoto Protocol.

Integral optimization of energy seems the most sensible scenario for a more efficient future. It has the very important appeal of advanced technology. At the same time, it has the drawback of expensive investments to deploy new technologies. The key advantage is that the system will be more effective in the long run, reduces fuel consumption, and reduction collateral side effects (as those produced by contamination of combustion by-products) All these will represent a major benefit which will likely overwhelm the initial drawback of higher expenses. This scenario follows the typical law of economics of engineering systems: the higher the investment cost (a proxy to higher quality) the smaller the running cost, especially fuel costs.

Last scenario shown in the picture is the most aggressive one, and corresponds to a general deployment of fracking. Local and global opposition to fracking can disturb this course of action. It is certainly not a well-accepted option within the European Union. It has been briefly explained that this is mainly a North-American story, and a rather fast one. The "fracking" technology has got such momentum that Obama's Administration has changed its general vision of Energy.

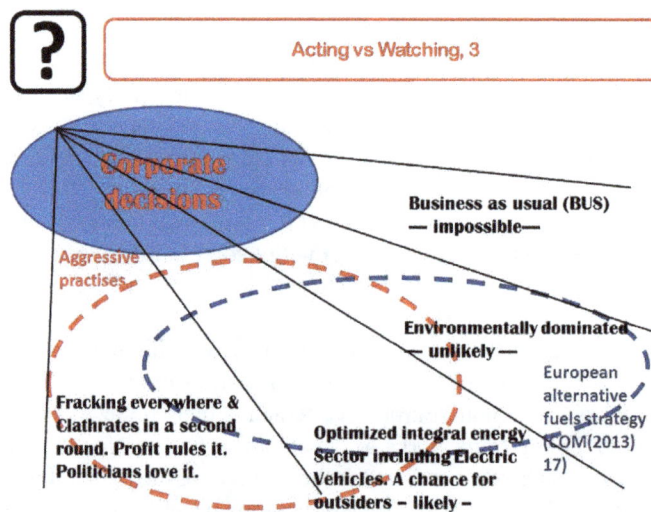

Figure 14. A map of pathways for arriving to different futures.

After considering that fourfold pathway towards a profitable future for the Energy corporation, what advice could be drawn from the contents of this paper? It seems the advice should be to invest and rely on technology. Further, to focus on the effects that the full development of an array of well coupled technologies would have on the market where the corporation works and the corporation itself.

From all papers studies under this scope, technologies that seem mandatory in this advice are: Electric Vehicles; High temperature fuel cells; Distributed and controlled electricity generation; and energy storage.

This can be counter intuitive and somehow a paradox, but oil can contribute towards those objectives (as well as natural gas), with higher efficiencies and lower CO_2 emissions per unit of energy applied to any given end use.

In other words, hydrocarbons can play a unique role in the context of Integral Energy Efficiency. Petroleum products can become the tools that guarantee the energy supply chain thanks to their suitable properties for storage and for feeding new electrochemistry devices.

This means that it is possible to make a corporation that is based on hydrocarbons operation compatible with the requirements for lowering the greenhouse gases emissions. In fact, it is possible because there are many roads for improving synergies among energy conversion mechanisms.

Hence, there seems to be a good agreement between the overall mandate of increasing the Integral Energy Efficiency as a general priority at global scale and the corporation objective of maximizing profits within a framework of commercial activity.

Corporations should be prepared for the higher complexity of future energy sector, with more degrees of freedom in a less segmented fueling structure.

Acknowledgements

This paper is part of the Ph. D work of J.M "Chema" Martinez-Val, at Madrid Technical University (UPM) in the School of Mines. The Ph. D work is carried out under the direction of professors Maldonado and Rodriguez-Pons. He also studied at The French Institute of Petroleum and the Colorado School of Mines. He is a Registered Engineer at Texas Board of Engineering.

References

[1]   International Energy Agency (2014) http://www.iea.org/etp/explore/

[2]   International Energy Agency (2014) World Energy Outlook 2014. www.worldenergyoutlook.com

[3]   U.S. Energy Information Administration (2014) Annual Energy Outlook 2014 with Projections to 2040. http://www.eia.gov/

[4]   IEEE (2014) IEEE National Energy Policy Recommendations. www.ieeeusa.org

[5]   International Monetary Fund (2012) Coping with High Debt and Sluggish Growth. http://www.imf.org/external/pubs/ft/weo/2012/02/

[6]   Chema Martínez-Val, J.M., Maldonado-Zamora, A. and Ramon Rodríguez Pons-Esparver, R.R. (2013) Adapting Business of Energy Corporations to Macro-Policies Aiming at a Sustainable Economy: The Case for New Powering of Automobiles. *Energy and Power Engineering*, **5**, 92-108. http://dx.doi.org/10.4236/epe.2013.51010

[7]   Chema Martínez-Val, J.M. (2013) Improving the Global Energy Industry by Integrating Macro-Technologies: Challenges and Opportunities for Corporations. *Energy and Power Engineering*, **5**, 604-621. http://dx.doi.org/10.4236/epe.2013.510067

[8]   Red Eléctrica de Espa—A Website. http://www.ree.es/sites/default/files/downloadable/sintesis_ree_2013_v1.pdf

[9]   Perez, Y. and Ramos-Real, F.J. (2009) The Public Promotion of Wind Energy in Spain from the Transaction Costs Perspective 1986-2007. *Renewable and Sustainable Energy Reviews*, **13**, 1058-1066. http://dx.doi.org/10.1016/j.rser.2008.03.010

[10] Martínez Montes, G., Serrano López, M.M., Rubio Gámez, M.C., Menéndez Ondina, A. (2005) An Overview of Renewable Energy in Spain. The Small Hydro-Power Case. *Renewable and Sustainable Energy Reviews*, **9**, 521-524.

[11] Hernández, F., Gual, M.A., Del Río, P. and Caparrós, A. (2004) Energy Sustainability and Global Warming in Spain. *Energy Policy*, **32**, 383-394. http://dx.doi.org/10.1016/S0301-4215(02)00308-7

[12] Foidart, F., Oliver-Solá, J., Gasol, C.M., Gabarrell, X. and Rieradevall, J. (2010) How Important Are Current Energy Mix Choices on Future Sustainability? Case Study: Belgium and Spain—Projections towards 2020-2030. *Energy Policy*, **38**, 5028-5037. http://dx.doi.org/10.1016/j.enpol.2010.04.028

[13] Batlle, C. and Rodilla, P. (2010) A Critical Assessment of the Different Approaches Aimed to Secure Electricity Generation Supply. *Energy Policy*, **38**, 7169-7179. http://dx.doi.org/10.1016/j.enpol.2010.07.039

[14] Moreno, F. and Martinez-Val, J.M. (2011) Collateral Effects of Renewable Energies Deployment in Spain: Impact on Thermal Power Plants Performance and Management. *Energy Policy*, **39**, 6561-6574. http://dx.doi.org/10.1016/j.enpol.2011.07.061

[15] European Union (2011) Materials Roadmap Enabling Low Carbon Energy Technologies. Commission Staff Working Paper, SEC (2011) 1609 Final.

[16] European Union (2013) Guidelines for Financial Incentives for Clean and Energy Efficient Vehicles. Commission Staff Working Document, SWD (2013) 27 Final.

[17] Klaassen, G. and Riahi, K. (2007) Internalizing Externalities of Electricity Generation: An Analysis with Message-Macro. *Energy Policy*, **35**, 815-827. http://dx.doi.org/10.1016/j.enpol.2006.03.007

[18] Intergovernmental Panel on Climate Change (2007) Fourth Assessment Report, 2007. www.ipcc.ch

[19] Terrell, H. (2011) US Gas Reserves Estimated at Record High. *World Oil*, **232**, 13.

[20] Hydrocarbon Processing (2011) Natural Gas Enters a New Era of Abundance. *Hydrocarbon Processing*, **90**.

[21] McIlvaine, R. and James, A. (2010) The Potential of Shale Gas. *World Pumps*, **7**, 16-18. http://dx.doi.org/10.1016/S0262-1762(10)70195-4

[22] Lebre, E., Borghetti, J., Basto, L. and Lauria, T. (2010) Sustainable Expansion of Electricity Sector: Sustainability Indicators as an Instrument to Support Decision Making. *Renewable and Sustainable Energy Reviews*, **14**, 422-429.

[23] Holsapple, C.V. and Singh, M. (2001) The Knowledge Chain Model: Activities for Competitiveness. *Expert Systems with Applications*, **20**, 77-98. http://dx.doi.org/10.1016/S0957-4174(00)00050-6

[24] Shina, M., Holden, R. and Schmidt, R.A. (2001) From Knowledge Theory to Management Practice: Towards an Integrated Approach. *Information Processing & Management*, **37**, 335-355. http://dx.doi.org/10.1016/S0306-4573(00)00031-5

[25] Liao, W., Heijungs, R. and Huppes, G. (2011) Is Bioethanol a Sustainable Energy Source? An Energy-, Exergy-, and Emergy-Based Thermodynamic System Analysis. *Renewable Energy*, **36**, 3479-3487. http://dx.doi.org/10.1016/j.renene.2011.05.030

[26] Luo, L., Van Der Voet, E. and Huppes, G. (2009) Life Cycle Assessment and Life Cycle Costing of Bioethanol from Sugarcane in Brazil. *Renewable and Sustainable Energy Reviews*, **13**, 1613-1619. http://dx.doi.org/10.1016/j.rser.2008.09.024

[27] Hanff, E., Dabat, M.-H. and Blin, J. (2011) Are Biofuels an Efficient Technology for Generating Sustainable Development in Oil-Dependent African Nations? A Macroeconomic Assessment of the Opportunities and Impacts in Burkina Faso. *Renewable and Sustainable Energy Reviews*, **15**, 2199-2209.

[28] Koh, M.Y. and Ghazi, T.I.M. (2011) A Review of Biodiesel Production from *Jatropha curcas* L. Oil. *Renewable and Sustainable Energy Reviews*, **15**, 2240-2251. http://dx.doi.org/10.1016/j.rser.2011.02.013

[29] Kinnaman, T.C. (2011) The Economic Impact of Shale Gas Extraction: A Review of Existing Studies. *Ecological Economics*, **70**, 1243-1249. http://dx.doi.org/10.1016/j.ecolecon.2011.02.005

[30] Arthur, J.D., Hochheiser, H.W. and Coughlin, B.J. (2011) State and Federal Regulation of Hydraulic Fracturing: A Comparative Analysis. *Proceedings of the SPE Hydraulic Fracturing Technology Conference*, The Woodlands, Texas, 24-26 January 2011.

[31] Lafollette, R.F. and Holcomb, W.D. (2011) Practical Data Mining: Lessons Learned from the Barnett Shale of North Texas". *Proceedings of the SPE Hydraulic Fracturing Technology Conference*, The Woodlands, Texas, 24-26 January 2011.

[32] BP Statistical Review of World Energy 2014. http://www.bp.com/en/global/corporate/about-bp/energy-economics/statistical-review-of-world-energy.html

[33] Eurostats (2014) Energy Price Statistics. http://epp.eurostat.ec.europa.eu/statistics_explained/index.php/Energy_price_statistics

[34] European Union Directive 2009/28 CE. http://eur-lex.europa.eu/legal-content/EN/ALL/?uri=CELEX:32009L0028

[35] Brundtland, G. (1987) Our Common Future. Report, United Nations World Commission on Environment and Development. Oxford University Press, UK.

[36] IPCC, Mitigation of Climate Change. http://report.mitigation2014.org/spm/ipcc_wg3_ar5_summary-for-policymakers_approved.pdf

[37] Gellings, P.J. and Bouwmeester, H.J. (1997) Handbook of Solid State Electrochemistry. CRC Press, Boca Raton.

[38] Brandon, N.P. and Thompsett, D. (2005) Fuel Cells Compendium. Elsevier, Oxford.

[39] Ference Weiker & Company Ltd. (2010) Assessment of the Economic Impact of the Canadian Hydrogen and Fuel Cell Sector. British Columbia Ministry for Technology and Economic Development.

[40] Godo, H., Nedrum, L., Rapmund, A. and Nygaard, S. (2003) Innovations in Fuel Cells and Related Hydrogen Technology in Norway—OECD Case Study in the Energy Sector. NIFU Report 35/2003.

[41] Wyld Group Pty Ltd. (2008) Hydrogen Technology Roadmap. Australian Government, Department of Resources, Energy and Tourism, ABN:53099078485.

[42] Ormerod, M. (2003) Solid Oxide Fuel Cells. *Chemical Society Reviews*, **32**, 17-28. http://dx.doi.org/10.1039/b105764m

[43] Park, S., Vohs, J.M. and Gorte, R.J. (2000) Direct Oxidation in a Solid-Oxide Fuel Cell. *Nature*, **404**, 265-267.

[44] Tse, L.K.C., Wilkins, S., McGlashan, N., Urban, B. and Martinez-Botas, R. (2011) Solid Oxide Fuel Cell/Gas Turbine Trigeneration System for Marine Applications. *Journal of Power Sources*, **196**, 3149-3162. http://dx.doi.org/10.1016/j.jpowsour.2010.11.099

[45] Turton, H. (2006) Sustainable Global Automobile Transport in the 21st Century: An Integrated Scenario Analysis. *Technological Forecasting & Social Change*, **73**, 607-629. http://dx.doi.org/10.1016/j.techfore.2005.10.001

[46] Thomas, C.E. (2009) Fuel Cell and Battery Electric Vehicles Compared. *International Journal of Hydrogen Energy*, **34**, 6005-6020. http://dx.doi.org/10.1016/j.ijhydene.2009.06.003

[47] Van Mierlo, J., Maggetto, G. and Lataire, Ph. (2006) Which Energy Source for Road Transport in the Future? A Comparison of Battery, Hybrid and Fuel Cell Vehicles. *Energy Conversion and Management*, **47**, 2748-2760.

[48] Moriarty, P. and Honnery, D. (2008) Low-Mobility: The Future of Transport. *Futures*, **40**, 865-872. http://dx.doi.org/10.1016/j.futures.2008.07.021

[49] Moriarty, P. and Honnery, D. (2008) The Prospects for Global Green Car Mobility. *Journal of Cleaner Production*, **16**, 1717-1726. http://dx.doi.org/10.1016/j.jclepro.2007.10.025

[50] Mullen, M. (2010) The State of Shale Plays in Europe. *World Oil*, **231**, D-79.

[51] Editorial News (2011) Quebec's Shale-Gas Moratorium. *Petroleum Economist*, **78**.

[52] Deluchi, M.A., Wang, Q. and Sperling, D. (1989) Electric Vehicles: Performance, Life Cycle Costs, Emissions and Recharging Requirements. *Transportation Research*, **23A**, 255-270.

[53] Schot, J., Hoogma, R. and Elzen, B. (1994) Strategies for Shifting Technological Systems: The Case of the Automobile System. *Futures*, **26**, 1060-1076. http://dx.doi.org/10.1016/0016-3287(94)90073-6

[54] Ford, A. (1994) Electric Vehicles and the Electric Utility Company. *Energy Policy*, **22**, 555-570. http://dx.doi.org/10.1016/0301-4215(94)90075-2

[55] Plowshare Project. https://www.osti.gov/opennet/reports/plowshar.pdf

19

Gedanken Experiment Examining How Kinetic Energy Would Dominate Potential Energy, in Pre-Planckian Space-Time Physics, and Allow Us to Avoid the BICEP 2 Mistake

Andrew Walcott Beckwith

Physics Department, Chongqing University Huxi Campus, Chongqing, China
Email: Rwill9955b@gmail.com, abeckwith@uh.edu

Abstract

We use Padmabhan's "Invitation to Astrophysics" formalism of a scalar field evolution of the early universe, from first principles, to show something which seems counter intuitive. How could, just before inflation, kinetic energy be larger than potential energy in pre-Planckian physics, and what physics mechanism is responsible for the Planckian physics result that Potential energy is far larger than kinetic energy. This document answers that question, as well as provides a mechanism for the dominance of kinetic energy in pre-Planckian space-time, as well as its reversal in the Planckian era of cosmology. The kinetic energy is proportional to $\rho_w \sim g^* T^4$, with g^* initial degrees of freedom, and T the initial temperature just before the onset of inflation. Our key assumption is the smallness of curvature, as given in the first equation, which permits adoption of the Potential energy and Kinetic energy formalism used, in the Planckian and pre-Planckian space-time physics. Interpretation of this result, if done correctly, will be able to allow a correct distinguishing of relic gravitational waves, as to avoid the BICEP 2 pickup of galactic dust as a false relic Gravitational wave signal, as well as serve as an investigative template as to if quantum gravity is embedded in a deterministic dissipative system, as cited in the conclusion.

Keywords

Geddanken Experiment, Kinetic Energy, Potential Energy

1. Introduction

We begin with a review from T. Padmanabhan [1] as to the foundations of a scalar field and a potential field, in terms of cosmological evolution. Following that, we are adding more detail as to a supposition by Handley *et al.* [2]. As to how one could invert the supposition of inflation, [1], that the Kinetic energy would be much larger than the Potential energy [2]. Here, we offer a mechanism for how this may happen. This is in lieu as to C. VAN DEN BROECK in [3] namely that we need to distinguish between multiple early universe sources of gravitational wave, or the onset of inflation generated early gravitational waves, as seen in the below quotation

> *"Omni-directional gravitational wave background radiation could arise from fundamental processes in the early Universe, or from the superposition of a large number of signals with a point-like origin. Examples of the former include parametric amplification of gravitational vacuum fluctuations during the inflationary era"*

In particular our approach is similar to the procedures outlined as to vacuum fluctuations in the (very early) inflationary era. And we will outline how we are avoiding multiple sources. Which would be done if there is a sharply focused peak frequency in magnitude and intensity, *i.e.* we wager that a sharply defined peak of GW amplitude and strain is about the only way to avoid the problems associated with Bicep 2, which was due to dust from galactic sources [3]-[5].

Our gedanken experiment is a thought experiment, that kinetic energy dominates potential energy in the initial phase of cosmological evolution. As a second component of our thought experiment is then the smallness of curvature, so to a good first approximation so that then the formulation of near flatness in the beginning of space-time is a starting point for surprisingly traditional looking Friedman cosmological equations, due to the minimal influence of curvature, which would be a consequence of [6].

Furthermore in our thought experiment is the supposition that the dominance of kinetic energy is in the regime of space-time before a Planck time interval has transpired. Implicit in this is the supposition that there is such a regime of space-time which the author has also worked out in [7]. If there is a regime of space-time just before Planck time, and this convention adheres to [2] that kinetic energy is, indeed greater than potential energy. In the pre-Planckian space time regime we will have, then a kinetic energy term with imaginary time, and that then the points just before Planck time will have a Potential energy as given by an imaginary time component as set by

$$\dot{\phi}^2 \left(\tau_{\text{Pre-Planckian-time}} \right) \Big|_{\tau_{\text{Pre-Planckian-time}} = \frac{\pm i \times 10^{-44} \text{ s}}{10^{-44} \text{ s}}} \tag{1}$$

Here, what has done, is to scale real to imaginary time, and to normalize it by a division of 10^{-44} seconds, effectively having a "rescaled" pre-Planckian time which would be written as an imaginary number with

$$\tau_{\text{Pre-Planckian-time}} = \frac{\pm i \times 10^{-44} \text{ s}}{10^{-44} \text{ s}} \tag{2}$$

This means that in an interval just before the Planck time, that we are postulating a pre-Planckian space-time which is purely imaginary, *i.e.* operationally Equation (2) is, effectively in the regime of analysis of Kinetic energy dominance given by

$$\tau_{\text{Pre-Planckian-time}} = \pm i \tag{3}$$

The change in time from imaginary to real time, is in line with a full evaluation of a transition from Equation (3) to the regime of space-time dominated by

$$t_{\text{real-time}} = \frac{t_{\text{measured-time}}}{t_{\text{Planck}} \sim 10^{-44} \text{ s}} \tag{4}$$

In particular what we are supposing is a transformation from the pre-Planckian to Planckian regime given by

$$\tau_{\text{Pre-Planckian-time}} = \left(\frac{\pm i \times 10^{-44} \text{ s}}{10^{-44} \text{ s}} \right) \xrightarrow{\text{Causal-barrier}} t_{\text{real-time}} = \left(\frac{t_{\text{measured-time}}}{t_{\text{Planck}} \sim 10^{-44} \text{ s}} \right) \tag{5}$$

The right hand side would represent non imaginary time, whereas the regime where there is imaginary structure would represent sub divisions in causal structure of space-time, as alluded to by Dowker [5] [8] with what is represented in Equation (5) as a division from pre-Planckian to Planckian space-time, and a defacto Causal structure with the imaginary time component one which is "tunneled" through, to reach Planckian Space-time physics.

What is conserved between the transference from a prior universe, to present universe would be at the Planckian regime, a minimum stress energy Tensor, which by [3] is given as a term with has the label of T_{00} for transferred space-time energy, and then a measure of curvature of space-time as given in [6] as, if

$$T_{00} = \rho_{\text{Energy-density}} = \frac{-(g_{00} = 1)}{16 \cdot \pi} \cdot \left(\frac{3\tilde{k}_{\text{Curvature-measure}}}{a_{\text{initial-scale-factor}}^2} + \Lambda_{\text{initial-value}} \right) \qquad (6)$$

The $\tilde{k}_{\text{Curvature-measure}}$ is defined in [6] and can be usually thought of as how gravity affects space-time geometry, as given in [9] which is part of the definition of a Riemann Scalar, for which as given in [6] we have that the Riemann scalar is given as

$$\Re = \text{Riemann-scalar} = \frac{-6 \cdot \tilde{k}_{\text{Curvature-measure}}}{\left(a_{\text{initial-value}} \right)^2} \qquad (7)$$

The term for a minimum scale factor is given in Equation (7) which is the smallest unit of evolving space-time which is accessible to analysis, which is akin to adopting the bounce definition of a non-singular starting point of a universe expansion as given in [7] [8] [10] [11] with a minimum initial scale value given in [3] [6] as

$$a_{\text{initial-value}} \sim 10^{-55} \qquad (8)$$

If so, then we find then that, if we are looking at a cosmological constant, as defined in [9] [10] [12] [13] with an initial value of this cosmological constant as written as $\Lambda_{\text{initial-value}}$ (leaving open the possibility that the cosmological constant could have changed over time, *i.e.* quintessence, as seen in [11] [14]), that then, as given in part by [3] [6], where $V^{(3)}$ is three space volume, not including the time dimension,

$$\begin{aligned} T_{00} = \rho_{\text{Energy-density}} &= \frac{-(g_{00} = 1)}{16 \cdot \pi} \cdot \left(\frac{3\tilde{k}_{\text{Curvature-measure}}}{a_{\text{initial-scale-factor}}^2} + \Lambda_{\text{initial-value}} \right) \\ \Leftrightarrow \tilde{k}_{\text{Curvature-measure}} &= -\frac{a_{\text{initial-scale-factor}}^2}{3} \times \left[\left(16 \cdot \pi \cdot \left(\rho_{\text{Energy-density}} = \left[\frac{S_{\text{initial-Entropy}} \cdot m_{\text{graviton}}}{V^{(3)}} \right] \right) \right) + \Lambda_{\text{initial-value}} \right] \end{aligned} \qquad (9)$$

Here in part we are partaking of the idea of a massive graviton, of about 10^{-62} grams, as given by [12] [15]. and we also, to those whom object to the idea of a massive Graviton refer to Maggiore [13] [16]

$$-3m_{\text{graviton}} h = \frac{\kappa}{2} \cdot T \qquad (10)$$

Here, the T is the trace of the Einstein Stress energy Tensor, whereas h is the trace of h_{uv}. In our case, if one took the trace of the massless graviton, h_{uv} would be identically zero in line with $T = 0$, whereas in the formulation given by Maggiorie [16] if one did not have the trace of h_{uv} not equal to zero, one is assuming a modification of the usual massless spin 2 graviton, whereas in terms of our treatment of T, we are effectively, due to the work we did in [7] restricting the T to be equivalent to T_{00} in the pre-Planckian treatment of space-time which is equivalent to looking at, if T_{00} is the same as the time component of the Stress energy tensor in the pre-Planckian regime of space time, that we have by [7] that

$$\begin{aligned} T\left(= trace T_{uv} \right) &\approx \Delta T_{tt} \sim \Delta \rho \sim \frac{\Delta E}{V^{(3)}} \\ \frac{\Delta E}{V^{(3)}} &= \frac{\hbar}{\delta t \cdot \delta g_{tt} \cdot V^{(3)}} \equiv \frac{\hbar}{\delta t \cdot V^{(3)} \cdot a^2(t) \cdot \phi} \\ \Leftrightarrow \Delta m_{\text{graviton}}^2 \left(\text{rest-mass} \right) &\sim \frac{\left(1 - \left[v(\text{velocity})_{\text{graviton}} / c \right]^2 \right) c^2}{72\pi^2 \hbar} \times \left[\frac{\kappa}{\delta t \cdot V^{(3)} \cdot a^2(t) \cdot \phi} \cdot \left(\frac{2\pi v(\text{velocity})_{\text{graviton}}}{\Delta \omega_{\text{graviton}}} \right) \right]^2 \end{aligned} \qquad (11)$$

The term of the ϕ inflaton scalar field factor shows up in the Equation (13) whereas we define the Graviton coupling term κ [16] via, if G is the gravitational constant, we define it as follows, namely

$$G = \frac{4}{3} \cdot \left(\frac{\kappa^2}{32\pi} \right) \tag{12}$$

If the curvature measure, above is almost zero, then we can use from [1], with $V(\phi)$ a cosmological potential energy term (usually $V(\phi)$ is more dominant in inflation, to the kinetic energy, but our thought experiment has that the kinetic energy we call $\dot{\phi}$, i.e. the time derivative of the inflaton field, is the dominant factor before the Planck time regime which starts about 10^{-44} seconds), and the term we will be watching as extremely significant, $\dot{\phi}$, the time derivative of the inflaton field. Here, we define the dynamical equation for H, below i.e. that

$$H = \frac{\dot{a}}{a} = \sqrt{\frac{8\pi G}{3}} \cdot \left[V(\phi) + \frac{\dot{\phi}}{2} \right]^{1/2} \tag{13}$$

If so, then using the Potential energy and Kinetic energy values from [1] we can write the following: Keep in mind that in Equation (14) that the H referred to is defined by Equation (13) above, and so we write below explicit entries for the Potential energy, $V(\phi)$, the Kinetic energy $\dot{\phi}$, and we define a value for Equation (13).

$$
\begin{aligned}
&V(t) = \frac{3H^2}{8\pi G} \cdot \left(1 + \frac{\dot{H}}{3H^2} \right) \\[2mm]
&\phi(t) = \int dt \cdot \sqrt{-\frac{\dot{H}}{4\pi G}} \\[2mm]
&H = \frac{2}{t \cdot \left(1 + \frac{p}{\rho} \right)} \sim \frac{2 \cdot t_{\text{REAL-TIME}}^{-1}}{\xi^+} \\[2mm]
&\Rightarrow \phi(t) \sim \sqrt{\frac{1}{4\pi G}} \cdot \frac{1}{\xi^+} \cdot \log\left[\frac{t_{\text{final-real-time}}}{t_{\text{initial-real-time}}} \right] \\[2mm]
&\Rightarrow \dot{\phi}^2(t) \sim \frac{1}{4\pi G} \cdot \left(\frac{1}{\xi^+} \right)^2 \cdot \left(\frac{1}{t_{\text{real-time}}} \right)^2 \\[2mm]
&\&\ V(t) \sim \frac{3}{2\pi G} \cdot \left(\frac{1}{\xi^+} \right)^2 \cdot \left[\left(\frac{1}{t_{\text{real-time}}} \right)^2 \cdot \left(1 + \left(\frac{1}{t_{\text{real-time}}} \right)^2 \right) \right]
\end{aligned}
\tag{14}
$$

The term $V(t)$ is for Potential energy, and it is by inspection $\gg \dot{\phi}^2(t)$ in the Planckian space-time regime which is the Kinetic energy component, provided that time here is a real co-ordinate.

We will, for now on, to keep this real time non dimensional, make the following identification with, once again, Equation (4), which has $t_{\text{real-time}} = \dfrac{t_{\text{measured-time}}}{t_{\text{Planck}} \sim 10^{-44} \text{ s}}$. For the sake of identification, we will be assuming that

Equation (14) and Equation (4) are in the present universe and that ξ^+ is extraordinarily small.

2. Re Examination of Equation (14) and Equation (4) in a Pre Universe Configuration

Our supposition is that Equation (2) to Equation (5) in the matter of pre-Planckian space time, say in a boundary of 2 times Planck time to buttress the repeating cyclical universe we are assuming as possibility given by Penrose [17], is changed then to take into the quantum bounce analogy we think should be looked at [11] as given by C. Rovelli and F. Vidotto. So then we get from Planckian space time, a real time evaluation which shrinks to imaginary time, via the following rule

$$t_{\text{real-time}} = \frac{t_{\text{measured-time}}}{t_{\text{Planck}} \sim 10^{-44} \text{ s}} \xrightarrow{\text{Pre-Planckian}} \tau_{\text{Pre-Planckian-time}} = i \cdot \frac{t_{\text{measured-time}}}{t_{\text{Planck}} \sim 10^{-44} \text{ s}} = i \cdot t_{\text{real-time}} \qquad (15)$$

i.e. what we are saying is that, there is a retime in the pre-Planckian regime of space-time

In a boundary of $\tau_{\text{Pre-Planckian-time}} \sim \dfrac{\pm i \times 10^{-44} \text{ s}}{10^{-44} \text{ s}}$ *i.e.* about a bounce area, of space time, then there would be this

switch, so then in this regime, we would re write the relevant evaluative time for the Potential and Kinetic ener-

gy as $\tau_{\text{Pre-Planckian-time}} \sim \dfrac{\pm i \times 10^{-44} \text{ s}}{10^{-44} \text{ s}}$.

Pick the following point of evaluation, namely at the transit point between the plus to the minus regions of

$\tau_{\text{Pre-Planckian-time}} \sim \dfrac{\pm i \times 10^{-44} \text{ s}}{10^{-44} \text{ s}}$ that we are looking at a Vanishing Potential energy, but a Kinetic energy which

would be very different from Zero.

$$\dot{\phi}^2 \left(\tau_{\text{Pre-Planckian-time}} \right) \Big|_{\tau_{\text{Pre-Planckian-time}} = \frac{\pm i \times 10^{-44} \text{ s}}{10^{-44} \text{ s}}} \sim \frac{1}{4\pi G} \cdot \left(\frac{1}{\xi^+} \right)^2 \cdot \left(\frac{1}{\tau_{\text{Pre-Planckian-time}}} \right)^2 < 0$$

$$\& \ V \left(\tau_{\text{Pre-Planckian-time}} \right) \Big|_{\tau_{\text{Pre-Planckian-time}} = \frac{\pm i \times 10^{-44} \text{ s}}{10^{-44} \text{ s}}} \sim \frac{3}{2\pi G} \cdot \left(\frac{1}{\xi^+} \right)^2 \cdot \left[\left(\frac{1}{\tau_{\text{Pre-Planckian-time}}} \right)^2 \cdot \left(1 + \left(\frac{1}{\tau_{\text{Pre-Planckian-time}}} \right)^2 \right) \right] = 0 \qquad (16)$$

The fact we have a very large non zero $\dot{\phi}^2 \left(\tau_{\text{Pre-Planckian-time}} \right) \Big|_{\tau_{\text{Pre-Planckian-time}} = \frac{\pm i \times 10^{-44} \text{ s}}{10^{-44} \text{ s}}}$ going into the

$\tau_{\text{Pre-Planckian-time}} \sim \dfrac{\pm i \times 10^{-44} \text{ s}}{10^{-44} \text{ s}}$ region, as a pre-Planckian bounce bubble, with this flipping to

$t_{\text{real-time}} = \dfrac{t_{\text{measured-time}}}{t_{\text{Planck}} \sim 10^{-44} \text{ s}}$ with the result that.

$$\frac{\dot{\phi}^2 \left(\tau_{\text{Pre-Planckian-time}} \right) \Big|_{\tau_{\text{Pre-Planckian-time}} = \frac{\pm i \times 10^{-44} \text{ s}}{10^{-44} \text{ s}}}}{\xrightarrow[\tau_{\text{Pre-Planckian-time}} \to t_{\text{real-time}}]{}} V \left(t_{\text{real-time}} \right) \sim \frac{3}{2\pi G} \cdot \left(\frac{1}{\xi^+} \right)^2 \cdot \left[\left(\frac{1}{t_{\text{real-time}}} \right)^2 \cdot \left(1 + \left(\frac{1}{t_{\text{real-time}}} \right)^2 \right) \right] \neq 0 \qquad (17)$$

In making this evaluation, we are assuming that there could be use of the following for relic Gravitational waves., *i.e.* for Equation (17) to hold we will be looking at a time interval which may be specified by [18] [19]

$$\left(\delta t \right)^2_{\text{emergent}} = \frac{\sum_i m_i l_i \cdot l_i}{2 \cdot \left(E - V \right)} \to \frac{m_{\text{graviton}} l_P \cdot l_P}{2 \cdot \left(E - V \right)} \qquad (18)$$

Initially, as postulated by Babour [18] [19] this set of masses, given in the emergent time structure could be for say the planetary masses of each contribution of the solar system. Our identification is to have an initial mass

value, at the start of creation, for an individual graviton. So If $\left(\delta t \right)^2_{\text{emergent}} = \delta t^2 \sim t_{\text{real-time}} = \dfrac{t_{\text{measured-time}}}{t_{\text{Planck}} \sim 10^{-44} \text{ s}}$ Then

there may be gravitons which are [18] [19]

$$m_{\text{graviton}} \geq \frac{2\hbar^2}{\left(\delta g_{tt} \right)^2 l_P^2} \cdot \frac{\left(E - V \right)}{\Delta T_{tt}^2} \qquad (19)$$

This would entail assuming relic gravitation generated by a massive graviton bounded below by

$$m_{\text{graviton}} \geq \frac{2\hbar^2}{\left(\delta g_{tt}\right)^2 l_P^2} \cdot \left.\frac{(E-V)}{\Delta T_{tt}^2}\right|_{\tau_{\text{Pre-Planckian-time}}=\frac{\pm i \times 10^{-44}\,\text{s}}{10^{-44}\,\text{s}}} \to \frac{2\hbar^2}{\left(\delta g_{tt}\right)^2 l_P^2} \cdot \frac{\left|\dot{\phi}^2\left(\tau_{\text{Pre-Planckian-time}}\right)\right|_{\tau_{\text{Pre-Planckian-time}}=\frac{\pm i \times 10^{-44}\,\text{s}}{10^{-44}\,\text{s}}}}{\Delta T_{tt}^2} \quad (20)$$

And the magnitude of K.E. as defined by

$$(E-V) \sim \left|\dot{\phi}^2\left(\tau_{\text{Pre-Planckian-time}}\right)\right|_{\tau_{\text{Pre-Planckian-time}}=\frac{\pm i \times 10^{-44}\,\text{s}}{10^{-44}\,\text{s}}} \quad (21)$$

If so, then if we use Equation (16) and Equation (20) and Equation (21) so as to obtain

$$m_{\text{graviton}} \geq \frac{2\hbar^2}{\left(\delta g_{tt}\right)^2 l_P^2} \cdot \frac{\left|\dot{\phi}^2\left(\tau_{\text{Pre-Planckian-time}}\right)\right|_{\tau_{\text{Pre-Planckian-time}}=\frac{\pm i \times 10^{-44}\,\text{s}}{10^{-44}\,\text{s}}} = \frac{1}{4\pi G} \cdot \left(\frac{1}{\xi^+}\right)^2 \cdot \left|\left(\frac{1}{\tau_{\text{Pre-Planckian-time}}}\right)^2\right|}{\Delta T_{tt}^2} \quad (22)$$

If so, then, we come to a conclusion, which uses a basic energy density result from Kolb and Turner [9] that the Kinetic energy, as defined in pre-Planckian physics, as defined in this document is decisively important, as given in the conclusion.

3. Re-Examining Relic Gravitational Wave Models as to What Relic Gravitational Waves Could Tell Us about the Origins of the Early Universe

It is very noticeable that in [21] we have that the following quote is particularly relevant to consider, in lieu of our results

"Thus, if advanced projects on the detection of GWs will improve their sensitivity allowing to perform a GWs astronomy (this is due because signals from GWs are quite weak) [16], one will only have to look the interferometer response functions to understand if General Relativity is the definitive theory of gravity. In fact, if only the two response functions (2) and (19) will be present, we will conclude that General Relativity is definitive. If the response function (22) will be present too, we will conclude that massless Scalar-Tensor Gravity is the correct theory of gravitation. Finally, if a longitudinal response function will be present, i.e. Equation (25) for a wave propagating parallel to one interferometer arm, or its generalization to angular dependences, we will learn that the correct theory of gravity will be massive Scalar-Tensor Gravity which is equivalent to f(R) theories. In any case, such response functions will represent the definitive test for General Relativity. This is because General Relativity is the only gravity theory which admits only the two response functions (2) and (19) [21]. Such response functions correspond to the two "canonical" polarizations h+ and h×. Thus, if a third polarization will be present, a third response function will be detected by GWs interferometers and this fact will rule out General Relativity like the definitive theory of gravity"

We argue that a third polarization in Gravitational waves from the early universe may be detected, if there is proof positive that in the pre-Planckian regime that the Corda conjecture [20] as given below, namely if the following analysis is part of our take on relic gravitational waves, is supported by the kinetic energy being larger than the potential energy, namely what if

"The case of massless Scalar-Tensor Gravity has been discussed in [22] [23] with a "bouncing photons analysis" similar to the previous one. In this case, the line-element in the TT gauge can be extended with one more polarization, labelled with $\Phi(t + z)$, i.e. ..."

i.e. the dominance of Kinetic energy over Potential energy, in pre-Planckian physics could serve as a template for verification for the existence of a third polarization along the lines brought up in [20] and its confirmation or falsification would yield foundational insight available nowhere else possible.

4. Conclusions

Our hypothesis, as to Equation (22), is equivalent to what is frequently postulated as an energy density as given by Kolb and Turner [12]. First of all the below is equivalent to T_{00}, *i.e.* the T_{00} is the same as

$$\rho_w \propto a^{-3(1-w)} \sim g^* T^4 \tag{23}$$

If so, then the lower bound to the graviton will be then as given by Equation (24) below, if we use Equation (23), then

$$m_{\text{graviton}} \geq \frac{2\hbar^2}{\left(\delta g_{tt}\right)^2 l_P^2} \cdot \frac{\dfrac{1}{4\pi G} \cdot \left(\dfrac{1}{\xi^+}\right)^2 \cdot \left|\left(\dfrac{1}{\tau_{\text{Pre-Planckian-time}}}\right)\right|^2}{\Delta T_{tt}^2 = \left(\rho_w \propto a^{-3(1-w)} \sim g^* T^4\right)^2} \tag{24}$$

i.e. if we have a comparatively low initial temperature, T, it will mean a large initial graviton mass. If the T is of the order of Planck temperature, say 10^{32} Kelvin, then the above will have a lower graviton mass value of about 10^{-66} grams. It goes up if there is what is called a (colder cosmology) about 1 order of magnitude lower initial temperature, leading to the mass of a graviton bounded below by 10^{-58} grams. Really cold initial temperatures far lower than 10^{32} degrees Kelvin for T would lead to maybe even 10^{-50} grams for the initial lower bound to the graviton mass.

For consistency with the 10^{-62} gram value as given by [15] we would probably be considering it desirable for 10^{32} degrees Kelvin for T. In all this we will be considering g^* initial degrees of freedom, of about 100, in terms of what was given by Kolb and Turner [12].

It is worth noting that Dr. Corda in [21] has extended the Maggiorie results [16] as given in the prior reference [21] section and that indeed Maggiore studied the detectability only for GWs having a wavelength very much longer than the interferometer's arms, while Corda [21] extended the results to all the GWs wavelengths. The importance of this contribution is, if we find out if there is a third polarization as indicated above, possibly due to a dominance of kinetic energy, *i.e.* the dominance in a pre-Planckian mode of space time may allow for settling the question given in [21], with an appropriately chosen magnitude, and frequency, and also allow for avoiding the mistake of Bicep 2, as given in references [21] [24].

We also state unequivocally, that confirmation of this result would give reality to the suppositions given in references [16]-[20], which would through analysis help toward falsifiable measurements which would allow us to determine if Quantum physics and quantum gravity are, indeed part of a larger non deterministic theory, as given in [25].

Acknowledgements

This work is supported in part by National Nature Science Foundation of China grant No. 11375279.

References

[1]   Padmanabhan, T. (2006) An Invitation to Astrophysics. World Scientific Co., Pte. Ltd., Singapore.

[2]   Handley, W.J., Brechet, S.D., Lasenby, A.N. and Hobson, M.P. (2014) Kinetic Initial Conditions for Inflation. http://arxiv.org/pdf/1401.2253v2.pdf

[3]   Van Den Broeck, C., *et al.* (2015) Gravitational Wave Searches with Advanced LIGO and Advanced Virgo. http://arxiv.org/pdf/1505.04621v1.pdf

[4]   Cowen, R. (2015) Gravitational Waves Discovery Now Officially Dead; Combined Data from South Pole Experiment BICEP2 and Planck Probe Point to Galactic Dust as Confounding Signal. http://www.nature.com/news/gravitational-waves-discovery-now-officially-dead-1.16830

[5]   Cowen, R. (2014) Full-Galaxy Dust Map Muddles Search for Gravitational Waves. http://www.nature.com/news/full-galaxy-dust-map-muddles-search-for-gravitational-waves-1.15975

[6]   Beckwith, A. (in press) Geddankenexperiment for Degree of Flatness, or Lack Of, in Early Universe Conditions. Accepted for publication, *JHEPGC*. http://vixra.org/abs/1510.0108

[7]   Beckwith, A. (2016) "Gedanken Experiment for Fluctuation of Mass of a Graviton, Based on the Trace of a GR Stress Energy Tensor-Preplanckian Conditions Lead to Gaining of Graviton Mass, and Planckian Conditions Lead to Graviton Mass Shrinking to 10^{-62} Grams. *Journal of High Energy Physics, Gravitation and Cosmology*, **2**, 19-24. http://vixra.org/abs/1510.0495

[8]   Dowker, F. (2005) Causal Sets and the Deep Structure of Space-Time. http://arxiv.org/abs/gr-qc/0508109

[9]   Katti, A. (2013) The Mathematical Theory of Special and General Relativity. Create Space Independent Publishing, North Charleston.

[10]   Haggard, H.M. and Rovelli, C. (2014) Black, Hole Fireworks: Quantum Gravity Effects Outside the Horizon Spark Black to White Hole Tunneling. http://arxiv.org/abs/1407.0989

[11]   Rovelli, C. and Vidotto, F. (2015) Covariant Loop Quantum Gravity: An Elementary Introduction to Quantum Gravity, and Spinfoam Theory. Cambridge University Press, Cambridge.

[12]   Kolb, E.W. and Turner, M.S. (1990) The Early Universe. The Advanced Book Program, Addison-Wesley Publishing Company, Redwood City.

[13]   Padmanabhan, T. (2010) Gravitation, Foundations and Frontiers. Cambridge University Press, Cambridge.

[14]   Ratra, P. and Peebles, L. (1988) Cosmological Consequences of a Rolling Homogeneous Scalar Field. *Physical Review D*, **37**, 3406. http://dx.doi.org/10.1103/PhysRevD.37.3406

[15]   Goldhaber, A. and Nieto, M. (2010) Photon and Graviton Mass Limits. *Reviews of Modern Physics*, **82**, 939-979.
        http://arxiv.org/abs/0809.1003
        http://dx.doi.org/10.1103/RevModPhys.82.939

[16]   Maggiorie, M. (2008) Gravitational Waves, Volume 1: Theory and Experiments. Oxford University Press, Oxford.

[17]   Penrose, R. (2010) Cycles of Time: An Extraordinary New View of the Universe. The Bodley Head, London.

[18]   Barbour, J. (2009) The Nature of Time. http://arxiv.org/pdf/0903.3489.pdf

[19]   Barbour, J. (2010) Shape Dynamics: An Introduction. In: Finster, F., Muller, O., Nardmann, M., Tolksdorf, J. and Zeidler, E., Eds., *Quantum Field Theory and Gravity, Conceptual and Mathematical Advances in the Search for a Unified Framework*, Birkhauser, Springer-Verlag, London, 257-297.

[20]   Beckwith, A.W. (2015) Gedankenexperiment for Refining the Unruh Metric Tensor Uncertainty Principle via Schwartzshield Geometry and Planckian Space-Time with Initial Nonzero Entropy and Applying the Riemannian-Penrose Inequality and Initial Kinetic Energy for a Lower Bound to the Graviton. http://vixra.org/abs/1509.0173

[21]   Corda, C. (2009) Interferometric Detection of Gravitational Waves: The Definitive Test for General Relativity. *International Journal of Modern Physics D*, **18**, 2275-2282. http://arxiv.org/abs/0905.2502
        http://dx.doi.org/10.1142/S0218271809015904

[22]   Capozziello, S. and Corda, C. (2006) Scalar Gravitational Waves from Scalar-Tensor Gravity: Production and Response of Interferometers. *International Journal of Modern Physics D*, **15**, 1119-1150.
        http://dx.doi.org/10.1142/S0218271806008814

[23]   Corda, C. (2007) The Virgo-Minigrail Cross Correlation for the Detection of Scalar Gravitational Waves. *Modern Physics Letters A*, **22**, 1727-1735. http://dx.doi.org/10.1142/S0217732307024140

[24]   Das, S., Mukherje, S. and Souradeep, T. (2015) Revised Cosmological Parameters after BICEP 2 and BOSS.
        http://arxiv.org/abs/1406.0857

[25]   t'Hooft, G. (1999) Quantum Gravity as a Dissipative Deterministic System.
        http://arxiv.org/PS_cache/gr-qc/pdf/9903/9903084v3.pdf

20

Influence of a Semiconductor Gap's Energy on the Electrical Parameters of a Parallel Vertical Junction Photocell

Nfally Dieme

Laboratory of Semiconductors and Solar Energy, Department of Physics, Faculty of Science and Technology, Cheikh Anta Diop University, Dakar, Senegal
Email: nfallydieme@yahoo.fr, fallydieme@gmail.com

Abstract

The present work is a theoretical study on a parallel vertical junction solar cell under a multi-spectral illumination in static regime. The density of the minority charge carriers was determined based on the diffusion equation. Photocurrent and photovoltage are deducted from such density. All these parameters are studied taking into account the influence of the gap energy (E_g).

Keywords

Vertical Junction, Energy Gap, Photocurrent Density, Photovoltage

1. Introduction

The operation of solar cells is basically dependent on photon-electron interaction. For an electron to be removed from the valence stripe to the conduction, the minimum value of the photon energy must at least equal E_g. The gap energy (E_g) is determined by the material and fluctuates according to temperature [1].

The aim of this work is to investigate the influence of gap energy (E_g) on electrical parameters such as photocurrent and photovoltage. Knowing the evolution of these two quantities based on the gap energy is a good indicator for us to comment on the performance of solar cells and types of semiconductors for use in the manufacture of solar cells able to run at high temperature.

The density of excess minority carriers, photocurrent and photovoltage will be determined from the diffusion equations. In the second part of this work we present our simulation results.

2. Theory

This study is based on a parallel vertical junction silicon solar cell [2] presented in **Figure 1**. The solar cell is illuminated along the z axis in steady state.

We assume that the following hypotheses are satisfied.

- The contribution of the emitter is neglected.
- Illumination is made with polychromatic light, and is considered to be uniform on the $z = 0$ plane.
- There is no electric field without space charge regions.

2.1. Density of Minority Charge Carriers

When the solar cell is illuminated, there are simultaneously three major phenomena that happen: generation, diffusion and recombination.

These phenomena are described by the diffusion-recombination equation obtained with:

$$\frac{\partial^2 n(x)}{\partial x^2} - \frac{n(x)}{L^2} = -\frac{G(z)}{D} \tag{1}$$

D is the diffusion constant and is related to the operating temperature through the relation [2]

$$D = \mu \cdot \frac{K}{q} \cdot T \tag{2}$$

with q as the elementary charge, k the Boltzmann constant and T temperature.

$G(z)$ is the carrier generation rate at the depth z in the base and can be written as [2] [3]:

$$G(z) = \sum a_i e^{-b_i z} \tag{3}$$

a_i and b_i are obtained from the tabulated values of AM1.5 solar illumination spectrum and the dependence of the absorption coefficient of silicon with illumination wavelength.

$n(x)$, L, τ, and μ are respectively the density of the excess minority carriers, the diffusion length, lifetime and mobility.

The solution to the Equation (1) is:

$$n(x) = A \sinh\left(\frac{x}{L}\right) + B \cosh\left(\frac{x}{L}\right) + \sum \frac{a_i}{D} L^2 e^{-b_i z} \tag{4}$$

Coefficients A and B are determined through the following boundary conditions:
at the junction ($x = 0$):

$$\left.\frac{\partial n(x)}{\partial x}\right|_{x=0} = \frac{S_f}{D} n(0) \tag{5}$$

This boundary condition introduces a parameter S_f which is called recombination velocity at the junction; S_f determines the flow of the charge carriers through the junction and is directly related to the operating point of the solar cell. The higher S_f is, the higher the current density will be.

In the middle of the base ($x = W/2$):

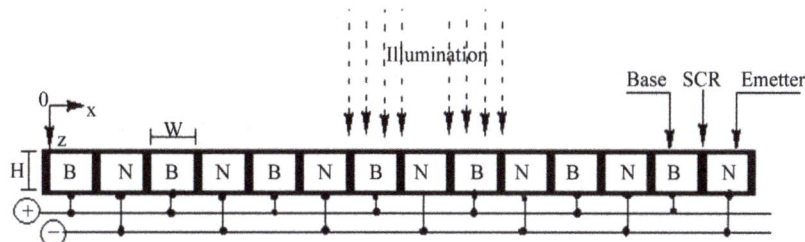

Figure 1. Parallel vertical junction solar cell ($H = 0.02$ cm; $W = 0.03$ cm).

$$\frac{\partial n(x)}{\partial x}\bigg|_{x=\frac{w}{2}} = 0 \qquad (6)$$

Equation (8) illustrates the fact that excess carrier concentration reaches its maximum value in the middle of the base due to the presence of junction on both sides of the base along x axis (**Figure 1**).

2.2. Photocurrent Density

The photocurrent J_{ph} is obtained from the following relation given that there is no drift current:

$$J_{ph} = qD\frac{\partial n(x)}{\partial x}\bigg|_{x=0} \qquad (7)$$

2.3. Photo-Voltage

The photo-voltage derives from the Boltzmann relation:

$$V_{ph} = \frac{k\cdot T}{q}\cdot\ln\left(N_B\cdot\frac{n(0)}{n_i^2}+1\right) \qquad (8)$$

with

$$n_i = A_n\cdot T^{\frac{3}{2}}\cdot\exp\left(\frac{E_g}{2KT}\right) \qquad (9)$$

n_i refers to the intrinsic concentration of minority carriers in the base,
A_n is a specific constant of the material ($A_n = 3.87\times10^{16}$ for silicon)
N_B is the base doping concentration in impurity atoms
E_g is the energy gap; it is given by [4] [5]:

$$E_g = E_{g0} - \frac{a\cdot T^2}{b+T} \qquad (10)$$

($E_{g0} = 1.170$ eV; $a = 4.9\times10^{-4}$ eV$\cdot$K^{-2}; $b = 655$ K for silicon).

3. Results and Discussion

In this section of our work, we present the results obtained from simulations.

3.1. Gaps Energy

When the solid temperature tends to absolute zero, two allowed energy bands play a special role. The last completely filled band is called "valence band: E_V". The allowed energy band is called following the "conduction band: E_C". It can be empty or partially filled. The energy between the valence band to the conduction band is called the "energy gap: E_g" [6]. **Figure 2** illustrates the band representation.

To have extraction of electrons under the influence of light, incident photon must have an energy greater than or equal to the energy of the gap. $E_g = E_C - E_V$. The gap energy is the energy that electron must absorb to be extracted.

3.2. Photocurrent Density

Figure 3 shows the photocurrent density profile versus junction recombination velocity for various values of energy gap.

It can be seen that photocurrent density quickly increases with the recombination velocity at the S_f junction until short circuit occurs. Given that the recombination velocity at the junction reflects the stream of carriers crossing the junction, an increase in this rate suggests an increase in the photocurrent density. S_f higher values represent a short circuit operation point and lower values are obtained in a situation of open circuit.

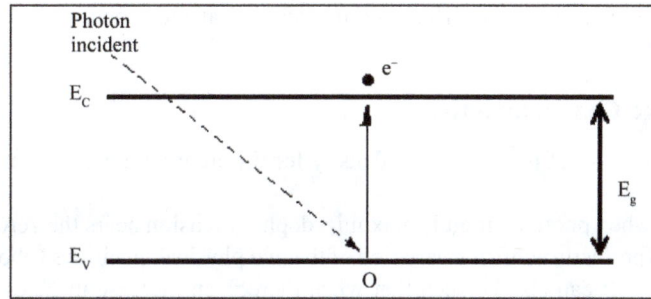

Figure 2. Gap's energy.

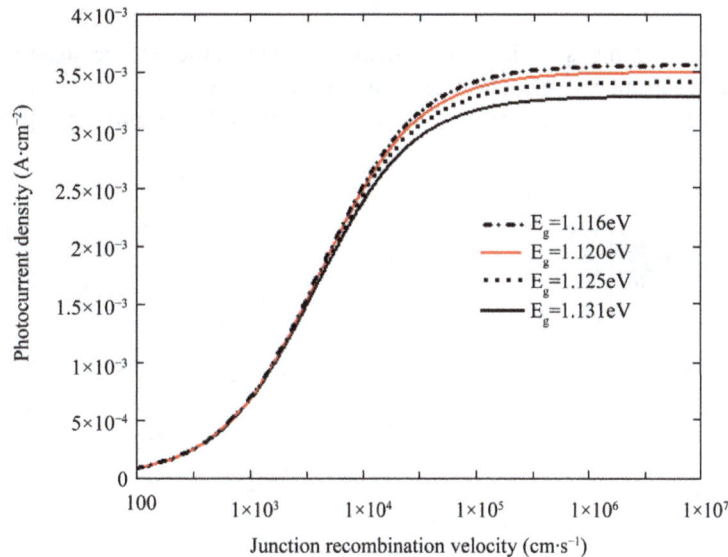

Figure 3. Photocurrent density versus junction recombination velocity ($z = 10^{-2}$ cm).

It can also be seen that the increase in the material's gap energy causes a decrease in the photo-courant density. This variation is much more visible in short circuit situations.

Indeed, photocurrent is produced by a movement of carriers photo-generated through the junction. When the height of the barred band increases, many electrons are extracted with low kinetic energy [7] [8]: this phenolmenon is called the photoelectric effect:

$$E_{\text{cinetik}} = E_{\text{photon}} - E_g.$$

Consequently the diffusion of carriers through the junction weakens as some carriers do not have enough kinetic energy to jump the depletion zone. It is said that photocurrent density decreases as the gag energy (E_g) increases.

3.3. Photo-Voltage

Figure 4 shows the evolution of photo-tension depending on recombination velocity at the junction regarding different values of the material's gap energy E_g.

Figure 4 shows that photovoltage decreases along with S_f junction recombination velocity. When S_f increases, the flow of charge carriers crossing the junction increases. Thus, fewer and fewer carriers are stored, which causes decrease in photovoltage at the junction.

Unlike photocurrent, it can also be seen that photovoltage increases as the gap energy increases. This should not be that surprising. Because the low kinetic energy possessed by some of the electrons and which is due to the growth of E_g, is not sufficient to make the charge carriers jump the depletion zone [9]. So they reach the junction and start piling up, thus increasing the difference in potential at the junction: it is said that photovoltage increases when E_g is high.

3.4. Current-Voltage Characteristics

Figure 5 shows the evolution of photo-courant density for different values of the gap energy and in relation to photo-tension.

Figure 4 shows that when photo-courant is maximized, photo-tension nears the zero level and vice versa. It can be noted that this figure perfectly confirms variation of the two physical quantities (photovoltage and photocurrent) in relation to gap energy. It can also be seen that when there is an increase in E_g of $\Delta E_g = 4 \times 10^{-3}$ eV, photovoltage increases by almost 10% while photocurrent decreases by about 2%.

4. Conclusion

In the simulation carried out in this work, we have demonstrated that the electric quantities of a solar cell such as photovoltage and photocurrent are very sensitive to the variation of a material's gap energy. Under the influence of temperature, an increase in the gap energy of $\Delta E_g = 4 \times 10^{-3}$ eV can prompt a growth in photovoltage of

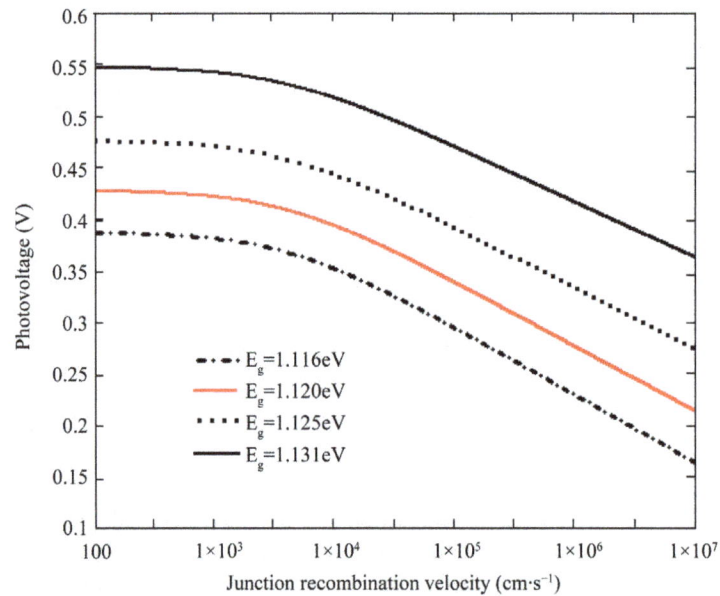

Figure 4. Photo-voltage versus junction recombination velocity ($z = 10^{-2}$ cm).

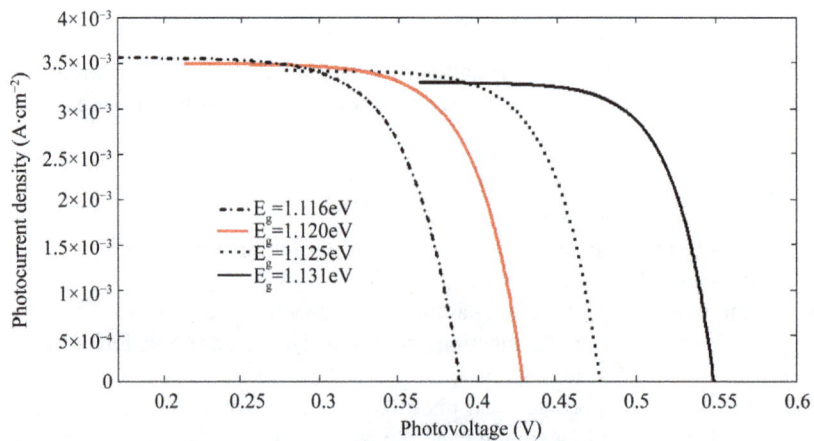

Figure 5. Photocurrent density versus photovoltage, $z = 10^{-2}$ cm.

almost 10% and a decrease in photocurrent of about 2%. We can estimate that according to experience, semiconductors have a great height of the band gap such as GaP, GaN, which are most suitable for high operating temperatures. This theoretical study can be confirmed by a comparative study of the conversion efficiency of semiconductor of different energy gap.

References

[1] Levy, F. (1995) Traité des matériaux 18: Physique et technologie des semi-conducteurs, Presses Polytechniques et Universitaires Romandes.

[2] Dieme, Nf., Zoungrana, M., Mbodji, S., Diallo, H.L., Ndiaye, M., Barro, F.I. and Sissoko, G. (2014) Influence of Temperature on the Electrical Parameters of a Vertical Parallel Junction Silicon Solar Cell under Polychromatic Illumination in Steady State. *Research Journal of Applied Sciences, Engineering and Technology*, **7**, 2559-2562.

[3] Furlan, J. and Amon, S. (1985) Approximation of the Carrier Generation Rate in Illuminated Silicon. *Solid-State Electronics*, 1241-1243. http://dx.doi.org/10.1016/0038-1101(85)90048-6

[4] Sze, S.M. and Kwok, K.Ng. (2007) Physics of Semiconductor Devices. Third Edition, John Wiley & Sons, Hoboken.

[5] Pässler, R. (2003) Semi-Empirical Descriptions of Temperature Dependences of Band Gaps in Semiconductors. *Physica Status Solidi*, 710-728.

[6] Valkov, S. (1994) Electronique Analogique. Edition Castéilla, Collection A. CAPLIEZ.

[7] Agroui, K. (1999) Etude du Comportement Thermique de Modules Photovoltaïquesde Technologie Monoverre et Biverre au Silicium Cristallin. *Revue des Energies Renouvelables: Valorisation*, 7-11.

[8] Mohammad, S.N. (1987) An Alternative Method for the Performance Analysis of Silicon Solar Cells. *Journal of Applied Physics*, 767-772. http://dx.doi.org/10.1063/1.338230

[9] Landis, G.A., Jenkins, P., Scheiman, D. and Rafaelle, R. (2004) Extended Temperature Solar Cell Technology Development. *AIAA 2nd International Energy Conversion Engineering Conference*, Providence, 16-19 August 2004.

21

Energy Analysis and Proposals for Sustainability from the Energy Transition

Eduardo Oliveira Teles[1]*, Marcelo Santana Silva[1], Francisco Gaudêncio M. Freires[2], Ednildo Andrade Torres[2]

[1]Federal Institute of Bahia (IFBA), Brazil and Federal University of Bahia (UFBA), Salvador, Brazil
[2]School of Industrial Engineering (PEI), Federal University of Bahia (UFBA), Salvador, Brazil
Email: *eoteles@yahoo.com.br

Abstract

The energy transition has become an increasingly attractive and necessary issue nowadays because of the tendency of scarcity and increased demand for fossil energy, and associated environmental impacts—for example, increased emission of greenhouse gases (GHGs), particularly CO_2, CH_4 and N_2O. From the study of several papers and reports from various international agencies like the World Energy Council (WEC), the International Energy Agency (IEA), the Policy Network for Renewable Energy (REN) and the World Organization of the United Nations (UN), this paper analyzes the global energy context, making a survey of what is being discussed under the theme "energy transition", and suggests ways to reduce the consumption of fossil fuels, aiming at sustainability.

Keywords

Energy, Sustainable, Energy Transition

1. Introduction

Worldwide, efficient, cheap and reliable energy resources are now increasingly being demanded. Certainly, this is the reason why oil and natural gas have dominated the energy market many years and will continue for some time, despite the change in oil price. Furthermore, most of current infrastructure depends on these fuels, meaning that they will be increasingly be disputed by various countries. However, the current debate is about the physical and economical oscillations of these resources, together with a strong political instability in regions

---
*Corresponding author.

with oil reserves, without considering awareness of the use of fossil fuels as contribution to global climate change.

There are several worries in the areas of Energy and Environment. Firstly, global warming, caused by emission of greenhouse gases (GHGs) as a result of the burning of fossil fuels. These forecasts about climate changes affect, especially, agriculture and many species of animals and plants. Secondly, the population growth in many countries of the world, particularly in China and India. Thirdly, the continued growth trend of supply and demand of energy in the world. Fourthly, the global dependence on oil and its derivatives, in particular from the countries of the Middle East.

The strong price instability promotes the increase of Biofuels use. In July 2008, the price of oil was $147 per barrel and within a period of six months showed a decline, reaching $30. In 2012, the average price of Brent crude oil was US $111.65 per barrel [1]. The picture of the period between 2008 and 2012 clearly shows the consecutive increases in the price, and soon, the decrease of price.

The energy transitions get more attention due to increasing concerns about the future availability of energy sourced from fossil fuels and the impact of their use on the environment [2]. The increase in oil and gas reserves is conditioned to the exploration in increasingly difficult environments (unconventional sources and pre-salt). However, these projects become unviable due to reduction of oil prices; the exploitation of these resources is more expensive, has higher emissions of greenhouse gases, and is more risky for the environment compared with conventional sources [3]-[5]. Therefore, there are increasing requests for fossil fuels to be replaced by alternative sources of energy cleaner and renewable. However, the production costs of the clean and renewable energy are still high and the profitability of this market is quite low.

The purpose of this article is to show the reality of the areas of Energy and Environment, through the analysis of the energy matrix in the world and checking the energy transition possibilities. Besides, this article mentions the participation of some emerging countries and shows the importance of increasing the studies on renewable energy.

2. Methodology

This research was divided into stages, namely, collection, description, tabulation and analysis of a wide range of current data on energy topics, sustainability and energy transition.

The analysis of the global energy sector (the current consumption and prospects for use of several energy sources) was held from reports of the World Energy Council (WEC), the International Energy Agency (IEA), the Renewable Energy Policy Network (REN), the Intergovernmental Panel on Climate Change (IPCC) of the United Nations (UN) and other international institutions such as the Institute the Pew Charitable Truts.

The analysis of these reports has details about the energy scene in the world, perspectives and evaluation of the current status in energy consumption. At the same time, there was a survey of information about the theme of "energy transition". In this survey, the collected material was classified into three research areas: 1) history of transition energy, 2) management of the energy transition, and 3) the transition in the oil and gas energy systems.

Based on information from the reports, projections of the energy matrix and study of several works on energy transition, is displayed the main study trends and best sustainable development paths. Finally, there is the analysis of the influence that the energy transition has on sustainability and the usage of energy sources, showing paths for this conversion.

3. Contextualization of Energy

The energy sector is strategic for the economy of nations and very important to social development [6] [7]. Approximately three years after the fossil fuel prices have suffered high increases due to the economic crisis of 2008 and 2009, the world faced a recession that did not happened for a long time. This recession has generated serious risks to political and economic stability, represented largely by the world's dependence on fossil fuels. Events like this one show the need to a change of the energy matrix in the world, converging to clean energy. The energy transition with a focus on sustainability would increase the energy security of the world, would promote economic growth to a long period and would face environmental challenges such as anthropogenic climate change. Therefore, this revolution will demand unprecedented investment in Research and Development (R & D) of clean technologies and low-carbon, in the next decades.

Despite the significant increase in global investment in renewable energy, led by the wind, solar and biomass, the growth of fossil fuels has increased in significant volumes too. This is not surprising, considering the fact that more than a quarter of the world's population, which is approximately 1.6 billion people still not having access to electricity, according to the report by the World Energy Council [8]. The largest growth in energy consumption comes from Asia, particularly China and India. These two countries, according to projections by the IEA, will represent more than 50% of total growth in global energy demand by 2030 [9]. By another hand, after the natural disaster happened in Japan, the uncertainty regarding some energy sources, for example the nuclear energy, increased significantly.

In accordance with data from the International Energy Agency [10], the consumption of primary energy in the world is increasing every year and the scenario is of growing in all sources. According to the same report, fossil fuels continue to be the main sources of energy in the world, but the share of oil will fall from 32% to 27% in 2035. Moreover, it is estimated incremental increases in the price of oil that will come in order to $125/barrel in 2035. This increase in demand is mainly supported by incentives to overcome market barriers, the decline in technology costs, increase in prices of fossil fuels, consumption mode of modern society and the increased purchasing power.

The International Energy Agency expects that between 2008 and 2030, due to the projected growth in 1.0% of the total world average, China and India will even get greater oil consumption averages [11]. Because of this short scenario, in case of any disruption of supply influenced by the monopoly power of the Oil Producers Exporting Countries Organization—OPEC, or even a terrorist attacks in countries with enormous quantities of reserves, the oil-importing countries would be significantly political and socio-economically vulnerable. In accordance with the World Energy Council, in their report called "Roadmap towards a Competitive European Energy Market", currently about 50% of all European energy is being imported, and this tends to grow to 70% in 2030.

The World Energy Council showed the growth of emerging economies, especially China and India, which raised and placed the availability of energy as a priority item on the global agenda. Moreover, it is assumed the occurrence of the doubling of energy demand on earth by 2050. Fossil fuels will continue to be the main component of the energy mix for at least another generation [12].

In 2010, the WEC mapped and developed future scenarios of global energy, in the report entitled "Logistics Bottlenecks", with a focus on fossil and renewable fuels most used in the world [13]. In this report, was made a projection of consumption and demand for 2020 and 2050, using data from the International Energy Agency—IEA and other sources, the Asia/Pacific was the largest producer and consumer of coal worldwide, with more than 3 billion tons, equivalent to 61% of the market.

Although the environmental unbalance caused from burning coal, it still plays a part of fundamental importance to the energy balance in the world because of their abundance and its well balanced wide geographic scope. According to a projection of the WEC coal production will increase by 20% in 2020 and 54% in 2050 compared with 2008 [13]. This way is observed growth in almost all regions of the world, with the exception of European region.

The situation in the oil market is far more complex than the coal, because it is more distant from the consumer markets and its production is limited in some areas, even if two-thirds of reserves are in the Middle East. There are big importers of oil, for example, Europe and Asia/Pacific, which imports about 70% and 68%, respectively. In the following regions a growth in consumption is expected: Latin America, Africa, Middle East and Asia/Pacific. The other regions showed a reduction in consumption [13].

In 2020, according to the report, the world oil consumption will increase 12% and the trend is that some regions will reduce its dependence about oil imports, together with the expansion of production rather than reduce the demand and control their consumption [13].

Natural gas also follows a growth path in the both production and their consumption, and is widely used internally, with only 12% of the export gas produced in 2008, while oil was on the order of 48%. Latin America will have the highest growth in consumption in the order of 4.2% annually until 2050, followed by the regions of Africa (2.8% per year), Asia/Pacific (2.7% per year) and the Middle East (2.5% per year) and North America (0.4% per year) (WEC, 2010).

The Biofuels Production

The Study of the Energy Information Administration—EIA with respect to electricity generation in the US, shows that the need for energy imports will be offseted by the increase in biofuels, linked to reductions in the demand resulted from the adoption of new fuel economy standards and increasing prices. This fact could provide a downturn in demand for fossil fuels, as the example of the United States (US) [14].

Figure 1 shows the increase of all renewable energy sources, especially biomass and wind power. The big reason for the increase of wind farms and biomass was the regulatory mark that drove a large investment in the sector supported by tributaries and tax credits granted to the respective companies producing these energies.

According to the EIA [14], the global biofuel production hit a record level of more than 34 million toe (tons of oil equivalent) in 2007, representing 1.5% of the total world fuel consumption. Furthermore, the EIA estimated an annual growth rate of 7%, which means that by 2030 biofuels will account for about 5% of total road transport, compared with about 2% in 2009 [15].

Currently, Brazil and the USA accounts for almost 80% of global production of biofuels. Both produce bioethanol, the American made from corn, and Brazilian from sugarcane. On the world scene, the USA is now the largest consumer of biofuels, with strong growth in demand in the coming years. Europe, on the other hand, is the largest producer of biodiesel, representing for about 90% of world production, due to subsidies and other incentives, with an increase in energy imports. China and India are also largest producers of biofuels, particularly bioethanol. Africa and Asia have great potential to become major producers and exporters of biofuels. Southeast Asian countries are big producers of palm oil and could develop a competitive biodiesel for the production and export [14].

In accordance with the American Institute, The Pew Charitable Trusts, in their report entitled "Who is Winning the Clean Energy Race", is shown the clean energy investments made in 2011, according to **Table 1**.

The United States leads with 48 billion dollars, followed by China with 45.5 billion dollars. Brazil appears in tenth place with $8 billion, equivalent to 16.7% this when compared to the amount of investments in the United States, with a 15% growth compared to 2010. The document points out that Brazil registered the third largest growth rate in alternative energy production in the last five years between the G20 countries.

In accordance with the same study, at the world level, investment in clean energy, without regard to research and development, has grown 600% since 2004 and represented a record investment of $263 billion, a 6.5% increase confronted with investments made in 2010. The production sources that grew most were solar and wind produced by the G20 group [16].

The success for the increasing development of biofuels will depend exclusively on global policies for the production and marketing performed on domestic and international markets, and the strong presence of the Government with the implementation of tax incentives and mainly of subsidies to offset the difference in values compared to other fossil fuels, considering that biofuel production costs are still very high. Another extremely important issue is the establishment of targets at the global level, for the use of biofuels, which could be blended with conventional fuel [17].

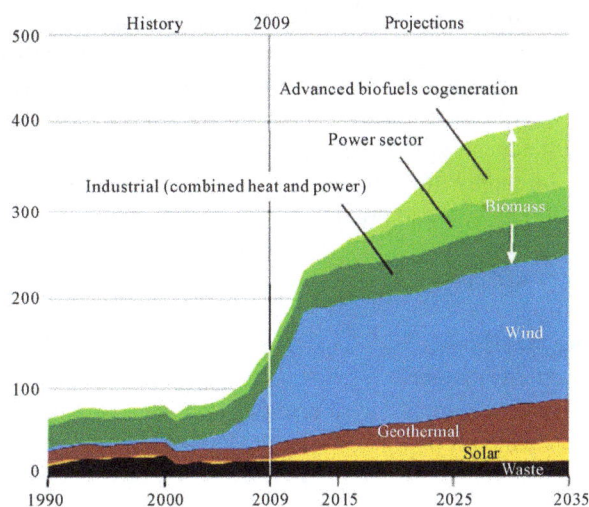

Figure 1. Renewable electricity generation in billion kilowatt per year. Source: EIA (2011).

Table 1. Countries with greater investment in clean energy in 2011.

Ranking	Countries/investment		
	Countries	Investment (billions- US $)	Investments in 2010 (billions- US $)
1	USA	48.0	33.7
2	China	45.5	45.0
3	Germany	30.6	32.1
4	Italy	28.0	20.2
5	The remaining EU - 27	11.1	15.2
6	India	10.2	6.6
7	United Kingdom	9.4	7.0
8	Japan	8.6	7.0
9	Spain	8.6	6.9
10	Brazil	8.0	6.9

Currently in the world there are about 10 countries that require oil companies to add a percentage of biofuels to the regular fuel, as example, European Union and Brazil. The Statistical Report produced by BP in 2014 shows, for example, Brazil as world's second largest producer of biofuels, with 24% of the world total. Ethanol production has risen about 18.5% and biodiesel, 7.6%. With these values, Brazil begins to bear a mark of 31% and 10% of production in the world, respectively.

The IEA has developed a biofuel scenario in road transportation sector, with projection by 2030, and the participation of all biofuels will be 6.8%, while in 2004 was only 1%. Brazil will control about 30% biofuels in the transport sector [15].

In this global context, Brazil is classified as a tropical and continental country, holder of a vast expansible land, has a historic opportunity to be qualified as a country with the largest energy reserves from renewable resources of the planet. Therefore, it is necessary to review and improve public policies, as well as its regulatory framework as a whole, so that the industry can emerge in this increasingly attractive market.

4. Energetic Transition

The interest about the energy transition has grown considerably in recent years due to increased preoccupation about the impact of the production and use of energy from fossil fuels to the environment.

Transition studies involve the understanding of the interaction between many factors (actors, social, institutional and technological) that influence and feedback resulting in the transformation of the energy system [18] [19]. Smil defines the energy transition as the time that happens between the introduction of a new primary energy source (oil, nuclear power, wind captured by large turbines) and its growth to the point of significant importance (20 - 30 percent) for the global market, or even to become the largest contributor or an absolute leader (over 50 percent) in the supply of national or global energy. The energy transition could also refer to the gradual diffusion of the technology used to produce and distribute energy for domestic, industrial and transport use [20].

The main studies on energy transition were classified in three research areas:
- Energy transition of history;
- Management of the energy transition;
- Transition in oil and gas energy systems.

The big focus of the studies on the transition of energy is the managing the transition to alternative energy or low carbon economy. The transition from traditional to unconventional oil is also happening today in the industry due to the decline of conventional sources [3] [21]. However, few researches have been conducted on this type of transition.

Research suggests a large unconventional oil transition starting before 2030 [4]. The increasing dependence on a few oil producers will continue, because of the fact that the largest reserves are concentrated in the Middle

East and North Africa. This represents a big problem because most of the countries of these regions are politically and economically unstable. The development of this debate is an indication of wider changes that are happening in the energy sector [22].

Therefore, competition for access to reserves will be even greater, even with the gradual increase of exploitation of unconventional sources. This factor may be one reason why many companies in the sector are involved in the development of alternative energy sources. Investments in this sense will not only help companies to get a competitive edge, but also will provide an opportunity for their green supply chains. It would be a strategic move that could help companies to stay in business and be compatible with a sustainable future.

A change for a renewable energy system will demand large investments and good planning and strategy in the medium and long-term. In addition, there is a socio-political appeal on the development of renewable energies, especially among those who are worried about excessive industrial growth and the domination of basic energy by some groups [23]. It is proposed that, in order to accelerate the transition to renewable energy system, the breaking of all energy systems is necessary. The intervention is necessary to encourage the development of renewable energy sources that are widely available [24]. However, the intense transformation is not an answer to get to a sustainable energy system, at the risk of causing a serious impact on the energy market, economy, environment and society. The solutions to these energy questions require the achievements of feasible actions to a long-term, to ensure that development can be truly sustainable [25] [26].

Therefore, effective policies promote the development of sustainable energy taking into consideration the limits of renewable energy in terms of technical and commercial feasibility. The adoption and abandonment of economic and market policies that provide little incentive for technological innovation and energy development does not help the energy transition in a view of sustainability.

5. Why Energetic Transition?

First two important topics can summarizes all the current set for the input of renewable energy: energy security and environmental impact. In 2009, it was estimated that 86.7% of energy demand in the world was originated from fossil fuels [27]. Present data show that two-thirds of global emissions in 2011 originated from just ten countries, especially China and the US, producing together approximately more than 12 Gt/CO_2, which was approximately 40% of global emissions of dioxide of carbon dioxide (CO_2), as shown in **Figure 2** [28].

This increase in greenhouse gas emissions—GHG took place with more intensity since the Industrial Revolution, the second half of the XVIII century, with a result of the implementation and modernization of the industrial sector, the farming projects and the transport sector.

These gases, among which stands out carbon dioxide (CO_2) let in the sunlight, but prevent part of the heat in which light is transformed back for space, holding the gas and causing the greenhouse effect. It is important to

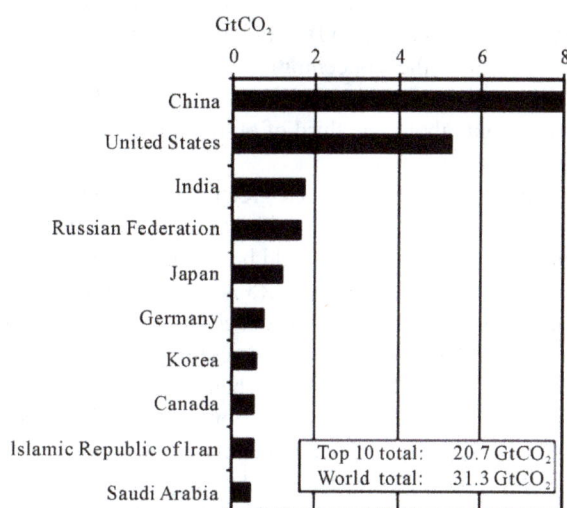

Figure 2. Top 10 emitting countries in 2011. Source: IEA (2013).

be noted that the greenhouse gases, including water vapor, do not cause harm to health and does not pollute the environment, as well as the greenhouse phenomenon is not a problem. If the effect was not produced, the average temperature of the earth would oscillate around negative 170˚C, making it impossible life on the planet as it is. Human activities that emit greenhouse gases—GHG, especially CO_2 and CH_4, are related to changes in levels of industrialization and use of land experienced by humanity since 1750, and a more intense way in the last decades of the last century and early of the XXI century.

In accordance with recent reports by the IPCC, the main sources for the increase in carbon dioxide concentration in the atmosphere are: the use of fossil fuels and the change of management and land use, although the last one, with a small contribution to these increase. The carbon dioxide emission includes all emissions due to the production, distribution and consumption of fossil fuel and also, for example, byproducts of the production of cement, while emissions originating from changes in land handling refer to emissions from deforestation, forest burning, and biomass decomposition resulting from logging [29] [30].

According to the report "Technology Roadmap: Biofuels for Transport" [31], if the pattern of energy consumption continues with the same current levels, the GHG emissions will double by 2050. Furthermore, by 2050 biofuels will account for about 27% fuel for the transportation sector, and will avoid about 2.1 Gt of CO_2 emissions a year when produced sustainably.

In accordance with the AR4, carbon dioxide (CO_2) annually increased by an average from 6.4 GtC per year in the 1990s to 7.2 GtC per year in 2000-2005. Although the data for the years 2004 and 2005 are approximate estimates, if the emissions associated with changes in land use are considered, must be added 1.6 GtC to these figures per year throughout the 1990 s, although the report point some uncertainties in these estimates. An emission of 1 GtC (1 GtC = 1 Giga-ton carbon = 1 billion tons of carbon) corresponds to 3.67 $GtCO_2$ [29]).

In AR5 was verified that emissions in a year of CO_2 from combustion of fossil fuels and cement production were 8.3 GtC per year (range 7.6 to 9.0 GtC per year), calculated over the period in 2002-2011. In 2011, it was announced 9.5 GtC (range 8.7 to 10.3 GtC per year), representing 54% above the 1990 level. The liquid CO_2 emissions in the year, due to changes in land use that has anthropogenic basis totaled 0.9 GtC per year (range 0.1 to 1.7 GtC per year), the average for the period of 2002-2011 [30].

In total context, between 1750 and 2011, the CO_2 emissions from combustion of fossil fuels and cement production sent about 365 GtC (335 - 395 GtC interval) into the atmosphere, while deforestation and other changes released 180 GtC (range 100 - 260 GtC). These result in accumulated anthropogenic emissions of 545 GtC (range 460 - 630 GtC) [30].

The concentration of methane gas in the global atmosphere follows the same growth trend. Also according to the IPCC, the concentration increased from a value from the pre-industrial period from about 735 parts per billion—ppb to 1765 ppb in the early 1990 s, and it was in 1774 ppb in 2005. According to the reflections published in the AR4 and AR5, it is very likely that the observed increase in methane gas concentration is due to anthropogenic activities, mainly as agriculture and the use of fossil fuels [29] [30].

Finally, the concentration of nitrous oxide (N_2O) in the global atmosphere. According to the information taken from the publication in question, the concentration of nitrous oxide increased from a value from the pre-industrial period from about 280 ppb to 340 ppb in 2005. The level of growth has been approximately constant since 1980. Out of this amount, about one third of all the nitric oxide emissions are classified as anthropogenic, primarily due to agriculture [29] [30].

The reports conclude warning that, with continued greenhouse gas emissions at the present rate or an even higher rate, an extra heating would occur and lead to several changes in the global climate system in the course of 21 st century, and very possibly these changes would be much more aggressive than those observed in the 20 th century. In addition, the main consideration of the AR5 is that the climate system warming is unequivocal.

It is important to show that the warming tends to reduce the capture of atmospheric carbon dioxide by the oceans and also the soil, which increases the anthropogenic emission fraction that remain in the atmosphere. In one of the scenarios developed by the IPCC [29], a very heterogeneous world is reported where the key is self-reliance and preservation of local identities, as well as economic development is primarily regionally oriented with more fragmented and slow technological development.

Collaborating with the result of the IPCC report, "Technology Roadmap: Biofuels for Transport" of the IEA also said that if the pattern of energy consumption continue with the same current levels, GHG emissions will double by 2050. The same report included that by 2050, biofuels will account for about 27% of fuel for the transport sector, and will avoid about 2.1 Gt of CO_2 emissions per year when produced in a sustainable way, worldwide [31].

Thus, it is clear the need for better action in conversion, transmission and distribution of energy. In addition, actions to improve the composition of fossil fuels (decarbonisation), including the diesel fuel, encourage increased renewable energy and bioenergy, enlarge the nuclear energy in a balanced way and invest in cost reduction for capture and storage carbon, are essential for better quality of life and economic stability in the coming years.

6. Conclusions

In a general way, the economical and population growth are the most important drivers of increases in CO_2 emissions from fossil fuels. The trajectory of population growth between 2000 and 2010 has been practically identical with respect to the past three decades, while economic growth has increased considerably, which leads to considerable increases in energy demand.

A large part of climate change has already become irreversible because of the past emissions. The current methods proposed for reducing the concentration of greenhouse gases are still preliminary, with large margin of doubt about its cost, its application and its probability to succeed.

The analysis of the various reports helps to understand the growth of energy demand and their sources. Enables understand the points of gradual energy transition: the biofuels development, with the projection to coming to 1/4 of the Brazilian matrix in the next decade, for example.

In addition, it became clear that fossil fuels have different representations in environmental instability. For instance, the availability of coal and its geographical dissemination has a very different view of oil. The latter has reserves concentrated in some regions that are politically unstable.

Thus, it is clear that the search for sustainability can be done through energy transition, with the development of effective policy, with good long-term planning, tax incentives and subsidies to reduce the costs of production.

Acknowledgements

The authors of this paper would like to thank PRPGI/Federal Institute of Education, Science and Technology of Bahia for providing the financial support to complete and publish this article (Edital 07/2014).

References

[1] Agência Nacional de Petróleo, Gás e Biocumbustíveis—ANP (2012) Boletim Anual de Preços: Preços do petróleo, gás natural e combustíveis nos mercados nacional e internacional. http://www.anp.gov.br/?dw=59757

[2] Fouquet, R. (2010) The Slow Search for Solutions: Lessons from Historical Energy Transitions by Sector and Service. *Energy Policy*, **38**, 6586-6596. http://dx.doi.org/10.1016/j.enpol.2010.06.029

[3] Farrell, A.E. and Brandt, A.R. (2006) Risks of the Oil Transition. *Environmental Research Letters*, **1**, Article ID: 014004. http://dx.doi.org/10.1088/1748-9326/1/1/014004

[4] Greene, D.L., Hopson, J.L. and Li, J. (2006) Have We Run Out of Oil Yet? Oil Peaking Analysis from an Optimist's Perspective. *Energy Policy*, **34**, 515-531. http://dx.doi.org/10.1016/j.enpol.2005.11.025

[5] Mohr, S.H. and Evans, G.M. (2010) Long TERM Prediction of Unconventional Oil Production. *Energy Policy*, **38**, 265-276. http://dx.doi.org/10.1016/j.enpol.2009.09.015

[6] Vera, I. and Langlois, L. (2007) Energy Indicators for Sustainable Development. *Energy*, **32**, 875-882. http://www.sciencedirect.com/science/article/pii/S0360544206002337 http://dx.doi.org/10.1016/j.energy.2006.08.006

[7] Matos, S. and Silvestre, B.S. (2012) Managing Stakeholder Relations When Developing Sustainable Business Models: The Case of the Brazilian Energy Sector. *Journal of CleanerProduction*, **45**, 61-73. http://dx.doi.org/10.1016/j.jclepro.2012.04.023

[8] World Energy Council—WEC (2010) Energy and Urban Innovation. United Kingdom. http://www.worldenergy.org/publications/2010/energy-and-urban-innovation-2010/

[9] World Energy Council—WEC (2010) Roadmap towards a Competitive European Energy Market. United Kingdom. http://www.worldenergy.org/publications/2010/roadmap-towards-a-competitive-european-energy-market/

[10] Internation Energy Agency—IEA (2012) World Energy Outlook 2012. OECD/IEA. http://www.worldenergyoutlook.org/publications/weo-2012/

[11] Internation Energy Agency (IEA) (2010) CO_2 Emissions from Fuel Combustion Highlights. OECD/IEA.

http://www.iea.org/media/training/presentations/statisticsmarch/co2highlights.pdf

[12] World Energy Council (WEC) (2008) Criando um Novo Impulso. Declaração do Conselho Mundial de Energia.
      https://www.yumpu.com/pt/document/view/31640436/criando-um-novo-impulso-world-energy-council

[13] World Energy Council (WEC) (2010) Logistics Bottlenecks. United Kingdom.
      https://www.worldenergy.org/publications/2010/logistics-bottlenecks-2010/

[14] Internation Energy Agency (IEA) (2011) Technology Roadmap: Biofuels for Transport. OECD/IEA.
      http://www.iea.org/publications/freepublications/ publication/name,3976,en.html./

[15] Energy Information Administration (EIA) (2015) Annual Energy Outlook 2015 with Projections to 2040. Forrestal
      Building, Washington, ABR, 2015. http://www.eia.gov/forecasts/aeo/

[16] The Pew Charitable Trusts (2012) Who Is Winning the Clean Energy Race.
      http://www.pewenvironment.org/news-room/reports/whos-winning-the-clean-energy-race-2011-edition-85899381106

[17] Renewable Energy Policy Network for the 21st Century (REN21) (2012) Renewables 2012 Global Status Report.
      REN21 Secretariat, Paris. www.ren21.net

[18] Geels, F.W. and Schot, J. (2007) Typology of Sociotechnical Transition Pathways. *Research Policy*, **36**, 399-417.
      http://dx.doi.org/10.1016/j.respol.2007.01.003

[19] Martens, P. and Rotmans, J. (2005) Transitions in a Globalising World. *Futures*, **37**, 1133-1144.
      http://dx.doi.org/10.1016/j.futures.2005.02.010

[20] Smil, V. (2008) Moore's Curse and the Great Energy Delusion. *American*, **2**, 34.

[21] Brandt, A.R., Plevin, R.J. and Farrell, A.E. (2010) Dynamics of the Oil Transition: Modeling Capacity, Depletion, and
      Emissions. *Energy*, **35**, 2852-2860. http://dx.doi.org/10.1016/j.energy.2010.03.014

[22] Wolf, C. (2009) Does Ownership Matter? The Performance and Efficiency of State Oil vs. Private Oil (1987-2006).
      Energy Policy, **37**, 2642-2652. http://dx.doi.org/10.1016/j.enpol.2009.02.041

[23] Lior, N. (2010) Sustainable Energy Development: The Present (2009) Situation and Possible Paths to the Future. *Energy*, **35**, 3976-3994. http://dx.doi.org/10.1016/j.energy.2010.03.034

[24] Verbruggen, A., Fischedick, M., Moomaw, W., Weir, T. and Nadaï, A. (2010) Renewable Energy Costs, Potentials,
      Barriers: Conceptual Issues. *Energy Policy*, **38**, 850-861. http://dx.doi.org/10.1016/j.enpol.2009.10.036

[25] Omer, A.M. (2008) Energy, Environment and Sustainable Development. *Renewable and Sustainable Energy Reviews*,
      12, 2265-2300. http://dx.doi.org/10.1016/j.rser.2007.05.001

[26] Markevicius, A., Katinas, V., Perednis, E. and Tamasauskiene, M. (2010) Trends and Sustainability Criteria of the Pro-
      duction and Use of Liquid Biofuels. *Renewable and Sustainable Energy Reviews*, **14**, 3226-3231.
      http://dx.doi.org/10.1016/j.rser.2010.07.015

[27] Brasil. Ministério de Minas e Energia (2012) Boletim mensal dos combustíveis Renováveis, Brasília, SPG, n. 49, fev.
      2012.

[28] Internation Energy Agency (IEA) (2013) CO$_2$ Emissions from Fuel Combustion Highlights. OECD/IEA.
      http://www.iea.org/publications/freepublications/publication/co2emissionsfromfuelcombustionhighlights2013.pdf

[29] Intergovernmental Panel on Climate Change (IPCC) (2007) Climate Change 2007: The Physical Science Basis—
      Summary for Policymakers. Working Group I. http://www.ipcc.ch/

[30] Intergovernmental Panel on Climate Change (IPCC) (2014) Climate Change 2014: Mitigation of Climate Change.
      Contribution of Working Group III to the Fifth Assessment Report of the Intergovernmental Panel on Climate Change.
      In: Edenhofer, O., Pichs-Madruga, R., Sokona, Y., Farahani, E., Kadner, S., Seyboth, K., Adler, A., Baum, I., Brunner,
      S., Eickemeier, P., Kriemann, B., Savolainen, J., Schlömer, S., von Stechow, C., Zwickel, T. and Minx, J.C., Eds.,
      Cambridge University Press, Cambridge and New York.

[31] Energy Information Administration (EIA) (2011) Annual Energy Outlook 2011 with Projections to 2035. Forrestal
      Building: Washington, ABR, 2011.
      http://www.columbia.edu/cu/alliance/documents/EDF/Wednesday/Heal_material.pdf

Permissions

All chapters in this book were first published by Scientific Research Publishing; hereby published with permission under the Creative Commons Attribution License or equivalent. Every chapter published in this book has been scrutinized by our experts. Their significance has been extensively debated. The topics covered herein carry significant findings which will fuel the growth of the discipline. They may even be implemented as practical applications or may be referred to as a beginning point for another development.

The contributors of this book come from diverse backgrounds, making this book a truly international effort. This book will bring forth new frontiers with its revolutionizing research information and detailed analysis of the nascent developments around the world.

We would like to thank all the contributing authors for lending their expertise to make the book truly unique. They have played a crucial role in the development of this book. Without their invaluable contributions this book wouldn't have been possible. They have made vital efforts to compile up to date information on the varied aspects of this subject to make this book a valuable addition to the collection of many professionals and students.

This book was conceptualized with the vision of imparting up-to-date information and advanced data in this field. To ensure the same, a matchless editorial board was set up. Every individual on the board went through rigorous rounds of assessment to prove their worth. After which they invested a large part of their time researching and compiling the most relevant data for our readers.

The editorial board has been involved in producing this book since its inception. They have spent rigorous hours researching and exploring the diverse topics which have resulted in the successful publishing of this book. They have passed on their knowledge of decades through this book. To expedite this challenging task, the publisher supported the team at every step. A small team of assistant editors was also appointed to further simplify the editing procedure and attain best results for the readers.

Apart from the editorial board, the designing team has also invested a significant amount of their time in understanding the subject and creating the most relevant covers. They scrutinized every image to scout for the most suitable representation of the subject and create an appropriate cover for the book.

The publishing team has been an ardent support to the editorial, designing and production team. Their endless efforts to recruit the best for this project, has resulted in the accomplishment of this book. They are a veteran in the field of academics and their pool of knowledge is as vast as their experience in printing. Their expertise and guidance has proved useful at every step. Their uncompromising quality standards have made this book an exceptional effort. Their encouragement from time to time has been an inspiration for everyone.

The publisher and the editorial board hope that this book will prove to be a valuable piece of knowledge for researchers, students, practitioners and scholars across the globe.

List of Contributors

Talal Masoud, Hesham Alsharie and Ahmad Qasaimeh
Civil Engineering Department, Jerash University, Jerash, Jordan

Miguel Edgar Morales Udaeta, Antonio Gomes dos Reis, José Aquiles Baesso Grimoni and Antonio Celso de Abreu Junior
GEPEA/EPUSP, Energy Group of the Department of the Electrical Energy and Automation Engineering/Polytechnic School of the University of São Paulo, São Paulo, Brazil

C. Nagarjuna Reddy and T. Harinarayana
Gujarat Energy Research and Management Institute, Gandhinagar, India

Talla Konchou Franck Armel
Environmental Energy Technologies Laboratory (EETL), University of Yaounde I, Yaounde, Cameroon
LISIE, University Institute of Technology Fotso-Victor, University of Dschang, Bandjoun, Cameroon

Aloyem Kazé Claude Vidal
LISIE, University Institute of Technology Fotso-Victor, University of Dschang, Bandjoun, Cameroon
HTTTC, Department of Electrical and Power Engineering/FSC, University of Bamenda, Bamenda, Cameroon

Tchinda René
LISIE, University Institute of Technology Fotso-Victor, University of Dschang, Bandjoun, Cameroon

Marc A. Rosen
Faculty of Engineering and Applied Science, University of Ontario Institute of Technology, Oshawa, Canada

V. Tsaplev
Department of Electroacoustics and Ultrasonic Engineering, Saint-Petersburg State Electrotechnical University, Saint-Petersburg, Russia
Department of Physics, North-West Open Technical University, Saint-Petersburg, Russia

R. Konovalov and K. Abbakumov
Department of Electroacoustics and Ultrasonic Engineering, Saint-Petersburg State Electrotechnical University, Saint-Petersburg, Russia

Angel Fierros Palacios
Instituto de Investigaciones Eléctricas, División de Energías Alternas, Mexico City, México

Meiyan Wang
Environmental Engineering, University of Kitakyushu, Kitakyushu, Japan
School of Landscape and Architecture, Zhejiang A&F University, Lin'an, China

Shenglan Huang
School of Landscape and Architecture, Zhejiang A&F University, Lin'an, China

Xinxin Lin, Didit Novianto and Weijun Gao
Environmental Engineering, University of Kitakyushu, Kitakyushu, Japan

Liyang Fan
Nikken Sekkei Research Institute, Osaka, Japan

Zu Wang
School of Architecture & Civil Engineering, Zhejing University, Hangzhou, China

Xiaoxia Wei
State Grid Energy Research Institute, Beijing, China

Jie Liu, Tiezhong Wei and Lirong Wang
Heilongjiang Grid Company, Harbin, China

Marc A. Rosen
Faculty of Engineering and Applied Science, University of Ontario Institute of Technology, Oshawa, Ontario, Canada

Didit Novianto, Weijun Gao and Soichiro Kuroki
The University of Kitakyushu, Kitakyushu, Japan

Mouhammad Alanfaf Mohamed Mladjao
University of Lorraine, IUT Henri Poincaré de Longwy, LERMAB Longwy, Nancy, France
ECAM-EPMI, LR2E-Lab, Cergy-Pontoise, France

El Abbassi Ikram
ECAM-EPMI, LR2E-Lab, Cergy-Pontoise, France

Darcherif Abdel-Moumen
ECAM-EPMI, Quartz-Lab, Cergy-Pontoise, France

El Ganaoui Mohammed
University of Lorraine, IUT Henri Poincaré de Longwy, LERMAB Longwy, Nancy, France

Abubakar Kabir Aliyu
Faculty of Electrical Engineering, Centre of Electrical Energy System, Johor Bahru, Malaysia

Abba Lawan Bukar
Department of Electrical Engineering, Faculty of Engineering, University of Maiduguri, Maiduguri, Nigeria

Jamilu Garba Ringim
Federal Airport Authority of Nigeria, Katsina, Nigeria

Abubakar Musa
Department of Electrical Engineering, Faculty of Engineering, Ahmadu Bello University, Zaria, Nigeria

Isiaka Adeyemi Abdul-Azeez
Department of Urban & Regional Planning, Modibbo Adama University of Technology, Yola, Nigeria

Chin Siong Ho
Faculty of the Built Environment, Universiti Teknologi Malaysia, Johor Bahru, Malaysia

Ashish Mishra and Nilay Khare
Department of Computer Science and Engineering, Maulana Azad National Institute of Technology, Bhopal, India

Mouaaz Nahas
Department of Electrical Engineering, College of Engineering and Islamic Architecture, Umm Al-Qura University, Makkah, KSA

M. Sabry
Department of Physics, College of Applied Sciences, Umm Al-Qura University, Makkah, KSA
Solar Research Department, National Research Institute of Astronomy and Geophysics, Cairo, Egypt

Saud Al-Lehyani
Department of Physics, College of Applied Sciences, Umm Al-Qura University, Makkah, KSA

Neilton Fidelis da Silva
Energy Planning Program (PPE), Coordination of Postgraduate Programs in Engineering at the Federal University of Rio de Janeiro (COPPE/UFRJ), Cidade Universitária, Rio de Janeiro, Brazil
International Virtual Institute of Global Change—IVIG, Centro de Tecnologia, Cidade Universitária, Rio de Janeiro, Brazil
Federal Institute of Education, Science and Technology of Rio Grande do Norte (IFRN), Natal, Brazil

Angela Oliveira da Costa, Rachel Martins Henriques and Marcio Giannini Pereira
Energy Planning Program (PPE), Coordination of Postgraduate Programs in Engineering at the Federal University of Rio de Janeiro (COPPE/UFRJ), Cidade Universitária, Rio de Janeiro, Brazil

Marcos Aurelio Freitas Vasconcelos
Energy Planning Program (PPE), Coordination of Postgraduate Programs in Engineering at the Federal University of Rio de Janeiro (COPPE/UFRJ), Cidade Universitária, Rio de Janeiro, Brazil
International Virtual Institute of Global Change—IVIG, Centro de Tecnologia, Cidade Universitária, Rio de Janeiro, Brazil

Jose M. "Chema" Martínez-Val Piera
ETSI Minas, Universidad Politécnica de Madrid, Madrid, Spain

Andrew Walcott Beckwith
Physics Department, Chongqing University Huxi Campus, Chongqing, China

Nfally Dieme
Laboratory of Semiconductors and Solar Energy, Department of Physics, Faculty of Science and Technology, Cheikh Anta Diop University, Dakar, Senegal

Eduardo Oliveira Teles and Marcelo Santana Silva
Federal Institute of Bahia (IFBA), Brazil and Federal University of Bahia (UFBA), Salvador, Brazil

Francisco Gaudêncio M. Freires and Ednildo Andrade Torres
School of Industrial Engineering (PEI), Federal University of Bahia (UFBA), Salvador, Brazil